New Emitters for OLEDs: The Coordination- and Photo-Chemistry of Mononuclear Neutral Copper(I) Complexes

Zur Erlangung des akademischen Grades eines

DOKTORS DER NATURWISSENSCHAFTEN

(Dr. rer. nat.)

Fakultät für Chemie und Biowissenschaften

Karlsruher Institut für Technologie (KIT) - Universitätsbereich

angenommene

DISSERTATION

von

Larissa Bergmann

aus Stühlingen

Dekan:	Prof. Dr. Peter Roesky
Referent:	Prof. Dr. Stefan Bräse
Korreferent:	Prof. Dr. Uli Lemmer
Tag der mündlichen Prüfung:	13. April 2015

Band 57
Beiträge zur organischen Synthese
Hrsg.: Stefan Bräse

Prof. Dr. Stefan Bräse
Institut für Organische Chemie
Karlsruher Institut für Technologie (KIT)
Fritz-Haber-Weg 6
D-76131 Karlsruhe

Bibliografische Information der Deutschen Bibliothek

Die Deutsche Nationalbibliothek verzeichnet diese Publikation in der
Deutschen Nationalbibliografie; detaillierte bibliografische Daten sind
im Internet über http://dnb.d-nb.de abrufbar.

ISBN 978-3-8325-4183-5
ISSN 1862-5681

Logos Verlag Berlin GmbH
Comeniushof, Gubener Str. 47,
10243 Berlin
Tel.: +49 030 42 85 10 90
Fax: +49 030 42 85 10 92
INTERNET: http://www.logos-verlag.de

Die vorliegende Arbeit wurde im Zeitraum November 2011 bis März 2015 am Institut für Organische Chemie des Karlsruher Instituts für Technologie (KIT) unter der Leitung von Herrn Prof. Dr. Stefan Bräse und bei der CYNORA GmbH unter Betreuung von Dr. Thomas Baumann durchgeführt.

Table of Contents

Abstract...1

Kurzzusammenfassung...2

1. **Introduction**...3

  1.1 Organic Light-Emitting Diodes 3

    1.1.1 The Route to a New Lighting and Display Technology.............................3

    1.1.2 Architecture and Working Principle of an Organic Light-Emitting Diode 6

  1.2 Copper(I) Complexes as Emitting Materials...17

    1.2.1 Structural Diversity of Copper(I) Complexes...18

    1.2.2 Thermally Activated Delayed Fluorescence...20

    1.2.3 Photophysical Characterization of Thermally Activated Delayed Fluorescence in Copper(I) Complexes 22

2. **Aims and Objectives**..26

3. **Results and Discussion**..28

  3.1 Development of Mononuclear Neutral Copper(I) Complexes based on various Anionic N^N Ligands 28

    3.1.1 Synthesis of Ligands and Complexes...29

    3.1.2 Structural studies..32

    3.1.3 Photophysical Characterization...52

  3.2 Mononuclear Neutral Copper(I) Complexes based on Tetrazolate Ligands 65

    3.2.1 The Model Complex [(PyrTet)Cu(DPEPhos)] 65

    3.2.2 Synthesis of Ligands and Complexes...71

    3.2.3 Copper(I) Complexes [(PyrTet)Cu(P^P)] with different Bis(phosphine) Ligands 85

    3.2.4 Copper(I) Complexes [(D-Tet)Cu(DPEPhos)] with different Tetrazolate Ligands 97

    3.2.5 Copper(I) Complexes [(R-PyrTet)Cu(P^P)] with different Substituents on the Pyridine-Tetrazolate Ligand..................115

  3.3 Investigation of the Thermally Activated Delayed Fluorescence by Time-resolved Spectroscopy 129

    3.3.1 Intersystem Crossing in the Copper(I) Complex [(PyrTet)Cu(DPEPhos)] 130

    3.3.2 The TADF Mechanism in Mononuclear Neutral Copper(I) Complexes based on Tetrazolate Ligands..................138

3.4   Optoelectronic Devices 146

   3.4.1   Quenching Processes in Copper(I) Complexes.......................................146

   3.4.2   Application of the Model Complex [(PyrTet)Cu(DPEPhos)] in an OLED..............148

4.   Conclusion **153**

5.   Experimental Part **158**

5.1   General...158

   5.1.1   Analytical Methods and Instrumentation...158

   5.1.2   Photophysical Methods and Instrumentation...160

   5.1.3   Cyclovoltammetry...163

   5.1.4   Thermogravimetric Analysis 163

   5.1.5   Differential Scanning Calorimetry 163

   5.1.6   Fabrication and IVL-Characterization of OLEDs..................................164

   5.1.7   Density-functional theory calculations..164

   5.1.8   Synthetic Methods and Materials..165

5.2   Synthetic Procedures and Analytical Characterization 166

   5.2.1   General Procedures 166

   5.2.2   Synthesis of Ligands 168

   5.2.3   Synthesis of Complexes 186

5.3   Supplemental Data..203

   5.3.1   Photophysical Characterization...203

   5.3.2   Theoretical Calculations 211

   5.3.3   Electrochemical Data 213

   5.3.4   Thermogravimetric Data...214

5.4   Crystallographic Data 215

6.   Abbreviations...**249**

7.   References...**253**

8.   Appendix..**267**

5.1   Curriculum vitae...267

5.2   Publications, conference contributions, patents...268

5.3   Acknowledgements..270

Abstract

Luminescent copper(I) complexes present a promising class of emitting materials for organic light-emitting diodes and have attracted enormous interest due to their high emission efficiencies, color tunability by ligand variation, and their abundancy compared to iridium(III) or platinum(II) compounds. Especially the assessment of thermally activated delayed fluorescence (TADF) to copper(I) compounds has accelerated the development and investigation of several complex classes. Herein, a large variety of highly luminescent mononuclear neutral copper(I) complexes of the type [(N^N)Cu(P^P)] has been developed and a deeper understanding of the structure-property relationships was achieved by comprehensive spectroscopical investigations.

In a systematic study on the coordination chemistry of copper(I) with bidentate pyridine-amines or their deprotonated amide analogues, respectively, and various phosphine ligands of different denticity and bite angle, a set of prediction rules for the complex's coordination motif with the given ligands could be derived. The bis(phosphine) DPEPhos has proven as a good candidate to yield mononuclear neutral copper(I) complexes of the type [(N^N)Cu(DPEPhos)] with all investigated pyridine-amides ligands. The neutral complexes show superior luminescence in the green to yellow region of the visible spectrum with photoluminescence quantum yields up to 78% compared to their cationic counterparts, which exhibit red-shifted emission of lower efficiencies (9–45%). The emission of the neutral complexes could be characterized as a (ML+IL)CT transition, which is strongly controlled by the electronic effects of the N^N ligand. The most promising candidate [(PyrTet)Cu(DPEPhos)] based on pyridyltetrazolates in terms of emission efficiency, low sensitivity against oxygen, and thermal stability was further varied by different modifications of its bis(phosphine) and tetrazolate ligands in order to tune the emission color and to understand the excited states scheme of this complex class. The investigation of a large variety of complexes by absorption and emission spectroscopy, supported by theoretical calculations and electrochemical measurements, enabled a thorough understanding of the steric and electronic effects of the ligands on the complexes' emission.

Furthermore, the mechanism of thermally activated delayed fluorescence could be illustrated by means of time-resolved emission spectroscopy in the pico-, nano-, and microsecond regime, and the intersystem crossing of the TADF complex [(PyrTet)Cu(DPEPhos)] determined to 27 ps in the solid state. Especially the determination of the intersystem crossing rate in such highly emissive compounds is challenging, but very important for the understanding of the excited state mechanisms in TADF materials that can be used in OLEDs.

Kurzzusammenfassung

Lumineszierende Kupferkomplexe stellen eine vielversprechende Klasse von Emittermaterialien für organische Leuchtdioden dar und sind aufgrund ihrer hohen Emissionsquantenausbeuten, der Variation der Emissionsfarbe durch Ligandenmodulation, und dem größeren Vorkommen von Kupfer gegenüber Iridium und Platin in den Fokus des Forschung gerückt. Insbesondere das Potential der thermisch aktivierten verzögerten Fluoreszenz (TADF) beschleunigte die Entwicklung und Untersuchung verschiedener Komplexklassen. In der vorliegenden Arbeit wurde eine große Bandbreite von stark lumineszierenden Kupferkomplexen der allgemeinen Form [(N^N)Cu(P^P)] entwickelt und ein tieferes Verständnis der Struktur-Eigenschafts-Beziehungen durch umfassende spektroskopische Untersuchungen erreicht.

Das Koordinationsverhalten von Kupfer mit bidentaten Pyridinaminen, beziehungsweise den entsprechenden deprotonierten Pyridinamiden, sowie von Phosphanen mit verschiedener Zähnigkeit und unterschiedlichen Bisswinkeln wurde systematisch untersucht, sodass Voraussagen über die bevorzugten Koordinationsmotive mit den entsprechenden Liganden gemacht werden konnten. Insbesondere das Bisphosphan DPEPhos hat sich als guter Kandidat für einkernige neutrale Kupferkomplexe der Form [(N^N)Cu(DPEPhos)] erwiesen. Diese zeigen herausragende Emissionseigenschaften mit Quantenausbeuten bis zu 78%, wohingegen ihre kationischen Pendants rot-verschobene Emissionen mit geringeren Effizienzen aufweisen (9–45%). Mithilfe von DFT Rechnungen konnte die Emission der neutralen Komplexe als (ML+IL)CT-Übergang charakterisiert werden, der stark von den elektronischen Eigenschaften des N^N-Liganden beeinflusst wird. Der vielversprechendste Kandidat, [(PyrTet)Cu(DPEPhos)], hinsichtlich Emissionsquantenausbeute, Oxidations- und thermischer Stabilität, wurde weiterhin durch verschiedene Modifikationen des Bisphosphans und des Tetrazolat-Liganden variiert. Anhand der Untersuchung zahlreicher Komplexe mithilfe von Absorptions- und Emissionsspektroskopie, unterstützt durch theoretische Berechnungen und elektrochemische Methoden, konnte so ein tieferes Verständnis der sterischen und elektronischen Eigenschaften der Liganden auf die Komplexemission erreicht werden.

Weiterhin konnte der Mechanismus der thermisch aktivierten verzögerten Fluoreszenz mithilfe von zeitaufgelöster Emissionsspektroskopie im Pico-, Nano- und Mikrosekundenbereich veranschaulicht und das Intersystem Crossing des Komplexes [(PyrTet)Cu(DPEPhos)] zu 27 ps im Festkörper bestimmt werden, was für diese stark lumineszierenden Systeme zwar anspruchsvoll aber umso wichtiger für das Verständnis der Prozesse in TADF-Materialien zur Anwendung in OLEDs ist.

1. Introduction

1.1 Organic Light-Emitting Diodes

1.1.1 The Route to a New Lighting and Display Technology

With the commercialization of the incandescent light bulb by THOMAS A. EDISON in 1879 the foundation stone for artificial light was laid. Around 130 years later one finds other illumination technologies such as halogen or fluorescent lamps, but only the ban of the incandescent light bulb by the European Union in 2009 ("EG-Verordnung 244/2009") – due to its low power efficiency of only 5% – led to a broadening of the lighting market (Figure 1).[1] Energy-saving lamps are used as short-term substitutes for the traditional incandescent light bulb, but light-emitting diodes (LEDs) based on inorganic semiconductors gain increasing market growth due to higher efficiencies and lifetimes (Figure 2). In 2014, the Nobel Prize was awarded to ISAMU AKASAKI, HIROSHI AMANO and SHUJI NAKAMURA for the invention of blue LEDs that made white light LED sources commercially available.[2] The impact of this invention is tremendous from the perspective that about 19% of the world's electricity consumption is used for lighting.[1] A still young lighting – and display – technology is based on the electroluminescence of organic materials, i.e. the generation of light by an organic molecule under an electric current, and already finds application as organic light-emitting diodes (OLEDs) in small displays, but is also dealt as an alternative illumination source.[3]

Figure 1. Changing illuminations sources: incandescent light bulb as patented by Thomas A. Edison, halogen light bulb, fluorescent tube, LEDs incorporated into a traditional light bulb design, WOLED panel.[4-8]

The advantages of white organic light-emitting diode (WOLED) panels are high efficiencies and the potential of low manufacturing costs with environmentally friendly and less expensive materials,[9] which is the main drawback of inorganic LEDs and fluorescent lamps. New designs and lighting concepts are possible with the ultrathin, light-weight and potentially flexible OLED devices, which emit areal light, such as large-area luminous wallpaper, automotive interior or dashboard light, and semitransparent illumination panels. However, further optimizations in increasing the lifetimes of organic light-emitting diodes to several ten thousand hours are necessary to meet the requirements for illumination sources. Nowadays, OLEDs are already used in displays for mobile phones or tablet computers in mass production, and the development of low-cost materials and processing techniques will lead to further applications in lighting as well as for smart packaging and e-papers.

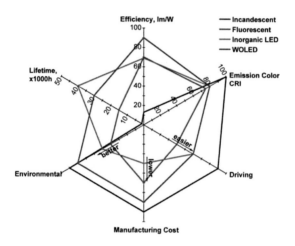

Figure 2. Qualitative comparison of the key-performance parameters of the illumination sources incandescent light, fluorescent tube, inorganic white LED and WOLED.[9] A symmetrical pentagon would reflect the preferred illumination source with high efficiency and lifetime, low environmental hazards and manufacturing costs (including material costs), easy driving and the emission of warm white light.

The working principle of OLEDs is based on the electroluminescence of organic materials, which was already found by ANDRÉ BERNANOSE in the 1950s. He applied an electric current to a polymer matrix doped with dies and observed a red luminescence at high voltages. [10] Thereby the die acts as semiconductor, which is excited by the recombination of electrons and holes in the electric current to generate light. Since in organic semiconductors electrical current is transported via distinct molecular orbitals instead of delocalized bands, charge carrier mobilities and conductivities are by several orders of magnitude lower than in inorganic semiconductors. Only 1987, when CHEN W. TANG and STEVEN VAN SLYKE developed an OLED with higher efficiency at lower driving voltage, the new field of organoelectronics started to grow.[11] A few years later, JEREMY BURROUGHES and coworkers discovered the electroluminescence of the polymer polyphenylene vinylene (PPV),[12] and the wish for thin and flexible displays emerged.[13] The significance of the discovery and development of conductive polymers as organic semiconductors was further awarded with the Nobel Prize to ALAN J. HEEGER, ALAN G. MACDIARMID and HIDEKI SHIRAKAWA in 2000.[14]

The research on organic electronics accelerated even more from the late 1990s on, when highly luminescent metal complexes based on iridium[15-21] and platinum,[22-27] – so-called triplet emitters – were developed, which made theoretical internal efficiencies of 100% in OLEDs possible.[28,29] Advanced photophysical studies on these materials have been conducted and a large spectrum of materials with desired properties designed for the application in optoelectronics. Recently, a new generation of OLED materials based on the emission concept of thermally activated delayed fluorescence (TADF), namely copper(I) complexes and particular organic materials, has emerged as low-cost alternative to triplet emitters.[30-33]

In the last few years, research has furthermore focused on the applicability of emitting materials in optoelectronic devices concerning efficiency, emission color, stability and processability, and the optimization and manufacturing of OLEDs. Nowadays, small displays based on OLEDs with polymer materials and iridium complexes as emitting materials are in mass production and show high color brightness even at flat viewing angles, short response times and high contrast ratios.[34,35] New challenges such as low-cost manufacturing by solution processing techniques are tackled.

In the introduction of this work, the setup and working principle of an OLED with basic electronic processes is explained, used materials and their requirements and major emission concepts presented, and further challenges to manufacture efficient, long-life OLEDs discussed (1.1). This section is followed by an introduction to luminescent copper(I) complexes as emitting material for OLEDs (1.2): A short overview on the structural diversity of copper(I) complexes is given in 1.2.1 and the characteristic emission concept thermally activated delayed fluorescence is explained and related to the application in OLEDs (1.2.2). A short summary on the photophysical methods to examine the TADF behavior of emitting materials is given in chapter 1.2.3. To conclude the introduction, the aims and objectives of this work are presented in chapter 2.

1.1.2 Architecture and Working Principle of an Organic Light-Emitting Diode

Setup and Fundamental Processes

An OLED is a multi-layer stack device with several organic layers in between two electrodes on a glass or plastic substrate. When applying an electric current to the device, light is generated in the emissive layer and emitted via the semi-transparent anode. In general, ITO – a mixed oxide of indium(III) and tin(IV) oxide – is used as transparent anode, whereas aluminum, calcium or magnesium is used as cathode. Further conductive layers for better electron or hole transport enable lower driving voltages and faster response times of the OLED. A schematic setup of an OLED device is shown in Figure 3.

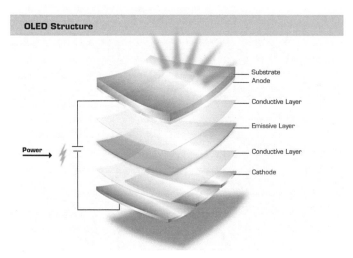

Figure 3 (by CYNORA GmbH). Device architecture of a multilayer OLED in bottom emission design. During operation, electrons from the cathode (Ca/Al) and holes from the anode (ITO) wander through the device and meet in the emissive layer forming an exciton, which excites an emitter molecule and deactivates by the emission of light.

During operation, electrons are injected from the cathode into an electron transport layer (ETL), while the anode withdraws electrons from the hole transport layer (HTL), thus generating holes.[1,36] On a molecular basis one can understand these processes as insertion of an electron in the lowest unoccupied molecular orbital (LUMO) of an organic molecule on the cathode side, i.e. formation of a radical anion

---

[1] Defect electrons are called "holes" herein.[39]

in a reduction, and the formal oxidation of adjacent organic molecules on the side of the anode. Due to the higher LUMO levels of the organic materials versus the Fermi level of the metal cathode, and the lower energy levels of the HOMOs (highest unoccupied molecular orbital) versus the ITO anode, high injection barriers result (see Figure 6a). To reduce the turn-on voltage caused by these high barriers, a thin layer of lithium fluoride is introduced on the aluminum cathode, whereas PEDOT/PSS (polyethylene dioxythiophene doped with polystyrene sulfonic acid) is usually used on the ITO, thereby adjusting the energy levels of adjacent layers in a stepwise design.[37-39]

In the electric field between the cathode and anode the charge carriers start to migrate towards the opposite electrodes via hopping processes (see Figure 4). The electrons "hop" from one molecule to the next one in a cascade of reductions in the direction of lower energy, while the holes are transported via formal oxidations of adjacent molecules in an energetic upward motion. As the injection barriers are usually high and charge carrier mobilities in the organic layers are very low, a relatively high driving voltage has to be applied until charge carriers start to wander through the device (~ 3–10 V).[37] The energy levels of the frontier orbitals of the organic materials adjust to the electric field as schematically depicted in Figure 4 and enable a stepwise transport of the charge carriers.

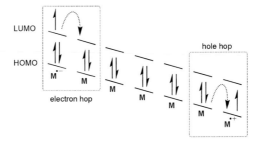

Figure 4. The transport of charge carriers in an organic semiconductor is understood as *hopping process*. Electrons move into the direction of lower energy via the LUMOs of the molecule and formally reducing it, while holes move energetically upwards by oxidation of the semiconductor molecule.

Further conductive layers such as electron and hole transport layers improve the transport of charge carriers in the device and thus reduce driving voltages. Organic materials with large, rigid and easily polarizable π-electron systems are used as conductive materials,[37] such as the materials N,N'-di(1-naphthyl)-N,N'-diphenyl-(1,1'-biphenyl)-4,4'-diamine (α-NPD) as HTL and 1,3,5-tris(1-phenyl-1H-benzimidazol-2-yl)benzene (TPBi) as ETL. Additionally, hole blocking layers are often introduced to minimize leakage currents at the cathode due to the intrinsically higher mobility of holes in organic materials.[40]

In Figure 5 an example of a monolayer (a) and a multilayer OLED (b) with typical electron and hole injection and transport layers is depicted with HOMO und LUMO energy levels of each material.

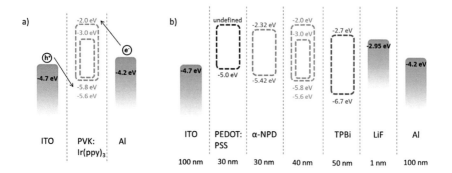

Figure 5. (by CYNORA GmbH) **a)** Monolayer OLED with the host poly(n-vinylcarbazole) doped with Ir(ppy)₃ as emitting material. Injection barriers are high and thus the OLED stack is optimized to a **b)** multilayer OLED with hole and electron injection layers (PEDOT:PSS, LiF) and hole and electron transport layers (α-NPD, TPBi). The electron transport layer TPBi with a very low HOMO further acts as hole blocking layer. The introduction of further conductive layers results in a stepwise increase of LUMO levels and decrease of HOMO levels, respectively, thus enabling better charge carrier transport and reducing turn-on and driving voltages.

The emission layer either consists of an electroluminescent conductive polymer as emitting material or of a small-molecule material. Thus, the two approaches are called polymer light-emitting diode (PLED) or small molecule light-emitting diode (SMOLED). Common polymer materials are based on the yellow-green emitting poly(1,4-phenylene vinylene) (PPV, **1**) or on the blue-emitting polyfluorene (PFO, **2**), which are shown in Figure 6.[41] First OLEDs were based on the small molecule tris(8-hydroxyquinolinato)aluminum (Alq₃, **3**) as neat emitting layer, whereas nowadays a host material is doped with a small-molecule emitting material in low concentrations (2–20%), e.g. tris(2-phenylpyridinato)iridium (Ir(ppy)₃, **4**). The host functions as diluter to prevent emission quenching from too high emitter concentrations, as well as semiconductor to transport the charge carriers and ensure good charge-balance. Typical host materials are based on the carbazole structure such as poly(n-vinylcarbazole) (PVK) or tris(4-carbazoyl-9-ylphenyl)amine (TCTA).

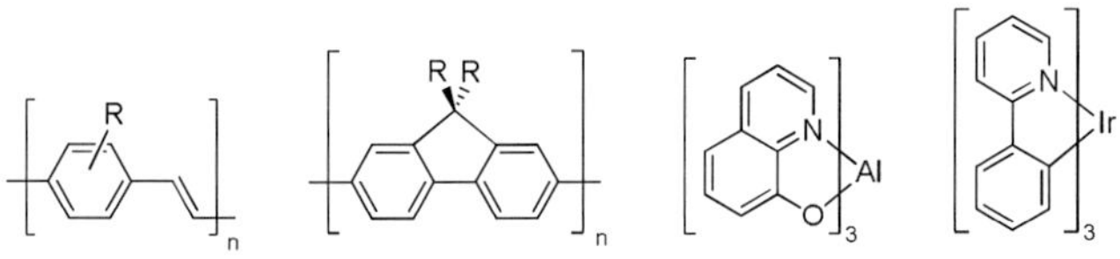

Figure 6. Commonly used polymer emitting materials poly(1,4-phenylene vinylene) (PPV, **1**), polyfluorene (PFO, **2**), and the small molecule emitting materials tris(8-hydroxyquinolinato)aluminum (Alq$_3$, **3**) and tris(2-phenylpyridinato)iridium (Ir(ppy)$_3$, **4**).

For the generation of light in OLEDs, an electron and hole have to meet and recombine in the emissive layer under the formation of an exciton – a bound electron and hole pair, which finally activates the emitter molecule to emit light. In PLEDs the exciton is directly formed on the polymer, while in host-guest systems the exciton resides on the host molecule and is then transferred to an adjacent emitter molecule. The recombination of electrons and holes is an electrostatically driven process, where one of the charges is localized on a host molecule (dependent on the host's polarizability) and the other one is attracted by it when migrating through the OLED in the applied electric field and coming close enough, so that the Coulomb interaction becomes effective. After formation of the exciton on the host – so called Frenkel exciton – it is transferred to the dopant by Förster or Dexter processes. Förster energy transfer occurs via overlap of the dipolar electrical fields of an excited matrix molecule and an emitter molecule, while in Dexter processes a direct electron exchange via overlapping molecule orbitals takes place and is thus range-limited.[42] In the less ideal case, excitons can also be formed on the emitter molecule itself, but this generally leads to degeneration of the emitting material. The exciton finally localized on the emitter molecule can be understood as excited emitter, which deactivates under emission of light. The efficiency of an emitting material is expressed in the ratio of radiant processes to all – also non-radiant – deactivation processes and is called photoluminescence quantum efficiency Φ_{PL} (PLQY). Both PLQY and emission color depend on the chosen emitting material and will be discussed in detail in the next section "Generation of light".

To conclude, the fundamental processes in an OLED under applied voltage are: 1) injection of charge carriers at the electrodes; 2) transport of the charge carriers via the HOMOs and LUMOs of the organic material to the oppositely charged electrode 3) recombination of electron and holes in the emission layer under formation of an exciton and subsequently 4) emission of light.

Generation of Light

The formation of an excited state of an emitter molecule can either be realized by charge carrier recombination as discussed for an OLED in the section before, or by optical excitation. To understand the fundamental processes after excitation, the case of photoluminescence (PL) is explained first and then later related to the excited state transitions in OLEDs (electroluminescence).

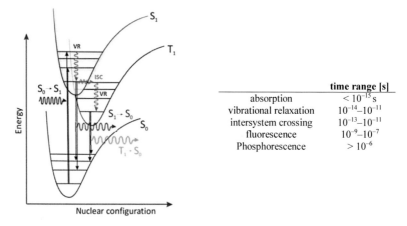

	time range [s]
absorption	$< 10^{-15}$ s
vibrational relaxation	$10^{-14}-10^{-11}$
intersystem crossing	$10^{-13}-10^{-11}$
fluorescence	$10^{-9}-10^{-7}$
Phosphorescence	$> 10^{-6}$

Figure 7. Jablonski diagram of an exemplary organic emitter with potential curves for the singlet ground state S_0 and excited states S_1 and T_1. The electronic transitions for absorption (black), vibrational relaxation (red arrow), intersystem crossing (green arrow), fluorescence (dark blue) and phosphorescence (light blue) are shown schematically.[43] The emission color of a radiative transition is given by its energy gap ($E = h\upsilon$). Typical time ranges for each electronic transition are given in the table.

Absorption of a photon occurs from the lowest vibronic ground state S_0 to higher lying vibronic states of the excited singlet state S_1 (see Figure 7). The probability of this transition is given by the Franck-Condon factor, which defines the overlap integral of the wave functions of the ground and respective excited state. After absorption, the photon relaxes quickly to the lowest vibronic state of the excited singlet state S_1 and can then either deactivate down to the ground state S_0 by emission of fluorescence or intersystem cross to the lower lying excited triplet state T_1. Transitions with spin flip such as intersystem crossing (ISC) $S_1 \rightarrow T_1$ and the deactivation path $T_1 \rightarrow S_0$ are usually forbidden in organic molecules due to low spin-orbit coupling (SOC). In particularly designed organic molecules or metal complexes with an intrinsically high SOC constant of the metal atom(s), intersystem crossing is allowed and after fast vibrational relaxation to the lowest vibronic state of the triplet T_1, the photon can deactivate to the singlet ground state S_0 by emission of phosphorescence. This is usually red-shifted compared to the fluorescent deactivation $S_1 \rightarrow S_0$ due to the intrinsically smaller energy gap between singlet ground state S_0 and excited triplet state T_1. Typical time ranges for each electronic transition are given in Figure 7 and can be understood/interpreted as a measure for the probability of the transitions, e.g. absorption occurs instantly and thus exhibits a very fast time constant of smaller than 10^{-15} s.

Besides fluorescence and phosphorescence, the photon can deactivate from excited states to the ground state by non-radiant transitions and conversion of its energy into heat. The photoluminescence quantum yield is defined by Equation 1 with k_r and k_{nr} as rates for radiant and competitive non-radiant processes (s^{-1}), and determines the efficiency of an emitting material:[39]

$$\Phi_{PL} = \frac{k_r}{k_{nr}+k_r} \qquad (1)$$

The lifetime or emission decay time of a specific state is further defined as the reciprocal sum of all potential deactivation (radiant and non-radiant) processes and is given in s:[44,45]

$$\tau = \frac{1}{k_{nr}+k_r} \qquad (2)$$

When the PLQY and lifetime of a specific state are available from experimental measurements, the radiative and non-radiative rates can be derived and thus, an estimation on the efficiency of the transition is obtained. Typical processes, which reduce the emission efficiency, found for phosphorescent metal complexes are dd-quenching and triplet-triplet annihilation (TTA). Especially in complexes based on Ir(III) (d^6) and Pt(II) (d^8), unoccupied metal-centered dd-stated can be populated during excitation, which weakens the metal-ligand bonds and results in non-radiative deactivation processes.[46] Triplet-triplet annihilation is an intermolecular process of two excited molecules in the triplet state T_1, which interact under formation of an excited singlet S_1 and ground state S_0, and thus the energy of one photon is lost. Since this process is range-limited (Dexter transfer), high concentrations of the emitting material lead to enhanced quenching.[31,42]

Both the efficiency of an emitting material as well as the emission color, are controlled by its molecular structure, electronic and steric substituents. As can be seen in the Jablonski diagram of Figure 7, the energy levels of the excited states have to be altered for a variation of emission color: The emission is blue-shifted by destabilization of the excited state energy levels, whereas a red-shift is obtained from stabilization of the respective excited states. If the population shift from the frontier orbital HOMO to the LUMO contribute strongly to the transition $S_0 \rightarrow S_1$, and furthermore the same trend for emission color as for the absorption is expected (although more red-shifted), the radiant transitions between particular states can roughly be visualized as HOMO-LUMO transitions. Several modifications of the molecular structure by enlargement of the aromatic system or introduction of substituents alter the energy gap between HOMO and LUMO and thus trends for the emission color can be predicted. In conjugated polymers and small molecules with conjugated electronic systems, the HOMO-LUMO energy gaps are narrowed the stronger the delocalization of the π-electrons,[37] i.e. in small molecules the emission color is red-shifted with an enlargement of the aromatic system. Introduction of electron withdrawing groups (–I) results in a lowering of the HOMO and LUMO, whereas electron donating substituents (+I) lead to a destabilization of the frontier orbitals. Depending on the localization of the frontier orbitals a blue- or red-shift can be obtained by modification of the ligands. These general rules for HOMO and LUMO adjustment can be used to design new emitting materials or alter emission colors of already known materials, but a correct description of the emission is only given by the transitions between particular states.

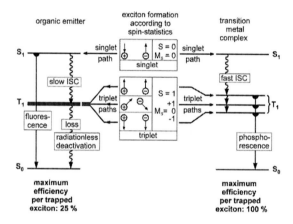

Figure 8. Comparison of light generation in OLEDs with purely organic emitters or metal complexes.[47] While in organic molecules only singlet excitons (25%) can be used for fluorescence and the energy of the triplet excitons (75%) is transferred to heat, in triplet emitters fast ISC populates the triplet state T_1, which enables phosphorescence. All formed excitons are used for the generation of light in triplet harvesting.

The generation of light in an OLED follows the same concept as discussed for photoluminescence, but further aspects such as spin statistics have to be taken into account. During operation of the OLED, two charge carriers recombine in the emissive layer and form an exciton. Both electron and hole have the same magnitude of spin, but oppositely directed ($s = +1/2, -1/2$). Linear combination of the wave functions of the hole and electron then gives four asymmetric wave functions corresponding to one singlet ($S = 0$, $M_s = 0$) and three triplet states ($S = 1$, $M_s = 0, 1, -1$). For all formed excitons this implies a population ratio of singlet to triplet states of 25% to 75%. As mentioned above, intersystem crossing $S_1 \rightarrow T_1$ as well as the deactivation path $T_1 \rightarrow S_0$ is very inefficient in organic molecules with small spin-orbit coupling, which means that only 25% of all excitons can be used for fluorescence. In transition metal complexes, a strong spin orbit coupling constant from the metal atom(s) induces mixing of the singlet and triplet states and thus enables fast intersystem crossing. All generated excitons are accumulated in the excited triplet state and can then be "harvested" for the emission of phosphorescence, i.e. so called triplet emitters can utilize 100% of all formed excitons for the generation of light.[28,48] This is further visualized in Figure 8.

The discussed spin statistics also play a role for the transfer of the formed exciton from the host to the emitter molecule in the emissive layer. While in Dexter processes the overall spin is retained and singlet and triplet excitons can be transferred to the dopant, Förster processes are only possible as energy transfer between singlet states.[42] So when using triplet emitters in OLEDs, a sufficient Dexter transfer by overlap of the host's emission and the dopant's absorption spectra as well as overlapping wave functions should be ensured.

Summing up the photophysical processes for the emission of light, two major emission principles – fluorescence and phosphorescence – were explained. Although both fluorescent organic molecules as well as triplet emitters can in principle reach photoluminescence quantum yields of 100%, in the OLED device only the triplet emitters can use all generated excitons for electroluminescence. Nevertheless, also fluorescent polymers are used for OLEDs in mass-production due to their good processability from solution and simpler stack design as well as better stability of blue-emitting polymers over small molecules (see also section "Challenges").

Characteristics of an Organic Light-Emitting Diode

There are two main characteristics for OLEDs: the luminance L, which physically describes the luminescence density of an areal light and is given in cd m^{-2}, and the efficiency η. The luminance is measured as luminous intensity (in cd) per area and depends on the applied voltage or current density. The efficiency of an OLED can be expressed as power efficiency in lm W^{-1} (Equation 3) or as current efficiency in cd A^{-1} (Equation 4), but both terms are also expressed as luminous efficacy. The efficiency depends on the luminance or applied voltage and thus should always be noted at a given luminance or voltage, respectively.[9,49]

$$\eta = \frac{\Phi_v}{P} \qquad (3)$$

Φ_v: luminous flux [lm], P: electrical power input [W]

$$\eta = \frac{L}{I_A} \qquad (4)$$

L: luminance [cd m^{-2}], I_A: current density [A m^{-2}]

The efficiency of an OLED strongly depends on the photoluminescence quantum efficiency of the emitting material, but further aspects such as charge balance and spin-statistics come into play in the electric device, as well as the outcoupling of the generated light in a glass body. The external quantum efficiency (EQE) expresses the quotient of light generated and outcoupled from the device per formed exciton. From Equation 5, the device-related parameters can be estimated at a given efficiency of emitting material:[47]

$$\eta_{ext} = \chi_{out} \cdot \beta \cdot \gamma \cdot \Phi_{PL} \qquad (5)$$

The outcoupling factor χ_{out} describes the percentage of actually emitted light, which is less than the generated light in the EML due to reflections at the interface transparent anode – air (refractive indices $n_{air} \sim 1$, $n_{device} \sim 1.7$). It is commonly given as 0.2 without any further outcoupling optimizations such as structured layers or outcoupling foils.[9]

The spin-statistics factor β expresses the percentage of excitons available for luminescent transitions relative to all excitons formed from the recombination of electrons and holes. As explained above, this factor depends on the chosen emitting material and is 0.25 for fluorescent materials and 1 for phosphorescent molecules.

The factor γ captures the recombination rate of electrons and holes and is a measure for a high charge carrier density as well as charge balance. A good adjustment of transport layers with respect to their energy levels and the introduction of blocking layers ensure high charge carrier densities and the EML as recombination zone.

A further parameter used in the field of OLEDs is the operating lifetime, which is especially important for displays, but also applications as illumination source require lifetimes of more than 10.000 h. The decrease of power and luminous efficiency over time is usually expressed as LT$_{50}$ or LT$_{70}$ and defines the time, after which the device shows only 50% or 70% of the initial luminous flux at a given current.

Challenges

In the last sections, the setup and working principle of OLEDs were explained, commonly used materials and material requirements introduced and efficiency and lifetime characteristics for OLEDs defined. Besides the development of highly, functionalized luminescent emitting materials with desired properties, many research groups recently focus on the challenges to process efficient OLED devices, which can be boiled down to the three main issues efficiency, device lifetime and cost-efficient manufacturing.

The efficiency of an OLED and device lifetime are usually interconnected and depend on many factors: A good alignment of the HOMO and LUMO levels of adjacent organic layers in the device prevent charge carrier trapping and unbalanced electron-hole transport, which further reduces electrochemical stress and finally degradation of the materials. The right choice of the host material is crucial to ensure sufficient energy transport to the emitter molecules and to control the morphology around the dopants. On the one hand, the band gap – the HOMO-LUMO energy gap – has to be larger than the emitter's to prevent re-absorption of the emitted light and on the other hand the triplet energy level of the host has to be higher to avoid quenching and energy back-transfer.[50,51] The concentration of the emitting material in the host can further play a role for emission quenching, especially when triplet emitters are used,[52] and the environment around the emitter molecule influences its emission color and efficiency by slight geometry distortions.[53] Multilayer devices with several supporting layers are designed and simulated (regarding layer thicknesses, interfaces, refractive indices etc.) to optimize the processes in the running device.

Material purity and stability are requirements, which include many different aspects, and can in some cases become only important after several hours of OLED operation.[54] Impurities or molecular fragments act as charge carrier traps and lead to a chain reaction of organic layer degradation in the device.[55] Thus, materials with high thermal stability temperatures are required with low tendency for molecule dissociation and subsequent diffusion of ionic fragments e.g. halides through the running device (chemical stability).[56-58] Even crystalline areas in an otherwise amorphous layer can act as charge carrier traps and so materials with high glass transition temperature are preferred over crystalline materials.[39] Water and oxygen impurities are excluded from the device by manufacturing in an inert atmosphere and subsequent encapsulation by a glass or polymer foil.[55,59,60] Photochemical stability defines the cycles, how often a molecule can be excited and emit light before breaking down (photobleaching) and is especially a problem in white light panels or displays due to different ageing of the colors.[3] And finally, the used materials have to be electrochemically stable, i.e. they should be easily oxidized or reduced, respectively, without undergoing side reactions or degradation.[61,62]

Besides these fundamental requirements of good energy level alignment of all layers and stability of the used materials, which control efficiency of the OLED in the short-term and device lifetime in the long run, the outcoupling of light influences mainly the efficiency of the device. Several approaches like outcoupling foils or patterned structures in the device have been studied to improve the outcoupling.[63-66] Recently, a horizontal orientation of the emitter molecules has been found to increase outcoupling as well, but the control of orientation is still difficult.[67-70]

OLED devices can be manufactured either by the sublimation technique at reduced pressure or by solution processing. While polymeric emitters can only be processed from the liquid phase, sublimable small molecules and especially organic materials can be deposited on a substrate by thermal evaporation or chemical vapor deposition (CVD). Although the deposition from gas phase ensures pure and homogenous layers,[39] it has some disadvantages such as high costs due to very high vacuum (10^{-6} mbar) and high material loss as well as the limitation to small devices and certain materials. The manufacturing from solution is favored as a cost- and material-efficient processing technique, but carries the challenge of depositing several layers on top of each other without dissolving the layer beneath. The orthogonal solvent route is used for PEDOT:PSS, which is not soluble in any organic solvent after deposition, but for other organic layers the approach of crosslinking (combined with sublimed layers) has proven as potential manufacturing technique for multilayer OLEDs.[71-75] Several printing techniques such as gravure and ink-jet printing and even knife-coating/doctor blading are examined (roll-to-roll) and inks with emitting materials developed to enable large-area printing at high quality and low costs.[76,77] So far, mainly thermal evaporation techniques are applied for the mass-production of displays (also due to its better resolution) and much research is still needed to bring solution processing to the market.

Nevertheless, many of the above discussed challenges have already been tackled by various approaches and OLED devices with EQEs of more than 20% and efficiencies of up to 107 lm/W at 1000 cd/A based on iridium complexes have been reported.[78] White light OLEDs with power efficiencies of 90 Lm/W and a brightness of 1000 cd/A already exceed commonly used fluorescent tubes with efficiencies of only 60–70 Lm/W.[79,80]

To address some further general considerations, namely abundance of materials and environmentally friendliness, in the following chapters copper(I) complexes will be introduced as emitting materials for OLEDs as alternative to iridium and platinum based materials.

1.2 Copper(I) Complexes as Emitting Materials

Efficient emitting materials for organic light-emitting diodes are a continued focus of academic and industrial research. Various transition metal complexes based on lanthanides, platinum, iridium and recently even copper have been developed and intensively studied for their application as emitters in optoelectronic applications. Since 2000, when MARC A. BALDO and STEPHEN R. FORREST proved that complexes based on transition metals with high spin-orbit coupling can use both singlet and triplet excitons in OLEDs,[28] the so-called triplet emitters gained increasing interest in research. While lanthanide complexes exhibit strong photoluminescence with narrow emission bands which makes them attractive for display applications, [81 - 84] OLED devices with these emitters only show low efficiencies.[85-87] Platinum(II) complexes are commonly applied as red emitters,[23,24,88] but their emission can be contributed by an excimeric band, which is taken advantage of in WOLEDs.[89,90] The seminal iridium(III) complexes, such as Ir(ppy)$_3$ (**4**), feature superior emission properties with photoluminescence quantum efficiencies near unity and short emission decay times of a few microseconds.[47,91] The transition metal complexes based on iridium(III) and platinum(II) utilize the strong spin-orbit coupling from the heavy metal to enable efficient triplet emission, however, they bring along high material costs due to their low abundance in the earth's crust (e.g. Eu: 1.1 ppm, Tb: 0.6 ppm, Pt: 0.0015 ppm, Ir: 0.000037 ppm) [92] and expensive production (Figure 9). Especially with material-intensive manufacturing techniques of OLEDs, such as sublimation, these high-cost although efficient emitting materials are not sustainable. The search for new emitting materials led to copper(I) complexes, which can show exceptional high emission efficiencies of nearly 1 despite small spin-orbit coupling.[93 , 94] Besides its higher natural abundance (27 ppm) and lower costs (Cu: 5 €/kg vs. Ir: 17.000 €/kg),[95,96] it can be easily produced from recycling of copper tubing, cables etc.

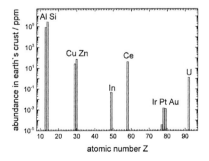

Figure 9. Natural abundance of selected metals in the earth's crust (given in logarithmic scale).[97] Copper is much more abundant (27 ppm) than iridium (0.000037 ppm) or platinum (0.0015 ppm).[92]

Although copper(I) exhibits only a small spin-orbit coupling constant when compared to iridium(III) (ξ_{Cu}=857 cm^{-1}, ξ_{Ir}=3909 cm^{-1}),[98] its complexes can harvest both triplet and singlet excitons for the emission of light and achieve internal quantum efficiencies of 100%[99,100] by utilizing the smaller spin-orbit coupling for thermally activated delayed fluorescence. This emission principle has proven to be an alternative strategy for high efficiencies based on more abundant starting materials with small spin-orbit coupling.[47,94] Furthermore, the concept of TADF enables naturally bluer emission and shorter emission decay times compared to phosphorescence and thus low roll-off in the running device at high current densities (see chapter 1.2.2). Remarkably, copper(I) complexes do not suffer from the quenching mechanisms found for triplet emitter, such as population of unoccupied dd* states in its d^{10}-configuration[101] or triplet-triplet annihilation. The metal copper(I) exhibits a broad structural diversity, which opens up a large field for the development of luminescent copper(I) complexes with desired properties, such as efficient luminescence at particular wavelengths regions (chapter 1.2.1).

1.2.1 Structural Diversity of Copper(I) Complexes

The structural diversity of copper(I) complexes ranges from mononuclear to polynuclear molecules, bridged or non-bridged by halides or ligands, trigonal to tetrahedral coordinated copper(I) centers (Figure 10). The flexible coordination numbers (2, 3, 4) and geometries (linear, trigonal, tetrahedral) of the copper(I) ions enable an especially high number of different coordination motifs with the halide ions of likewise flexible coordination numbers (left part of Figure 10).[103]

A plethora of copper(I) complexes based on imine (N), phosphine (P) or imine-phosphine (N^P) ligands surrounding a copper(I) halide core is available as emitting materials for OLEDs.[103-106] In particular, highly luminescent dinuclear complexes of the type [Cu$_2$X$_2$(P^N)(P)$_2$] (structure 9) with desired emission colors, solubilities and functional groups, e.g. for crosslinking, have been developed in the last few years,[72,94,107] but also dinuclear complexes [Cu$_2$X$_2$(P^N)$_2$] (structure 6) exhibit very promising photophysical properties with luminescence in the blue wavelength region and high quantum yields.[108] Tetranuclear cubane-like and octahedral structures (structures 13 and 14) hold interesting photophysical properties with a cluster-centered low-energy band and a high-energy band of halide-to-ligand charge-transfer character, whose relative intensities are temperature-dependent, but have not been applied to OLEDs yet.[105, 109 - 112] Despite their simple structures, mononuclear copper(I) complexes with halides (structures 1–3) have commonly been less considered as emitting materials for OLEDs due to their high tendency of dissociation reactions with monodentate ligands or cluster formation with bi- and tridentate ligands.[113-117] The emission properties of all the presented copper(I) complexes can easily be modulated by ligand variations, which influence the metal-to-ligand charge-transfer (MLCT), halide-to-ligand charge-transfer (XLCT), and metal- and halide-to-ligand charge-transfer ((ML+LX)CT) transitions responsible for emission from the excited states.

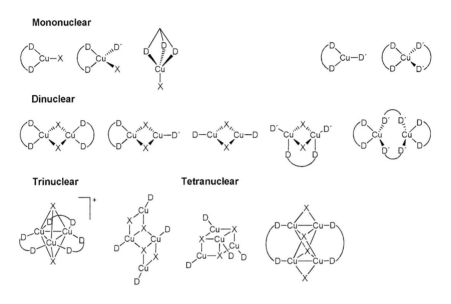

Figure 10. Common coordination modes of copper(I) complexes, where X denotes a halide ion and D, D′ are coordinating atoms of mono- or polydentate ligands (D, D′ = N, P). Non-halide copper(I) complexes can be cationic or neutral depending on the ligands' charge.

Besides the copper(I) halide complexes, a second class of copper(I) complexes based on non-halide counterions or, more recently, on anionic ligands has been developed. The mononuclear complexes of the type $[Cu(N{\wedge}N)_2]^+$, $[Cu(P{\wedge}P)_2]^+$ or $[Cu(N{\wedge}N)(P{\wedge}P)]^+$ (structure 5 in Figure 10) with symmetrical bis(imine) (N∧N) or bis(phosphine) (P∧P) ligands have been intensively studied by steady-state and time-resolved spectroscopy techniques (transient absorption, fluorescence upconversion, X-ray transient absorption) to understand the excited state processes and the influence of steric substituents on the ligands for applications in dye-sensitized solar cells (DSSC) and light-emitting electrochemical cells (LEECs).[118-124] By particular choice of anionic organic ligands, such as borates or deprotonated amines, and phosphines, neutral copper(I) complexes $[Cu(N{\wedge}N)(P{\wedge}P)]$ can be obtained, which show promising luminescence properties such as high efficiencies up to 90% and emission colors ranging from blue to red.[93,101,125-130] Further cluster-type coordination modes are also possible for copper(I) complexes without halides and usually realized by bis(phosphines) with small bite angles (e.g. structure 10),[131,132] but less common than for the halide analogues.

The huge range of coordination modes of copper(I) stimulates the search for new copper(I) complex classes with superior properties, but also challenges synthesis since the energetically favored coordination species might not be the desired one, and furthermore thermodynamic equilibria in solution can result in the formation of more than one complex species.[131-134] Thus, it is of high importance to understand the coordination chemistry of copper(I) and choose suitable reaction conditions for the formation of a sole complex species. It was found, that the stoichiometry plays an important role in the synthesis of polynuclear copper(I) halide complexes, but also the steric demand of applied ligands controls the formation of a particular cluster species.[105,134-137] Steric hindrance is the main parameter for the control of thermodynamic equilibria between heteroleptic $[Cu(N^\wedge N)(P^\wedge P)]^+$ and homoleptic $[Cu(N^\wedge N)_2]^+$ and $[Cu(P^\wedge P)_2]^+$ complexes with symmetrical phenanthrolines and bis(phosphines).[131] Especially when structure-property relationships are discussed, such as photoluminescence versus complex-type, the prediction of the coordination chemistry of copper(I) salts with particular ligands is crucial for the design of an emitter.

1.2.2 Thermally Activated Delayed Fluorescence[II]

As described above and in detail in chapter "Generation of light" (1.1.2), triplet emitters such as iridium(III) or platinum(II) complexes utilize the strong spin-orbit coupling of the heavy metal for efficient phosphorescence. The transition from the excited triplet to the ground state $T_1 \rightarrow S_0$ is more allowed and emission decay times are the shorter, the stronger the spin-orbit coupling induced by the metal ions. The coordination geometry, the ligand field splitting and the number of ligands control the extent of the SOC, but even more the atomic number of the metal contributes to the spin-orbit coupling by the fourth potency (SOC $\sim Z^4$).[48]

In contrast, luminescent copper(I) complexes with a small spin-orbit coupling (ξ_{Cu}=857 cm^{-1} versus ξ_{Pt}= 4481 cm^{-1}, ξ_{Ir}=3909 cm^{-1}),[98] also show short emission decay times of a few microseconds despite a largely forbidden $T_1 \rightarrow S_0$ transition. Due to small energy gaps ΔE_{ST} between the excited singlet and triplet states S_1 and T_1 and stable triplet states with few non-radiative processes, the copper(I) complexes can overcome the spin-forbidden phosphorescence by thermally activated delayed fluorescence from

---

[II] Parts of the introductory chapters "Thermally Activated Delayed Fluorescence" and "Intersystem Crossing in the Copper(I) Complex [(PyrTet)Cu(DPEPhos)]" have been published previously.[138]

the excited singlet S_1 (see Figure 11).[31] After excitation of the molecule or formation of an exciton in an OLED, the TADF emitter can either deactivate via spontaneous fluorescence (also called prompt fluorescence) or intersystem-cross to the triplet. The extent of these transitions depend on the ratio of their decay rates to each other, i.e. if the intersystem-crossing rate is high and thus the $S_1 \rightarrow T_1$ transition is faster than spontaneous fluorescence, no such fluorescence will be observed, as is the case for copper(I) complexes. At sufficiently high temperature the small energy gaps ΔE_{ST} can be overcome by thermal activation $k_B T$ and the triplet excitons transfer back to the excited singlet by reverse intersystem crossing (RISC) giving rise to thermally activated delayed fluorescence. This effect was first proposed for metal complexes by MCMILLIN in the early 1980s,[139] who studied the temperature-dependent luminescence and emission decay times of copper(I) bis(phenanthroline) complexes in solution, but is only recently used by several research groups to design highly efficient emitting materials based on the plentiful metal copper.[47,94,101,104,117,125,128,129,140]

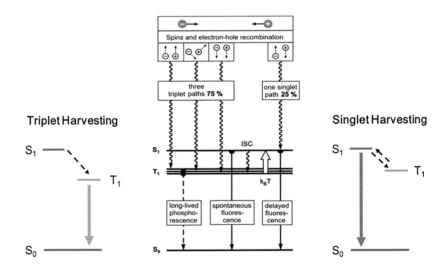

Figure 11. Comparison of the concepts of triplet and singlet harvesting: While for triplet emitters the phosphorescent transition $T_1 \rightarrow S_0$ is very effective, in singlet harvesting emitters a thermal up-conversion from the triplet T_1 back to singlet state S_1 is observed at ambient temperature giving rise to thermally activated delayed fluorescence.[141] Singlet harvesting enables copper(I) complexes to emit delayed fluorescence at shorter wavelengths (compared to phosphorescence), as the excited singlet is naturally higher in energy than the triplet, and ensures short emission decay times despite small spin-orbit coupling. It was shown, that copper(I) complexes can use both singlet and triplet excitons for the generation of light in OLEDs.[99,100]

Particular organic materials designed to have narrowed energy gaps between the excited singlet and triplet states have also been found to show thermally activated delayed fluorescence.[31,33,142,143] In all efficient TADF emitters a distinct charge-transfer (CT) character of the emission, which can be rationalized as a spatial separation of HOMO and LUMO orbitals on a donor and acceptor part of the molecule, lead to a small ΔE_{ST} and faster intersystem crossing. The major difference of copper(I) complexes versus organic TADF emitters is higher spin-orbit coupling and thus intrinsically faster ISC so that prompt fluorescence is not observed. In addition to the generally faster emission decay times, the ease of emission color tunability by ligand variation makes copper(I) complexes an attractive alternative approach to organic emitters.

When a TADF emitter is applied in an OLED device, the spin statistics lead to a lower ratio of spontaneous fluorescence (if any) to delayed fluorescence due to a population of the singlet of only 25% after the recombination of an electron and hole (see Figure 11). By utilizing the small spin-orbit coupling and narrow energy gaps ΔE_{ST}, luminescent copper(I) complexes can harvest both singlet and triplet excitons via the excited singlet state,[99,100] thus calling these materials singlet harvesting emitters.[47] The emission lifetimes can be as low as 1 or 2 microseconds and thus lie in the range of iridium and platinum complexes,[94] and further developments to reach much lower emission decay times by a combination of phosphorescence and singlet harvesting are undergoing.[144] Short emission decay times are crucial for the efficiency of an OLED to prevent the accumulation of excitons on the emitter, which can lead to emission quenching at high current densities (roll-off).

1.2.3 Photophysical Characterization of Thermally Activated Delayed Fluorescence in Copper(I) Complexes

Photophysical methods to investigate the thermally-activated delayed fluorescence behavior of luminescent materials are of high importance to gain a deeper understanding of the structure-property relationships and draw conclusions for the design of highly efficient emitters. The methods applied for luminescent copper(I) complexes are based on the two-state model proposed by DAVID R. MCMILLIN in 1983, where the singlet and triplet excited states – here denoted as 1 and 2 – are in thermal equilibrium, with the long-lived triplet state serving as exciton reservoir (Figure 12) for delayed fluorescence.[139]

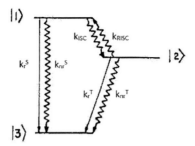

Figure 12. Two-state model in thermal equilibrium as proposed by McMillin.[139] He investigated the luminescence of copper(I) bis(imine) complexes of the type [Cu(N^N)$_2$]$^+$ in the temperature range –28–24 °C and found a red-shift of the emission maxima accompanied by an decrease of emission intensity at lower temperatures, which was ascribed to the emission from a lower-lying state of less radiative rate, namely the triplet. This triplet state is in thermal equilibrium with the higher-lying singlet state, so that delayed fluorescence from the singlet state is observed at sufficient high temperature.

A first hint on the TADF behavior of a material is given by a red-shift of the emission and a strong increase of the emission decay times when cooling down, until reverse ISC is completely disabled and phosphorescence occurs. Further confirmation is given by time-resolved photoluminescence measurements at ambient temperature, which show the same emission spectra for the spontaneous and delayed fluorescence, thus arising from the same excited state.[99] The common method to investigate TADF recently is temperature-dependent emission decay time measurements as reported by HARTMUT YERSIN.[30,93,108,145-148]

Two states of different energy, which are in fast thermal equilibrium (i.e. $k_{ISC} \gg k_r^S + k_{nr}^S$ and $k_{RISC} \gg k_r^T + k_{nr}^T$), can be understood as steady-state (after an initial time period) and thus described by Boltzmann statistics:[139]

$$K = \frac{N_1}{N_2} = \frac{g_1}{g_2} \cdot \exp\left(-\frac{\Delta E}{k_B T}\right) \approx \frac{k_{RISC}}{k_{ISC}} \qquad (6)$$

K: equilibrium constant, N: population of the respective states 1 or 2, g: degree of degeneracy of the excited state 1 or 2, ΔE: energy gap between the excited singlet and triplet states 1 and 2, k_B: Boltzmann constant, T: absolute temperature.

The temperature-dependent averaged decay-time of this system can be expressed by the following term (obtained from the rate equation for the total population N of the system):[149,150]

$$\tau_{ave} = \frac{g_2 + g_1 \cdot exp\left(-\frac{\Delta E}{k_B T}\right)}{K_2 + K_1 \cdot exp\left(-\frac{\Delta E}{k_B T}\right)} \qquad (7)$$

wherein $K_1 = k_r^S + k_{nr}^S$ and $K_2 = k_r^T + k_{nr}^T$. If a thermal population of the three substates of the triplet is assumed, on can use $g_2 = 3$ and $g_1 = 1$. By introduction of $\tau_1 = (K_1)^{-1}$ and $\tau_2 = (K_2)^{-1}$ as the lifetimes of the excited singlet and triplet states S_1 and T_1, Equation 7 can be simplified to:

$$\tau_{ave} = \frac{3 + exp\left(-\frac{\Delta E}{k_B T}\right)}{\frac{3}{\tau_2} + \frac{1}{\tau_1} \cdot exp\left(-\frac{\Delta E}{k_B T}\right)} \qquad (8)$$

The experimental data of emission decay times versus temperature can be fitted by Equation 8, as shown in Figure 13, and the two fit parameters ΔE and $\tau_1(S1)$ are obtained. This fit procedure is valid as long as the lifetimes of the excited states are temperature-independent, which might be problematic for the triplet state in some cases. Due to very weak prompt fluorescence, the lifetime of the excited singlet state cannot be measured directly for copper(I) complexes, but typical fit values range from 30–230 ns.[93,33] The energy splitting between the excited singlet and triplet state as well as the triplet decay rates control the lifetime of thermally activated delayed fluorescence, thus for efficient emitting materials ΔE values below 1000 cm^{-1} are favored.

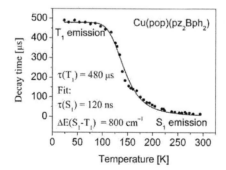

Figure 13. Emission decay times of the neutral copper(I) complex [Cu(pop)(pz$_2$Bph$_2$)] plotted against temperature in the range 25–300 K.[93] A strong increase of the emission lifetime from 13 μs to 480 μs is observed below 175 K, corresponding to emission of different origins, namely delayed fluorescence from the singlet and phosphorescence from the triplet state. By fitting Equation 8 to the experimental data, one obtains a lifetime of the singlet of 120 ns and an energy gap of 800 cm^{-1} as fitting parameters.

To further understand the interplay of all radiative and non-radiative transitions in the copper(I) TADF system, examinations on the non-emissive transitions intersystem crossing and reverse intersystem crossing are of high importance. So far, ISC rates have been reported only for the non-luminescent copper(I) bis(imine) complexes in solution, which were studied by time-resolved emission spectroscopy and transient absorption, determining the ISC rates to 7–15 ps.[124,151-155] Since in solution also other transitions occur such as a flattening distortion of the tetrahedral coordination around the copper(I) ion,[115,123,124,151,152,156,157] assignments of the observed decay dynamics are difficult.[154] Due to the higher degree of freedom in solution in general, the reported processes and attributed rate constants might not be transferable to the solid state in OLEDs, besides the fundamentally different emission properties of luminescent copper(I) complexes.

For organic TADF materials both the intersystem crossing and reverse rate constants can be obtained by comparison of the individual fractions of photoluminescence intensity by the spontaneous and delayed fluorescence and determination of their quantum yields. The derivation of the respective equations was reported by MÁRIO N. BERBERAN-SANTOS, who studied the thermally activated delayed fluorescence in fullerenes,[158] by ALFONS PENZKOFER for fluorescent dies,[159] and was recently revisited by CHIHAYA ADACHI for organic TADF materials.[31,32,160] However, the application of this method for copper(I) complexes is not possible due to the lack of spontaneous fluorescence in a system with very fast intersystem crossing rates.

2. Aims and Objectives

In the introductory chapters 1.1 and 1.2 the setup and working principle of an organic light-emitting diode have been presented, and material requirements and challenges for the design of highly efficient OLEDs explained. Focus was laid on the generation of light by an emitter molecule, which can be fluorescent or phosphorescent in origin, and thus defines the upper limit of exciton conversion into light to 25% or 100% in an OLED. Luminescent copper(I) complexes have been introduced as more abundant alternative to triplet emitters such as iridium(III) or platinum(II) complexes. Due to their superior emission properties such as high photoluminescence quantum yields close to unity and color tunability by ligand variation, research has focused on the development and photophysical examination of highly luminescent copper(I) compounds since 2004.[161] Furthermore, short emission decay times and internal efficiencies of theoretically 100%, when applied in an OLED, are facilitated by the utilization of thermally activated delayed fluorescence. Despite a broad knowledge on the cationic copper(I) bis(imine) complexes of the type $[Cu(N^{\wedge}N)]^+$, only few mononuclear neutral copper(I) complexes have been reported up to the beginning of this research work.[93,101,125,126,162,163] The aim of this work is thus to develop luminescent, mononuclear neutral copper(I) complexes and to study their coordination behavior, stability, and photoluminescence properties.

As first step, a synthetic route with appropriate anionic ligands has to be developed to obtain mononuclear neutral copper(I) complexes of the type $[(N^{\wedge}N)Cu(P^{\wedge}P)]$ or $[(N^{\wedge}N)Cu(P)_2]$. This is especially challenging in the light of oxygen sensitivity, which was found to be a problem for most of the neutral complexes being reported so far.[162-164] Focus will be laid on copper(I) complexes based on anionic, bidentate $N^{\wedge}N$ ligands with phosphines or bis(phosphines), respectively, since this combination induced strong luminescence in cationic, mononuclear complexes due to the energetic suppression of non-radiative deactivation.[165] To learn more about the complexes' chemical and photophysical behavior, different anionic ligands will be investigated and furthermore, the obtained complexes compared to their cationic counterparts $[(N^{\wedge}N)Cu(P^{\wedge}P)]^+$.

For the most promising complex class of mononuclear neutral complexes in view of luminescence and stability, a deeper understanding of the structure-property relationships is desired. While the emission color of the complexes can be tuned by ligand modulation of different electronic character, as was seen for dinuclear copper(I) complexes,[94,166] ligand substituents can also induce steric constraints, which influence the luminescence behavior.[157] One aim is to introduce various substituents on the most promising N^N ligand or vary its aromatic character and investigate both the steric as well as electronic effects on the complex's properties, and thus understand the excited state scheme of the complexes.

Besides the development of mononuclear neutral copper(I) complexes and the systematic examination of structure-property relationships for a broad range of ligands with different substituents, the thermally activated delayed fluorescence of copper(I) complexes will be focus of this work. This concept can either be studied by temperature-dependent measurements,[30] but further spectroscopic methods and setups have to be developed to also determine the non-radiative transitions occurring directly after excitation, namely intersystem crossing and reverse intersystem crossing. Since the ISC rate and the energy splitting between the excited singlet and triplet state control the efficiency of TADF, intersystem crossing will be examined for a model compound in different aggregation phases. This provides a deeper understanding of the interplay of all radiative and non-radiative transitions in the copper(I) TADF system, and furthermore conclusions can be drawn for the design of highly efficient emitting materials.

And finally, the herein developed and studied copper(I) complexes shall be applied to an organic light-emitting diode as emitting material as a proof-of principle experiment.

3. Results and Discussion

3.1 Development of Mononuclear Neutral Copper(I) Complexes based on various Anionic N^N Ligands[III]

Only few mononuclear neutral copper(I) complexes have been reported so far and they generally exhibit emission of low efficiency or are not luminescent, and suffer from a high sensitivity towards oxygen.[101,126,162-164] The recently published neutral complex [Cu(pop)(pz$_2$Bph$_2$)] based on the anionic bis(pyrazolyl)borate ligand exhibits blue emission and an exceptionally high photoluminescence quantum yield of 90%,[93] which stimulated the search for further mononuclear neutral copper(I) complexes. In this chapter, a systematic investigation of several anionic N^N ligands with various phosphines and their influence on the complexes' chemical and photophysical properties is reported.

To develop a synthesis route to mononuclear neutral copper(I) complexes, experimental results from cationic copper(I) complexes are considered: Bidentate ligands ensure low dissociation tendency in the complexes and thus improve their formation and stability. The combination of N^N and P^P ligands has shown to be promising for cationic complexes [(N^N)Cu(P^P)]$^+$, as it prevents non-radiative deactivation pathways from energetically high-lying MLCT states responsible for emission, and thus leads to higher emission efficiencies compared to the complexes of only bis(imines).[165] Steric bulkiness of the N^N ligands further hinders geometric changes after excitation, such as the proposed flattening distortion.[157] Thus, bidentate pyridine-amine ligands are chosen for the complexation with copper(I) salts in combination with neutral phosphines or bis(phosphines), respectively. Regarding their synthetic accessibility as well as easier introduction of substituents, variability of the aromatic character, or position of the coordination sites, anionic N^N ligands are advantageous over anionic N^P or P^P ligands (3.1.1).

As described in the introduction, copper(I) complexes exhibit a broad structural diversity, which can be challenging for the synthesis. Since the coordination motif controls the photoluminescence behavior of a complex, the formation of a particular complex species is desired, but this might not be the thermodynamically or kinetically favored one. Additionally, equilibria between different complex species are present in solution, so that a mixture of various complexes might be obtained. To chose the

---

[III] Parts of the chapter „Development of Mononuclear Neutral Copper(I) Complexes based on various Anionic N^N Ligands with Phosphines" have been published previously.[167]

right combination of ligands and reaction conditions, the coordination chemistry of copper(I) with these ligands has to be understood. Herein, both copper(I) complexes with the neutral and the deprotonated form of the N^N ligands with phosphines are structurally studied to derive general predictions for the formed coordination motifs and even more, design rules for mononuclear neutral complexes. From the comparison of both complex classes, differences in the coordination chemistry, the steric and electronic effects of the ligands for the formation of specific coordination motifs can be understood (3.1.2).

In chapter 3.1.3, the photoluminescence properties of all synthesized complexes are investigated and correlated to their molecular structures. DFT calculations are applied to explain their electronic transitions and to get a deeper understanding of the photophysics of mononuclear neutral copper(I) complexes and the influence of the ligands. And to conclude, their emission behavior is compared to the respective mononuclear cationic complexes based on neutral N^N ligands.

3.1.1 Synthesis of Ligands and Complexes

The simplest pyridine-imines, which can also act as anionic ligands in their deprotonated form, are pyridine-amines such as 2-(pyridin-2-yl)benzimidazole (PyrBimH, **5-H**), 2-(1,2,3-triazol-5-yl)pyridine (PyrTriH, **6-H**), 2-(tetrazol-5-yl)pyridine (PyrTetH, **7-H**) (Figure 14). Varying the number of N-atoms in these ligands changes the electron richness and thus donor ability, but also the acidity of the amine which can lead to spontaneous deprotonation during complexation. Besides the ligands' stereochemical and electronic influence on the complex coordination chemistry, a different photoluminescence behavior of the complexes is expected with increasing nitrogen ratio within a given coordination motif.

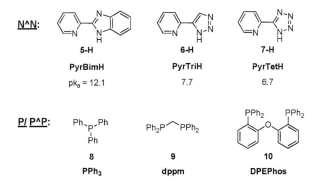

Figure 14. Pyridine-amine ligands (N^N) and their corresponding pk_a values, and phosphine (P) or bis(phosphine) (P^P) ligands, respectively.

The phosphines triphenylphosphine (PPh₃, **8**), bis(diphenylphosphino)methane (dppm, **9**) and bis(2-(diphenylphosphino)phenyl)ether (DPEPhos, **10**) are chosen to vary a) the denticity, b) the steric hindrance and c) the distance between the two coordinating P-atoms and thus, the bite angle. While the monodentate PPh₃ (**8**) exhibits a stronger tendency to undergo ligand exchange and dissociation reactions in the complex than chelating bis(phosphines), its large steric demand – reflected in the Tolman cone angle of 145°[168] – can be advantageous for the formation of particular complex species. The ligand dppm (**9**) usually acts as bridging ligand between two copper(I) centers due to its narrow natural bite angle of 72°, which would lead to a highly constrained coordination when acting as a bidentate ligand.[169-171] In contrast, DPEPhos (**10**) exhibits a flexible bite angle of 86–120° (natural bite angle of 102°) making it a good candidate for mononuclear copper(I) complexes to shield the metal center from further coordination.[172]

Scheme 1. Synthesis of the ligands 2-(1,2,3-triazol-5-yl)pyridine (**6-H**) and 2-(tetrazol-5-yl)pyridine (**7-H**) in [3+2] cycloaddition reactions as reported in literature.[173,174,175]

While the ligands 2-(1,2,3-triazol-5-yl)pyridine (**6-H**) and 2-(tetrazol-5-yl)pyridine (**7-H**) are obtained in [3+2] cycloaddition reactions from 2-ethynylpyridine (**11**) and 2-pyridinecarbonitrile (**12**) with an azide in moderate to good yields (Scheme 1),[173,174,175] the ligands 2-(pyridin-2-yl)benzimidazole (**5-H**), triphenylphosphine (**8**), bis(diphenylphosphino)methane (**9**) and bis[2-diphenylphosphino)phenyl]-ether (**10**) are commercially available.Synthesis of the cationic copper(I) complexes with PyrBimH (**5-H**), PyrTriH (**6-H**) or PyrTetH (**7-H**) is accomplished by dissolving the pyridine-amine and the bis(phosphine) ligands with the commercially available copper(I)salts [Cu(CH₃CN)₄]BF₄ (**13**) or [Cu(CH₃CN)₄]PF₆ (**14**) in the ratio of 1:1:1 or 1:2:1 for PPh₃, respectively, in dichloromethane/ethanol under nitrogen atmosphere and stirring for 6–12 h at room temperature (Scheme 2). The solvent mixture dichloromethane/ethanol has to be applied to dissolve all starting materials.

Copper(I) complexes with the anionic form of the N^N ligands, namely 2-(pyridin-2-yl)benzimidazolate (PyrBim, **5**), 5-(pyridin-2-yl)1,2,3-triazolate (PyrTri, **6**), 5-(pyridin-2-yl)tetrazolate (PyrTet, **7**), and the phosphines PPh₃ (**8**), dppm (**9**), DPEPhos (**10**) are obtained by deprotonation of the N^N ligand by a base (either *n*-butyllithium in the case of PyrBimH or potassium hydroxide for PyrTriH, PyrTetH) and subsequent complexation as described for the cationic complexes (Scheme 2).

After the reaction, inorganic salts and excess ligands are removed by evaporation of all solvents, extraction with few dichloromethane and precipitation of the filtrate in diethyl ether. For some complexes further purification is achieved by crystallization from a dichloromethane solution layered by diethyl ether. All complexes are obtained in high yields, as shown in Table 1.

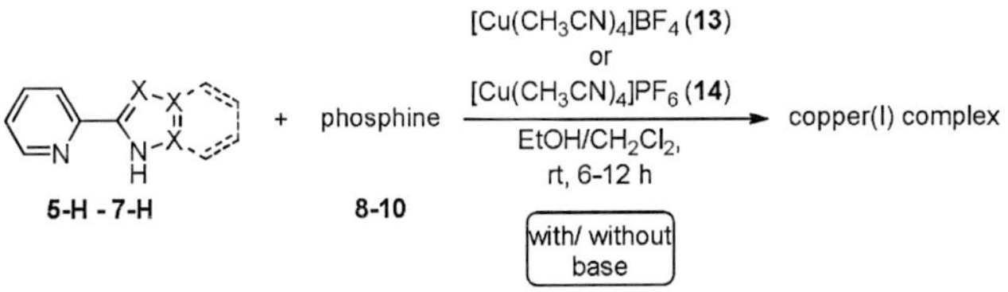

Scheme 2. General synthesis scheme for copper(I) complexes either with neutral (**5-H–7-H**) or anionic N^N ligands (**5–7**) and phosphines **8–10** (x denotes C or N).

All synthesized complexes are characterized by fast-atom bombardment mass spectrometry (FAB-MS), which usually gives the molecule, as well as fragmentations and clusters in the gas phase. The chemical structure is given by elemental analysis and X-ray diffraction. In particular for compounds obtained from precipitation, solvent residuals have to be taken into account for elemental analysis, which can also be reflected in X-ray crystal structures as crystal solvent. Crystals for X-ray diffraction structure analysis are usually obtained by layering diethyl ether upon a dichloromethane solution of the complex.

The complexes are denoted as *X-Cu-Y*, where X means the neutral pyridine-amine (**5-H, 6-H, 7-H**) or the deprotonated pyridine-amide (**5, 6, 7**) ligand, and Y is the phosphine ligand (**8, 9, 10**). Since [Cu(CH$_3$CN)$_4$]BF$_4$ (**13**) is used as standard copper(I) source, the counterion is only added to the numbering in the case of PF$_6^-$, e.g. **c-(5-Cu-8-PF$_6$)**. If the compound exhibits a cluster-type structure, *c-* is used as prefix. Thereby, the coordination motif (mononuclear cationic or neutral vs. cluster-type) as well as the applied ligands (and the counterion) can be deduced from the designation. However, in the case of cluster complexes, this numbering does not reflect the ratios of copper(I) versus ligands in the molecular structure or the charge of the complexes.

Table 1. Reaction yields of the synthesized complexes based on pyridine-amines **5-H–7-H** or pyridine-amides **5–7** from [Cu(CH₃CN)₄]BF₄ (**13**).

Entry		Complex	Yield	Entry		Complex	Yield
PyrBimH (5-H)	PPh₃ (8)	5-H-Cu-8	84%	PyrBim (5)	PPh₃ (8)	c-(5-Cu-8-PF₆)	82%
PyrTriH (6-H)		6-H-Cu-8	61%	PyrTri (6)		c-(6-Cu-8-PF₆)	62%
PyrTetH (7-H)		7-H-Cu-8	70%	PyrTet (7)		7-Cu-8	70%
PyrBimH (5-H)	dppm (9)	c-(5-H-Cu-9)	81%	PyrTri (6)	dppm (9)	c-(6-Cu-9)	78%
PyrBimH (5-H)	DPEPhos (10)	5-H-Cu-10	90%	PyrTet (7)		c-(7-Cu-9)	78%
PyrTriH (6-H)		6-H-Cu-10	84%	PyrBim (5)	DPEPhos (10)	5-Cu-10	34%
PyrTetH (7-H)		7-H-Cu-10	57%	PyrTri (6)		6-Cu-10	73%
				PyrTet (7)		7-Cu-10	68%

ª Complexes obtained from [Cu(CH₃CN)₄]PF₆ (**14**).

As the coordination chemistry of copper(I) strongly depends on the nature of the N^N ligand, in the next section first the complexes with neutral pyridine-amine ligands are discussed, followed by the section on copper(I) complexes with the anionic ligand form. When discussing N^N ligands the neutral form is meant in chapter 3.1.2.1, but the deprotonated species in chapter 3.1.2.2.

3.1.2 Structural studies

3.1.2.1 Copper(I) Complexes with Pyridine-amine Ligands

From reaction of 2-(pyridin-2-yl)benzimidazole (**5-H**) with two equivalents of PPh₃ (**8**) and [Cu(CH₃CN)₄]BF₄ (**13**), a yellow compound is obtained for which elemental analysis as well as single crystal X-ray diffraction give the molecular structure [(PyrBimH)Cu(PPh₃)₂]BF₄ (**5-H-Cu-8**). The crystal structure is shown in Figure 15 (already published but without data on the benzimidazole-(N)*H*)[176,177] revealing a distorted tetrahedral coordination of the central copper(I) atom by the bidentate 2-(Pyridin-2-yl)benzimidazole and two PPh₃ molecules. The Cu-N and Cu-P bonds (Cu-N1: 2.098 Å, Cu-N15: 2.086 Å, Cu-P1: 2.253 Å and Cu-P2: 2.258 Å) are comparable to other mononuclear copper(I) complexes,[123,125,126] see also Table 2 for details. Whereas the two bulky PPh₃ ligands span up a large P-Cu-P angle, the bite angle of the N^N ligand is much smaller due to its steric constraint (N-Cu-N: 80.4°, P-Cu-P: 127.0°). The pyridine and benzimidazole rings lie almost coplanar to each other with an angle of only 6.5° between the two ring planes (torsion angle N1-C9-C10-N15). A comparable complex based on a 2-(2′-quinolyl)benzimidazole ligand exhibits a larger torsion angle of 9.1° due to stronger steric repulsion of the quinoline with the phosphine ligands.[130] The N-Cu-N and P-Cu-P planes are perpendicular to each other, determined by a dihedral angle of 88.7°. The counterion BF_4^- interacts with the N^N ligand via hydrogen bridge bonds to the hydrogen (N)H of the imidazole ring and two hydrogens of the pyridine ring (F1-H(N8): 1.91 Å), and is even detectable in the mass spectrum of complex **5-H-Cu-8**.

Applying 2-(1,2,3-triazol-5-yl)pyridine (**6-H**) with $[Cu(CH_3CN)_4]BF_4$ (**13**) and two equivalents of PPh₃ (**8**) also gives a mononuclear complex of the composition $[(PyrTriH)Cu(PPh_3)_2]BF_4$ (**6-H-Cu-8**) as a brownish powder. When using the copper(I) salt $[Cu(CH_3CN)_4]PF_6$ (**14**) instead, the analogous compound **6-H-Cu-8-PF₆** with the counterion PF_6^- instead of BF_4^- is obtained. Figure 15 shows the molecular structure of **6-H-Cu-8-PF₆**. In compound **6-H-Cu-8-PF₆** the copper(I) atom is tetrahedrally coordinated by the N^N ligand with a bite angle of 79.7°, similar to compound **5-H-Cu-8**, and by two PPh₃ molecules with an P-Cu-P angle of 123.3° (Figure 15, Table 2). The sterically less bulky triazole compared to the afore-discussed benzimidazole results in the copper(I) atom moving slightly closer to the triazole ring, but further away from the pyridine ring (Cu-N1: 2.064 Å, Cu-N11: 2.130 Å). The Cu-P bond lengths are in a similar range as for compound **5-H-Cu-8**, but the phosphines lie closer to each other (**5-H-Cu-8**: P1-P2: 4.038 Å, **6-H-Cu-8-PF₆**: P1-P2: 3.952 Å) and thus exhibit a smaller bite angle (**5-H-Cu-8**: P1-Cu-P2: 127.0°, **6-H-Cu-8-PF₆**: P1-Cu-P2: 123.3°). The triazole and pyridine ring of the N^N ligand lie almost perfectly coplanar to each other (N1-C5-C6-N11: 2.8°) and span up a plane which is perpendicular to the P-Cu-P plane (angle: 88.57°). Also in this case, a hydrogen bridge interaction is found between the counterion PF_6^- with the acidic (N)H of the triazole ring, but difficult to examine in detail due to disorder of the counterion.

Compound **7-H-Cu-8** is obtained by complexation with 2-(tetrazol-5-yl)pyridine (**7-H**) and two equivalents of PPh₃ (**8**) as a white powder.[167] Elemental analysis and crystal structure analysis reveal a cationic, mononuclear $[(PyrTetH)Cu(PPh_3)_2]BF_4$ composition, where the central copper(I) atom is tetrahedrally coordinated by the N^N ligand **7-H** and two phosphine ligands **8** (Figure 15, Table 2). In comparison to the before discussed mononuclear complexes **5-H-Cu-8** and **6-H-Cu-8-PF₆**, the copper(I) center moves even slightly further away from the pyridine moiety of the N^N ligand, although steric demands of the PyrTriH and PyrTetH are quite similar (**6-H-Cu-8-PF₆**: Cu1-N11: 2.130 Å, **7-H-Cu-8**: Cu1-N11: 2.140 Å). This also results in two different Cu-P bond lengths, more pronounced in **7-H-Cu-8** than for compound **6-H-Cu-8-PF₆** (**6-H-Cu-8-PF₆**: Cu-P1: 2.239 Å, Cu-P2: 2.252 Å, **7-H-Cu-8**: Cu-P1: 2.234 Å, Cu-P2: 2.272 Å). In contrast to **5-H-Cu-8** and **6-H-Cu-8-PF₆**, the tetrazole and pyridine rings of PyrTetH are even more coplanar to each other, but the tetrahedral coordination of the N^N and P^P ligands is slightly more distorted due to the less symmetric coordination (angle: 85.2°).

As expected from analysis of the mass spectrum of complex **7-H-Cu-8**, a strong electrostatic interaction of the counterion with the acidic (N)*H* atom of the tetrazole moiety is observed in the crystal structure (F1-H(N4): 1.94 Å).

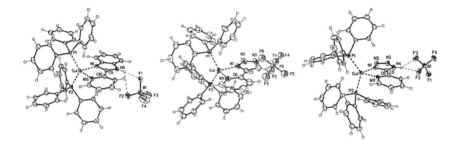

Figure 15. Molecular structures of the mononuclear complexes [(PyrBimH)Cu(PPh$_3$)$_2$]BF$_4$ (**5-H-Cu-8**), [(PyrTriH)Cu(PPh$_3$)$_2$]PF$_6$ (**6-H-Cu-8-PF$_6$**) and [(PyrTetH)Cu(PPh$_3$)$_2$]BF$_4$ (**7-H-Cu-8**) from left to right. Displacement parameters are drawn at 50% probability level.[IV]

Table 2. Selected bond lengths (Å) and angles (°) of structurally characterized mononuclear complexes [(pyridine-amine)Cu(PPh$_3$)$_2$]X (with X = BF$_4$ or PF$_6$).

	5-H-Cu-8	**6-H-Cu-8-PF$_6$**	**7-H-Cu-8**
Cu1-N1	2.0977 (15)	2.064 (3)	2.0794 (14)
Cu1-N11	2.0857 (14)[a]	2.130 (3)	2.1396 (14)
Cu1-P1	2.2531 (5)	2.2385 (9)	2.2342 (5)
Cu1-P2	2.2582 (6)	2.2521 (9)	2.2724 (5)
P1-P2	4.0377 (8)	3.952 (1)	4.082 (7)
F1-H(N)	1.91 (2)[b]	1.97 (3)[c]	1.94 (2)[d]
F2-H(N)	2.55 (2)[b]	1.91 (4)[c,e]	2.53 (2)[d]
		2.63 (4)[c,f]	
N1-Cu-N11	80.40 (6)g	79.69 (12)	79.01 (5)
P1-Cu-P2	127.02 (2)	123.28 (3)	129.93 (2)
φ[h]	88.72 (4)	88.57 (9)	85.21 (4)
N1-C5-C6-N11	6.5 (2)[i]	2.8 (5)	−2.01 (2)

[a] Cu1-N15; [b] F-H(N8); [c] F-H(N3); [d] F-H(N4); [e] F2′-N(H); [f] F5′-H(N); [g] N1-Cu-N15; [h] angle between N–Cu–N and P–Cu–P planes; [i] N1-C9-C10-N15.

[IV] The herein shown crystal structures were obtained in collaboration with Dr. Martin Nieger, University of Helsinki. Thereby the crystal growth was performed by the author of this thesis, while X-ray diffraction was measured by Dr. Martin Nieger (see also 5.1.1).

To sum up, complexation of the pyridine-amine ligands PyrBimH (**5-H**), PyrTriH (**6-H**) and PyrTetH (**7-H**) with two equivalents of PPh$_3$ (**8**) and copper(I) salt results in the formation of cationic mononuclear complexes with a distorted tetrahedral geometry around the copper(I) center where the angle between the N-Cu-N and P-Cu-P planes is close to 90° and the two aromatic rings of the N^N ligand lie almost coplanar. Additionally, for all three complexes interactions of the counterions with the acidic (N)*H* atoms of the pyridine-amine are found and thus the counterions are even detectable by mass spectroscopy.

Going further to copper(I) complexes with the bis(phosphine) ligand dppm (**9**), it is found that only with the pyridine-amine ligand PyrBimH (**5-H**) a complex is formed based on neutral N^N ligands, whereas the ligands PyrTriH (**6-H**) and PyrTetH (**7-H**) are deprotonated during reaction giving complexes, which will be discussed in detail in section 3.1.2.2. In Figure 16 the crystal structure of **c-(5-H-Cu-9-PF$_6$)** – the analogue to compound **c-(5-H-Cu-9)** with PF$_6^-$ counterions – is shown. Complex **c-(5-H-Cu-9)** is obtained by reaction of PyrBimH (**5-H**), copper(I) salt **13** and dppm (**9**) in the ratio of 1:1:1 in good yields and is characterized as [(PyrBimH)$_2$Cu$_2$(dppm)$_2$](BF$_4$)$_2$. The crystal structure shows two complex units, where each copper(I) atom is surrounded by the 2-(pyridin-2-yl)benzimidazole and a phosphorous atom of the two dppm ligands in a distorted tetrahedral coordination. The complex units are bridged via the bis(phosphine) ligands. This is a common structural motif for copper(I) complexes with dppm.[131,132,169,170] The natural bite angle of dppm is too narrow to coordinate to one copper(I) center and rather acts as bridging ligand of two metal atoms. In **c-(5-H-Cu-9-PF$_6$)** the dppm ligand is spanned up to a quasi-tetrahedral P1-C1-P2 angle of 116.5° (tetrahedral angle: 109°) with a distance between the phosphine atoms of only 3.121 Å (compared to P1-P2 distances around 4 Å in the case of PPh$_3$ complexes). The two PyrBimH ligands itself are virtually coplanar regarding the pyridine versus benzimidazole ring (torsion angle N26-C34-C35-N40: –3.3°), and furthermore lie parallel to each other (Table 3). The two dppm ligands lie perpendicular to this N-Cu-N plane (above and below) and interconnect the two complex units (angle: 89.2°). As in all other complexes **5-H-Cu-8**, **6-H-Cu-8-PF$_6$**, **7-H-Cu-8** with pyridine-amine ligands so far, the counterion PF$_6^-$ interacts with the (N)*H* and hydrogens of the pyridine or benzene ring, respectively, via hydrogen bridge bonds.

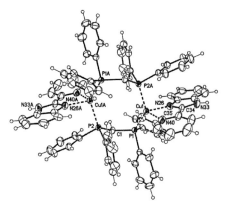

Figure 16. Molecular structure of the complex [(PyrBimH)$_2$Cu$_2$(dppm)$_2$](PF$_6$)$_2$ (**c-(5-H-Cu-9-PF$_6$)**). Displacement parameters are drawn at 50% probability level (solvent and anion omitted for clarity).[IV]

Table 3. Selected bond lengths (Å) and angles (°) of the structurally characterized dinuclear complex [(PyrBimH)$_2$Cu$_2$(dppm)$_2$](PF$_6$)$_2$ (**c-(5-H-Cu-9-PF$_6$)**).

Bond lengths [Å]		Angles [°]	
Cu1-N26	2.083 (3)	N26-Cu1-N40	79.05 (11)
Cu1-N40	2.118 (3)	P1-Cu1-P2A	132.45 (3)
Cu1-P1	2.2087 (8)	φ_1[a]	89.21 (9)
Cu1-P2	2.2540 (8)	N26-C34-C35-N40	−3.3 (5)
P1-P2	3.121 (1)	P1-C1-P2	116.47 (15)

[a] Dihedral angle between N–Cu–N and P–Cu–P planes.

Continuing with the coordination chemistry of copper(I) with bidentate bis(phosphines), the complexes based on pyridine-amines PyrBimH (**5-H**), PyrTriH (**6-H**) and PyrTetH (**7-H**) with the sterically more flexible, but bulkier DPEPhos (**10**) will be discussed in the next paragraph. From reaction of 2-(pyridin-2-yl)benzimidazole (**5-H**) with [Cu(CH$_3$CN)$_4$]BF$_4$ (**13**) and one equivalent of DPEPhos (**10**) a yellow powder is obtained, which is characterized as mononuclear complex [(PyrBimH)Cu(DPEPhos)]BF$_4$ (**5-H-Cu-10**) by single crystal X-ray diffraction and elemental analysis (Figure 17, Table 4). PyrBimH coordinates the copper(I) atom with a similar bite angle as in the PPh$_3$ complex **5-H-Cu-8** (N1-Cu-N11: **5-H-Cu-8** 80.4°, **5-H-Cu-10** 80.5°), but interestingly with a Cu-N bond length slightly shorter to the benzimidazole than to the pyridine ring (Cu1-N1: 2.063 Å, Cu1-N15: 2.098 Å). The bidentate DPEPhos moves closer to the central copper(I) atom with Cu-P bond lengths of around 2.24 Å, and exhibits a small bite angle of 116.1° due to the less steric demand compared to two PPh$_3$ molecules, which is also reflected in the short phosphorous atoms distance of 3.800 Å (in **5-H-Cu-8**, **6-H-Cu-8-PF$_6$**, **7-H-Cu-8**: 3.95–4.08 Å). In the crystal structure, the benzene and pyridine ring of the PyrBimH ligand of two adjacent complex molecules overlap each other, which is also observed for the PPh₃ complex **5-H-Cu-8**. This is further stabilized by electrostatic interactions of the BF_4^- counterion with hydrogens of the two adjacent PyrBimH ligands in the crystal.

Complexation of $[Cu(CH_3CN)_4]BF_4$ (**13**) with 2-(1,2,3-triazol-5-yl)pyridine (**6-H**) and DPEPhos (**10**) in the ratio 1:1:1 gave a brownish powder. Further purification by recrystallization of its dichloromethane solution layered by diethyl ether afforded crystals of the composition $[(PyrTriH)Cu(DPEPhos)]BF_4$ (**6-H-Cu-10**), which is proven by single-crystal X-ray diffraction and elemental analysis (Figure 17). The molecular structure shows a mononuclear, tetrahedral copper(I) coordination, where the PyrTriH ligand coordinates the central copper(I) atom with a bite angle of 79.4°, and with equal Cu-N bond lengths (Cu-N1: 2.092 Å, Cu-N1: 2.087 Å) in contrast to its PPh₃ complex **6-H-Cu-8-PF₆** (Cu-N1: 2.064 Å, Cu-N1: 2.130 Å). The smaller steric hindrance of the DPEPhos ligand in **6-H-Cu-10** results in a narrower bite angle of 115.3° and a shorter distance of the two coordinating phosphorous atoms (P1-P2: 3.833 Å). As expected from the analysis of the mass spectrum, a strong interaction of the counterion BF_4^- with the acidic (N)*H* atom of the PyrTriH ligand is found for **6-H-Cu-10** as well, but at a longer distance than in **6-H-Cu-8-PF₆**, which might also be a reason for the less symmetric coordination of the N^N ligand in the PPh₃ complex **6-H-Cu-8-PF₆** (see Table 4). When comparing the coordination geometry of the complex $[(PyrTriH)Cu(DPEPhos)]BF_4$ (**6-H-Cu-10**) with its analogous complex **5-H-Cu-10** based on PyrBimH (**5-H**), a slight distortion away from the copper(I) atom is observed for the triazole moiety (Cu-N1: **5-H-Cu-10** 2.063, **6-H-Cu-10** 2.092 Å), while the pyridine maintains similar coordination bond lengths (Cu-N11: **5-H-Cu-10** 2.098, **6-H-Cu-10** 2.087Å).

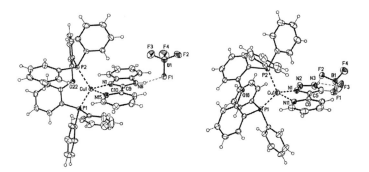

Figure 17. Molecular structures of the mononuclear complexes $[(PyrBimH)Cu(DPEPhos)]BF4$ (**5-H-Cu-10**) and $[(PyrTriH)Cu(DPEPhos)]BF4$ (**6-H-Cu-10**). Displacement parameters are drawn at 50% probability level (solvent molecules omitted for clarity).[IV]

To complete this series, also complex **7-H-Cu-10** is discussed here, which is obtained from reaction of copper(I) salt **13** with DPEPhos (**10**) and 2-(tetrazol-5-yl)pyridine (**7-H**).[167] Surprisingly, the compound shows a dimeric structure, where two individual complex units are linked via a hydrogen atom shared by two tetrazole ligands (Figure 18). The composition of the compound is denoted as [(PyrTetH)Cu(DPEPhos)][(PyrTet)Cu(DPEPhos)]PF$_6$ as the hydrogen is closer to one of the complex units than to the other (N46-H46: 0.78 Å, N46A-H46: 1.78 Å) (Table 4). In both units the copper(I) atom is coordinated by the DPEPhos and PyrTetH ligand in a distorted tetrahedral geometry, with bite angles comparable to the beforehand discussed copper(I) complexes (N-Cu-N: 80.0°, P-Cu-P: 115.4°). The two hydrogen-bridged PyrTet ligands lie parallel to each other, but are slightly twisted in themselves due to strong steric constraints in this coordination motif (torsion angle 7.3°). The orientation of the two N^N ligands towards each other further leads to a slightly longer Cu-N(tetrazole) versus Cu-N(pyridine) bond (Cu-N47: 2.095 Å, Cu-N37: 2.046 Å) and unequal coordination to the phosphorous atoms of DPEPhos (Cu-P1: 2.245 Å, Cu-P2: 2.200 Å). Since both tetrazole rings are shielded by copper(I) and a DPEPhos shell, interaction of the counterion with the N^N ligand or especially (N)*H* is not possible. In contrast, a mononuclear cationic complex is reported for similar pyridine-tetrazole ligands with DPEPhos, where the tetrazole (N)*H* is substituted by a *tert*-butyl group in 3-position. Thus no deprotonation or further coordination to other N-atoms in the tetrazole ring are possible (due to steric hindrance), leading to the formation of cationic monomers.[132] In the case of PyrTetH (**7-H**) the deprotonation of half of the complexes is favored and the quasi-monomeric complex molecules are linked via a hydrogen bond bridge.

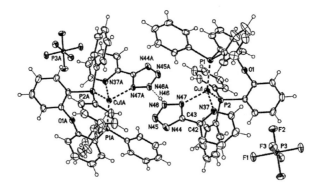

Figure 18. Molecular structure of the dimeric complex [(PyrTetH)Cu(DPEPhos)][(PyrTet)Cu(DPEPhos)]PF$_6$ (**7-H-Cu-10-PF$_6$**), here with PF$_6^-$ counterion. Displacement parameters are drawn at 50% probability level (solvent molecules omitted for clarity).[IV]

Table 4. Selected bond lengths (Å) and angles (°) of the structurally characterized mononuclear complexes [(pyridine-amine)Cu(DPEPhos)]X (with X = BF_4^- or PF_6^-).

	5-H-Cu-10	6-H-Cu-10	7-H-Cu-10-PF6
Cu1-N1	2.0626 (16)	2.0915 (14)	2.095 (2)[a]
Cu1-N11	2.0981 (15)[b]	2.0865 (14)	2.046 (3)[c]
Cu1-P1	2.2372 (5)	2.2611 (5)	2.2451 (8)
Cu1-P2	2.2413 (5)	2.2758 (5)	2.2002 (9)
P1-P2	3.8004 (8)	3.8331 (6)	3.7580 (11)
N1-Cu-N11	80.47 (6)[d]	79.40 (6)	79.99 (10)[e]
P1-Cu-P2	116.12 (2)	115.31 (2)	115.43 (3)
φ[f]	89.68 (5)	87.22 (4)	87.65 (6)
N1-C5-C6-N11	−0.8 (2)[g]	4.0 (2)	7.32 (4)[h]

[a] Cu1-N47; [b] Cu1-N15; [c] Cu1-N37; [d] F-H(N8); [d] N1-Cu-N15; [e] N47-Cu1-N37; [f] angle between N–Cu–N and P–Cu–P planes; [g] N1-C9-C10-N15; [h] N47-C43-C42-N37.

A mononuclear cationic complex structure is found for the reactions of copper(I) salt **13** with the bidentate DPEPhos (**10**) and PyrBimH (**5-H**) or PyrTriH (**6-H**), and a quasi-mononuclear motif for PyrTetH (**7-H**). In all compounds the copper(I) atom is coordinated by the pyridine-amine ligand with a narrow bite angle close to 80° and by DPEPhos with bite angles around 115° in a quasi tetrahedral coordination. Both ligands lie perpendicular to each other, with the N^N ligand slightly tilted when steric constraints are high (e.g. **7-H-Cu-10-PF6**). Compared to the respective complexes with PPh3 **5-H-Cu-8**, **6-H-Cu-8-PF6**, **7-H-Cu-8**, the steric demand of DPEPhos is smaller, determined by shorter Cu-P bond lengths on average and smaller bite angles (115–116° vs. 123–130°).

To sum up the coordination chemistry of copper(I) with pyridine-amine ligands **5-H–7-H** and different phosphines **8–10**, it can be found that steric over electronic factors control the coordination motifs of the formed complexes. Varying the number of N-atoms of the amine does not significantly influence the coordination behavior towards copper(I) but moreover the acidity of the ligand resulting in spontaneous deprotonation, as was seen for the combination of PyrTriH (**6-H**) and PyrTetH (**7-H**) with dppm (**8**), and partially in **7-H-Cu-10-PF6**. Since the steric demand of the applied pyridine-amine ligands PyrBimH (**5-H**), PyrTriH (**6-H**) and PyrTetH (**7-H**) is similar, same coordination geometries with comparable bond lengths and angles are expected for the same phosphine type, as shown for **5-H-Cu-8**, **6-H-Cu-8-PF6**, **7-H-Cu-8** and **5-H-Cu-10**, **6-H-Cu-10**, **7-H-Cu-10-PF6**. Benzannulation in the case of PyrBimH (**5-H**) gives same coordination motifs but changes the packing in the crystal structure. In contrast to the small steric influence of the pyridine-amine ligands, the phosphine has a major steric impact on the formed complex species. The following general statements for the coordination geometry of copper(I) can be made (Scheme 3): a) With two equivalents of bulky PPh3 (**8**) mononuclear cationic complexes are obtained. b) With the bidentate, but very rigid dppm (**9**) cluster-type structures are obtained, where dppm acts as a bridging ligand. For copper(I) complexes with the ligands PyrTriH (**6-H**) and PyrTetH (**7-H**) deprotonation is observed with dppm (**9**), which is either an intrinsic electronic effect of dppm, or might be a result of cluster formation when more coordinating N-atoms are present in the N^N ligand leading to an acidification of the amine moiety by copper(I) coordination. c) Monomeric or quasi-monomeric cationic coordination modes are accessible with the flexible DPEPhos (**10**) ligand, which can adapt its geometry to the actual steric demand around the copper(I) center.

Denticity of the phosphine ligand does not seem to play a role for the coordination modes, but controls the extent of dissociation processes, e.g. copper(I) complexes with monodentate PPh₃ are prone to dissociation and ligand exchange reactions.[113,133,134] DPEPhos is a chelating ligand and additionally, its homoleptic complex [Cu(DPEPhos)₂]⁺ is destabilized due to strong steric constraints, which leads to the formation of energetically favored heteroleptic complexes with DPEPhos.[178] How this general Scheme 3 compares to copper(I) complexes with the anionic ligands PyrBim (**5**), PyrTri (**6**), PyrTet (**7**) and phosphines **8–10** will be examined in the following section 3.1.2.2.

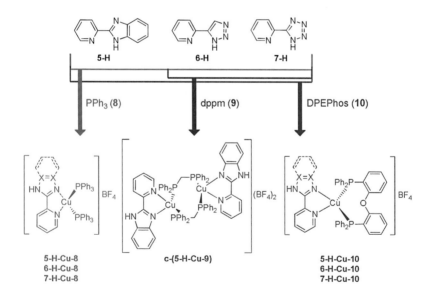

Scheme 3. The coordination chemistry of copper(I) with the pyridine-amine ligands PyrBimH (**5-H**), PyrTriH (**6-H**), PyrTetH (**7-H**) and monodentate or bidentate phosphines **8–10**.

3.1.2.2 Copper(I) Complexes with Anionic Pyridine-amide Ligands

Reaction of [Cu(CH3CN)4]PF6 (13) with the anionic ligand PyrBim (5) and two equivalents of PPh3 (8), followed by crystallization by layering a dichloromethane solution of the yellow compound with diethyl ether gave yellow crystals of the composition [(PyrBim)3Cu4(PPh3)6]BF4 (c-(5-Cu-8-PF6)), as shown in Figure 19. The crystal structure reveals a highly symmetric, star-shaped structure, where three copper(I) atoms are tetrahedrally coordinated by a PyrBim ligand and two PPh3 molecules, and a fourth copper atom interconnects the three units. Bond lengths and angles in the tetrahedrally coordinated molecule parts are comparable to similar copper(I) complexes (see Table 5, Cu1-N: 2.04–2.14 Å, Cu1-P: 2.26 Å, N-Cu1-N: 80.1°, P-Cu1-P: 118.4°, see also before discussed compounds). Compared to compound 5-H-Cu-8 with neutral PyrBimH, the two phosphine molecules slightly move away from the copper(I) atom resulting in surprisingly small bite angles of 118.4° (5-H-Cu-8: 127.0°) at exactly same Cu-P bond lengths. The fourth copper(I) atom acting as tie point is symmetrically surrounded by the PyrBim ligands of the tetrahedral copper(I) units. It is trigonally planar coordinated to the second N-atom of the benzimidazolate moiety, which is available as coordination site in the deprotonated form of the ligand (N-Cu2-N: 119.7°). Due to high steric constraints, the benzimidazole and pyridine rings are twisted away from each other with a torsion angle of 12.0°.

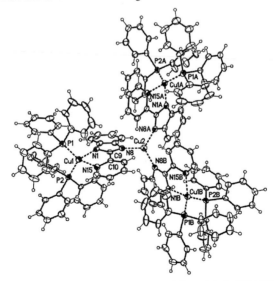

Figure 19. Molecular structure of the tetranuclear compound [(PyrBim)3Cu4(PPh3)6]PF6 (c-(5-Cu-8-PF6)). Displacement parameters are drawn at 50% probability level, 30% for PF6, (solvent and anions omitted for clarity).[IV]

Table 5. Selected bond lengths (Å) and angles (°) of the structurally characterized tetranuclear complex [(PyrBim)₃Cu₄(PPh₃)₆]PF₆ (**c-(5-Cu-8-PF₆)**).

	Bond lengths [Å]		Angles [°]
Cu1-N1	2.036 (6)	N1-Cu1-N15	80.1 (2)
Cu1-N15	2.138 (5)	P1-Cu1-P2	118.37 (7)
Cu1-P1	2.263 (2)	$\varphi_1{}^a$	83.1 (2)
Cu1-P2	2.263 (2)	N1-C9-C10-N15	12.0 (9)
P1-P2	3.887 (2)	N8-Cu2-N8A	119.70 (3)
Cu2-N8	1.980 (5)	N8A-Cu2-N8B	10.9 (2)
Cu2-N8A	2.60	N8-Cu2-N8B	80.1 (2)
Cu2-N8B	3.01	$\varphi_1{}^b$	118.37 (7)
Cu2-H11	2.036 (6)		
Cu2-H6	2.138 (5)		

[a] Dihedral angle between N–Cu1–N and P–Cu1–P planes; [b] angle between N–Cu2–N and N–Cu2–N planes.

Copper(I) complex **c-(6-Cu-8-PF₆)** based on the pyridine-amide PyrTri (**6**) with PPh₃ (**8**) is obtained by deprotonation of its cationic precursor [(PyrTriH)Cu(PPh₃)₂]PF₆ (**6-Cu-8-PF₆**) by potassium hydroxide in methanol, and precipitation from diethyl ether. It is further purified by crystallization from dichloromethane layered with diethyl ether giving transparent crystals. Elemental analysis and single crystal X-ray diffraction reveal a trinuclear coordination motif of the composition [(PyrTri)₂Cu₃(PPh₃)₆]PF₆ (Figure 20). Similar to the above discussed benzimidazolate based complex **c-(5-Cu-8-PF₆)**, two quasi-tetrahedral coordinated units are found, which are interconnected by a third copper(I) atom. In contrast to **c-(5-Cu-8-PF₆)**, the copper(I) atom acting as tie point does not exhibit a trigonal planar coordination from three N^N ligands, but saturates its coordination by two PyrTri ligands and two PPh₃ molecules in a tetrahedron. This can be explained by the smaller size of PyrTri versus PyrBim and thus a threefold coordination of the interconnecting copper(I) atom would not shield it enough but leave space for a fourth coordination. Also the different position of the third coordinating N-atom might influence the formed coordination motif. While Cu-P and Cu-N bond lengths and angles in the tetrahedral units are comparable to the before discussed complexes (see Table 6; Cu1-P: 2.2–2.3 Å, Cu1-N: 2.0–2.1Å, N-Cu1-N: 80.2°, P-Cu1-P: 125.9°), the bite angle of the two phosphine ligands is very small for the interconnecting copper(I) atom (P-Cu2-P: 112.4°, and P3-P4: 3.79 Å) and is together with a bite angle N-Cu2-N of 107.4° close to the perfectly tetrahedral angle (109°). The three copper(I) atoms and pyridine-triazolate ligands lie in the same, although slightly twisted, plane, with three PPh₃ molecules each above and below. Throughout the crystal, the complex molecules are orientated in layers with the copper planes parallel to each other.

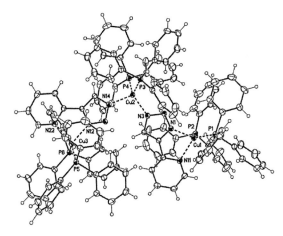

Figure 20. Molecular structure of the trinuclear compound [(PyrTri)$_2$Cu$_3$(PPh$_3$)$_6$]PF$_6$ (**c-(6-Cu-8-PF$_6$)**). Displacement parameters are drawn at 50% probability level (solvent and anions omitted for clarity).[IV]

Table 6. Selected bond lengths (Å) and angles (°) of the structurally characterized trinuclear complex [(PyrTri)$_2$Cu$_3$(PPh$_3$)$_6$]PF$_6$ (**c-(6-Cu-8-PF$_6$)**).

Bond lengths [Å]		Angles [°]	
Cu1-N1	2.046 (2)	N1-Cu1-N11	80.16 (9)
Cu1-N11	2.125 (2)	P1-Cu1-P2	125.90 (3)
Cu1-P1	2.2201 (8)	φ$_1$[a]	85.72 (7)
Cu1-P2	2.2805 (8)	N1-C5-C6-N11	9.0 (4)
P1-P2	4.008 (1)		
Cu2-N3	2.055 (2)	N3-Cu2-N14	107.41 (9)
Cu2-N14	2.067 (2)	P3-Cu2-P4	112.37 (3)
Cu2-P3	2.2768 (7)	φ$_2$[a]	89.35 (6)
Cu2-P4	2.2823 (7)	C4-N3-Cu2-N14	10.9 (2)
P3-P4	3.788 (1)		
Cu3-N12	2.056 (2)		
Cu3-N22	2.094 (2)	N12-Cu3-N22	80.58 (8)
Cu3-P5	2.2537 (7)	P5-Cu3-P6	125.46 (3)
Cu3-P6	2.2249 (7)	φ$_3$[a]	87.91 (7)
P5-P6	3.981 (1)	N12-C16-C17-N22	-7.4 (3)

[a] Dihedral angle between N–Cu1–N and P–Cu1–P planes.

The crystal structure of compound [(PyrTet)Cu(PPh$_3$)$_2$] (**7-Cu-8**) based on 5-(pyridin-2-yl)-tetrazolate (**7-H**) and PPh$_3$ (**8**) can be found in Figure 21. It is synthesized by reaction of its cationic analogue [(PyrTetH)Cu(PPh$_3$)$_2$]BF$_4$ (**7-H-Cu-8**) with potassium hydroxide.[167] Purification of the complex is difficult due to dissociation processes in solution. During precipitation or when suspending the compound **7-Cu-8** in diethyl ether, PPh$_3$ dissociated off the complex and dissolved in ether, giving cluster-compounds of the composition [(PyrTet)$_2$Cu$_2$(PPh$_3$)$_2$], proven by elemental analysis. Unfortunately no single crystals are obtained for X-ray diffraction to study this coordination motif, but one can assume a bridging behavior for the PyrTet (**7**) ligand as observed for PyrBim (**5**) and PyrTri (**6**) in **c-(5-Cu-8-PF₆)** and **c-(6-Cu-8-PF₆)** before. The mononuclear neutral complex [(PyrTet)Cu(PPh₃)₂] (**7-Cu-8**) motif is obtained by evaporation of all solvents after the reaction, extraction of the residue with dichloromethane and evaporation of the solvent, without further precipitation in diethyl ether, thus taking into account small phosphine impurities. A tetrahedral coordination of the central copper(I) atom by the PyrTet ligand with a bite angle of 79.8° and by two PPh₃ molecules with a bite angle P-Cu-P of 121.2° is found, which is much smaller compared to the cationic species [(PyrTetH)Cu(PPh₃)₂]BF₄ (**7-H-Cu-8**). Further differences arise from shorter Cu-N bond lengths, which hint at a stronger coordination of the anionic PyrTet ligand to Cu(I) compared to the neutral ligand PyrTetH and is especially effective for the tetrazolate moiety (Cu-N(tetrazolate): **7-H-Cu-8**: 2.079 Å, **7-Cu-8**: 2.043 Å; Cu-N(pyridine): **7-H-Cu-8**: 2.140 Å, **7-Cu-8**: 2.118 Å). The less symmetric coordination in the neutral complex **7-Cu-8** is also noticed in a less perpendicular orientation of the N^N ligand versus P-Cu-P plane (angle 83.4°) and a distortion of the PyrTet ligand (N1-C5-C6-N11: 5.6°). In the crystal structure, an adjacent molecule interacts with the PyrTet ligand via hydrogen bridge bonds.

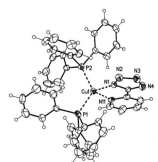

	Bond lengths [Å]		Angles [°]
Cu1-N1	2.0428 (4)	N1-Cu1-N11	79.816 (16)
Cu1-N11	2.1184 (4)	P1-Cu1-P2	121.21 (6)
Cu1-P1	2.2397 (16)	φ₁[a]	83.379 (1)
Cu1-P2	2.2508 (15)	N1-C5-C6-N11	5.561 (7)
P1-P2	3.912 (2)		
N4-H(C15A)	2.50		

[a] Dihedral angle between N–Cu–N and P–Cu–P planes.

Figure 21. Molecular structure of the mononuclear complex [(PyrTet)Cu(PPh₃)₂]BF₄ (**7-Cu-8**) and selected bond lengths and angles.[IV]

For copper(I) complexes based on the anionic pyridine-amide ligands PyrBim (**5**), PyrTri (**6**), PyrTet (**7**) with the monodentate phosphine PPh₃ (**8**), cluster-type structures are found, where the N^N ligand acts in a bridging manner usually interconnecting tetrahedral complex units. A third (potentially even a fourth) coordination site of the N^N ligand after deprotonation enables a bridging behavior, and this together with the monodentate phosphine PPh₃ results in formation of polynuclear coordination modes. The phosphine PPh₃, which is prone to dissociation reactions, can even favor the formation of more than one complex species dependent on the reaction and purification conditions. Which specific coordination structure is obtained, depends on the size or the steric demand of the N^N ligand (as seen for **c-(5-Cu-8-PF₆)**) and the position of the third coordinating N-atom in the 5-membered ring. For compound **7-Cu-8** also the mononuclear species could be obtained, but only by careful adjustment of the purification conditions. In general, it's difficult to compare complexes of different structure types because bond lengths and angles are also determined by intermolecular steric constraints or interactions, which differ for different coordination motifs. Nevertheless, for all complex types a stronger coordination of the deprotonated versus the neutral N^N ligand is observed, which is reflected by shorter Cu-N bond lengths. Furthermore, in the anionic form the Cu-N(amide) bond is usually shorter than the Cu-N(pyridine) bond in the tetrahedral units. This especially applies to the copper(I) complexes **5-H-Cu-8** and **c-(5-Cu-8-PF₆)** with the PyrBim(H) ligand: The Cu-N(benzimidazole) bond shortens by about 0.06 Å, but also the bond lengths difference of Cu-N(benzimidazole) versus Cu-N(pyridine) increases from 0.01 to 0.10 Å. For the ligands PyrTri (**6**) and PyrTet (**7**) a similar shortening is observed, but less pronounced due to their less basic character.

When going from PPh₃ to dppm copper(I) complexes, an acidification of the pyridine-amine ligands PyrTriH (**6-H**) and PyrTetH (**7-H**) occurs and complexes $[(N^\wedge N)_2Cu_4(dppm)_4](X)_2$ (X denotes the counterion BF_4^- or PF_6^- depending on the used copper(I) salt) with anionic PyrTri (**6**) and PyrTet (**7**) are obtained even from reactions without a base. Both compounds $[(PyrTri)_2Cu_4(dppm)_4](BF_4)_2$ (**c-(6-Cu-9)**) and $[(PyrTet)_2Cu_4(dppm)_4](BF_4)_2$ (**c-(7-Cu-9)**) show a tetranuclear coordination motif, where two copper(I) atoms are tetrahedrally coordinated by a N^N ligand and two phosphine atoms, and two copper(I) atoms connect these complex units (Figure 22). Both the pyridine-amide as well as the bis(phosphine) act as bridging ligands. Interestingly, the fourth potentially coordinating N-atom in the tetrazolate ring of PyrTet (**7**) does not influence the coordination chemistry of the formed complexes. Bond lengths and angles in the tetrahedral units do not differ significantly for PyrTri and PyrTet complexes except smaller P-Cu-P bite angles for the PyrTri complexes, which results from slightly longer Cu-P bonds (P-Cu1-P: **c-(6-Cu-9)**: 109.3°, **c-(7-Cu-9)**: 132.4°, P-Cu1: **c-(6-Cu-9)**: 2.222–2.253 Å, **c-(7-Cu-9)**: 2.195–2.2235 Å, Table 7). The two interconnecting copper(I) atoms bridge two pyridine-amide ligands by coordination to the N(2) and N(3A) atoms of the triazolate/tetrazolate moiety, and thus span up a plane of all copper(I) atoms and pyridine-amide ligands. Two bis(phosphine) ligands each lie above and below this plane and bridge one copper(I) atom of the tetrahedral unit and one interconnecting copper(I) atom. In the crystal, the complex molecules of **c-(6-Cu-9)** are orientated in (slightly distorted) layers with the copper planes parallel to each other. When going from the PyrTri to the PyrTet complex **c-(7-Cu-9)**, the interconnecting copper(I) atoms move closer to the copper(I) center of the tetrahedral unit and thus result in unequal Cu2-N bonds lengths (**c-(7-Cu-9)**: Cu2-N2: 2.204 Å, Cu2-N3A: 2.058 Å) and a longer Cu2-Cu2A distance of 4.406 Å (**c-(6-Cu-9)**: 3.946 Å). From comparison of the complexes **c-(6-Cu-9)**, **c-(7-Cu-9)** with their PF_6^--analogues no major differences are observed except a more distorted plane of copper(I) atoms and pyridine-amide ligands for **c-(6-Cu-9)** than for **c-(6-Cu-9-PF₆)** and thus, a slightly different geometry around the interconnecting copper(I) atoms (e.g. φ: **c-(6-Cu-9)**: 83.0°, **c-(6-Cu-9-PF₆)**: 88.7°) and a stronger distortion of the PyrTri ligand. This might be a result of the high symmetry in complex **c-(6-Cu-9-PF₆)** with a parallel orientation of the dppm ligands to each other. The counterions and cocrystallized solvents might have an impact on the exact orientation of the ligands, but less on the coordination geometry at the complex center.

a)

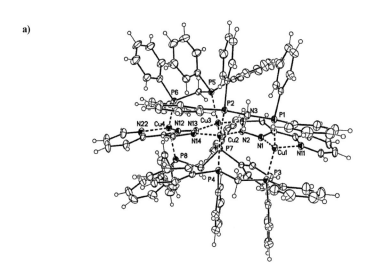

b)

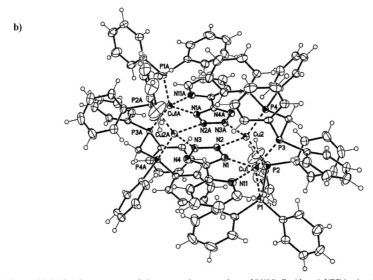

Figure 22. Molecular structures of the tetranuclear complexes [(N^N)₂Cu₄(dppm)₄](BF₄)₂ a) **c-(6-Cu-9)** and b) **c-(7-Cu-9)**. Displacement parameters are drawn at 50% probability level (solvent molecules and anions omitted for clarity).[IV]

Table 7. Selected bond lengths (Å) and angles (°) of the structurally characterized tetranuclear complexes [(N^N)₂Cu₄(dppm)₄](X)₂ (with X = BF₄⁻ or PF₆⁻).

	c-(6-Cu-9)	c-(6-Cu-9-PF₆)	c-(7-Cu-9)	c-(7-Cu-9-PF₆)
Cu1-N1	2.012 (2)	2.039 (3)	2.007 (2)	1.993 (3)[a]
Cu1-N11	2.139 (2)	2.156 (3)	2.169 (2)	2.160 (3)[b]
Cu1-P1	2.2528 (8)	2.2555 (10)	2.2354 (8)	2.2176 (11)[c]
Cu1-P3	2.2223 (8)	2.2558 (10)	2.1946 (8)	2.1811 (11)[d]
Cu2-N2	2.079 (2)	2.039 (3)	2.204 (2)	2.151 (3)[e]
Cu2-N3A	2.065 (2)[f]	2.052 (2)	2.058 (2)	2.045 (3)[g]
Cu2-P2	2.2722 (8)	2.2681 (9)	2.2940 (8)	2.2748 (11)[h]
Cu2-P4	2.2810 (8)	2.2523 (9)	2.3037 (8)	2.2933 (11)[i]
P1-P2	3.108 (1)	3.081 (1)	3.154 (1)	3.152 (2)
P3-P4	3.106 (1)	3.139 (1)	3.096 (1)	3.089 (1)
Cu1-Cu2	3.5974 (5)	3.6149 (6)	3.5232 (6)	3.536 (1)
Cu2-Cu2A	3.9456 (6)[j]	3.8485 (10)	4.4065 (9)	4.294 (1)[k]
N1-Cu1-N11	78.66 (9)	78.07 (10)	78.69 (9)	78.29 (13)[l]
P1-Cu1-P3	109.28 (3)	109.55 (4)	132.43 (3)	130.85 (4)[m]
N2-Cu2-N3A	99.46 (9)[n]	104.01 (10)	87.38 (8)	89.61 (12)[o]
P2-Cu2-P4	110.20 (3)	116.75 (3)	114.12 (3)	112.78 (5)[p]
φ[q]	86.59 (5)/82.97 (5)	87.20 (7)/88.67 (7)	87.96 (6)/89.76 (6)	88.47 (9)/89.33 (8)
N1-C5-C6-N11	−9.5 (4)	−4.6 (4)	1.9 (4)	1.1 (5)[r]
P1-C24-P2	114.74 (15)[s]	113.11 (17)	117.20 (14)[t]	118.3 (2)[u]

[a] Cu2-N61; [b] Cu2-N51; [c] Cu2-P2; [d] Cu2-P4; [e] Cu1-N60; [f] Cu2-N14; [g] Cu1-N59A; [h] Cu1-P1; [i] Cu1-P3; [j] Cu2-Cu3; [k] Cu1-Cu1A; [l] N51-Cu2-N61; [m] P2-Cu2-P4; [n] N2-Cu2-N14; [o] N60-Cu1-N59A; [p] P1-Cu1-P3; [q] angle between N1–Cu1–N11 and P1–Cu1–P3 planes/ N2-Cu2-N3A and P2-Cu2-P4 planes; [r] N51-C56-C57-N61; [s] P1-C35-P2; [t] P1-C12-P2; [u] P2-C1-P4.

Similarly to the case of compound **c-(5-H-Cu-9)**, the rigid structure of the bis(phosphine) dppm (**9**) with a narrow bite angle leads to cluster-type structures with the ligands PyrTri (**6**) and PyrTet (**7**), but here also the anionic pyridine-amide ligands with a third and fourth coordination site act as bridging ligand. The deprotonation of the ligand happens spontaneously during reaction even without a base. Probably, first clusters are formed when more potential coordinating N-atoms are present such as in PyrTriH (**6-H**) and PyrTetH (**7-H**), and the coordination of more than one copper(I) atom acidifies the amine moiety and then leads to deprotonation. In contrast, PyrBimH (**5-H**) only acts as bidentate ligand for one copper(I) center in **c-(5-Cu-9)** and additionally has much less acidic character by nature. Minor differences in the bond lengths and angles of the complexes **c-(6-Cu-9)** and **c-(7-Cu-9)** compared to their PF_6^--analogues arise from a different orientation of the phosphine ligands in the crystal.

From reaction of $[Cu(CH_3CN)_4]BF_4$ (**13**) with the pyridine-amine PyrBimH (**5-H**) and DPEPhos (**10**) in a ratio of 1:1:1 and potassium hydroxide a yellow compound is obtained. Elemental analysis determines a mononuclear neutral compound of the composition [(PyrBim)Cu(DPEPhos)] (**5-Cu-10**), for which the crystal structure is shown in Figure 23. The coordination to copper(I) acidifies the PyrBimH ligand so that potassium hydroxide is sufficient enough for its deprotonation. Selected bond lengths and angles for compound **5-Cu-10** are given in Table 8. The ligand PyrBim coordinates the central copper(I) atom in a bidentate manner with a typical bite angle of 80.9°, where benzimidazolate- and pyridine-rings lie nearly coplanar to each other (angle: 1.9°). The DPEPhos ligand on the other hand coordinates the copper(I) with a bite angle of 112.3°. Compared to its cationic analogue, the benzimidazolate moiety moves slightly closer to the copper(I) atom, while the pyridine shifts away (Cu-N(benzimidazole) **5-H-Cu-10**: 2.063 Å, **5-Cu-10**: 2.036 Å; Cu-N(pyridine): **5-H-Cu-10**: 2.098 Å, **5-Cu-10**: 2.107 Å). The less symmetric coordination is also reflected in unequal Cu-P bond lengths (Cu-P: **5-H-Cu-10**: 2.237, 2.241 Å, **5-Cu-10**: 2.228, 2.259 Å), and a slightly distorted tetrahedron with an angle of 80.2° between the P-Cu-P and N-Cu-N planes.

Using the pyridine-amide PyrTri (**6**) with $[Cu(CH_3CN)_4]BF_4$ (**13**) and one equivalent of DPEPhos (**10**) gives an off-white compound of the molecular composition [(PyrTri)Cu(DPEPhos)] (**6-Cu-10**) as proposed by elemental analysis, but no single crystals are obtained for X-ray diffraction analysis. A tetrahedral coordination of the copper(I) atom by the chelating ligands PyrTri and DPEPhos with similar bond lengths and angles as in **5-Cu-10** can be expected.

To complete the series of copper(I) complexes with DPEPhos, compound **7-Cu-10** based on the pyridine-amide PyrTet (**7**), is synthesized either from reaction of its cationic complex **7-H-Cu-10** with a base, or by direct complexation of copper(I) with PyrTet (**7**) and DPEPhos (**10**).[167] As in **5-Cu-10** and

6-Cu-10, a mononuclear neutral compound of the type [(PyrTet)Cu(DPEPhos)] is obtained and the crystal structure shown in Figure 23. Compared to the before discussed mononuclear PyrBim complex **5-Cu-10,** the effect of deprotonation on the Cu-N bond lengths is even more significant in **7-Cu-10:** While the Cu-N(tetrazole) bond shortens by 0.055 Å, the Cu-N(pyridine) bond is elongated by 0.036 Å (**5-Cu-10:** –0.027 Å, +0.009 Å). This indicates a stronger coordination of the copper(I) atom by the less electron rich PyrTet ligand than for PyrBim (5). In contrast, the Cu-P bond lengths and P-Cu-P angle remain similar to the cationic complex (compare Table 8 and 4). The coordination of copper(I) to the anionic PyrTet ligand results in a stronger distortion of the tetrahedral geometry, reflected in the angle of 84.3°. Here, not only electronic effects of the anionic pyridine-amide ligand influence the coordination geometry when compared to the cationic complex **7-H-Cu-10,** but also steric constraints are smaller in the monomeric structure in contrast to the dimeric arrangement of **7-H-Cu-10** with a hydrogen-bridge bond. Thus, derivation of general rules from this comparison is difficult.

Furthermore, it is observed during synthesis that oxidation stabilities of the mononuclear neutral complexes **5-Cu-10, 6-Cu-10, 7-Cu-10** increase in the order PyrBim, PyrTri, PyrTet. The highest occupied molecular orbitals (HOMO) of the complexes are destabilized due to higher electron density of the N^N ligand after deprotonation, while for the ligand PyrTet (**7**) the anionic charge is delocalized over the tetrazolate moiety, i.e. copper(I) complexes based on 5-(pyridin-2-yl)tetrazolate are less sensitive to oxygen.[163,164]

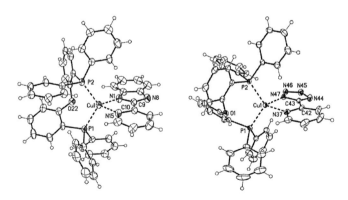

Figure 23. Molecular structures of the mononuclear complexes [(PyrBim)Cu(DPEPhos)] (**5-Cu-10**) and [(PyrTet)Cu(DPEPhos)] (**7-Cu-10**). Displacement parameters are drawn at 50% probability level.[IV]

Table 8. Selected bond lengths (Å) and angles (°) of the structurally characterized mononuclear complexes [(N^N)Cu(DPEPhos)] **5-Cu-10** and **7-Cu-10**.

	5-Cu-10	**7-Cu-10**
Cu1-N1	2.036 (6)	2.0400 (15)[a]
Cu1-N15	2.107 (6)	2.0818 (16)[b]
Cu1-P1	2.259 (19)	2.2262 (5)
Cu1-P2	2.228 (18)	2.2493 (5)
P1-P2	3.726 (2)	3.762 (1)
N1-Cu-N15	2.036 (6)	2.0400 (15)[a]
P1-Cu-P2	2.107 (6)	2.0818 (16)[b]
φ^{d}	2.259 (19)	2.2262 (5)
N1-C9-C10-N15	2.228 (18)	2.2493 (5)

[a] Cu1-N47; [b] Cu1-N37; [c] N47-Cu-N37; [d] angle between N–Cu–N and P–Cu–P planes; [e] N37-C42-C43-N47.

In a nutshell, copper(I) complexes with the bis(phosphine) DPEPhos (**10**) and the anionic pyridine-amide ligands PyrBim (**5**), PyrTri (**6**) and PyrTet (**7**) exhibit a mononuclear structure with a distorted tetrahedral coordination around the metal center. The coordination geometries with different N^N ligands are very similar except a stronger twisted PyrTet ligand compared to PyrBim. Since that effect was already observed for the complexes **5-H-Cu-10** and **7-H-Cu-10-PF₆** with the neutral forms of the pyridine-amine ligands, it can be attributed to an intrinsic property of PyrTet(H) instead of an electronic feature of the anionic species.

To sum up, the coordination chemistry of copper(I) with phosphines **8–10** and the deprotonated pyridine-amide ligands PyrBim (**5**), PyrTri (**6**), PyrTet (**7**) is controlled by steric but even more by electronic effects. A stronger coordination of the anionic N^N ligand to the copper(I) atom than in its neutral form and an increased electron density of the pyridine-amide ligand lead to different coordination modes with mono- or bidentate phosphine ligands compared to section 3.1.2.1 (see Scheme 3). Increasing the number of N-atoms in the 5-membered ring of the N^N ligand facilitates more potential coordination sites for copper(I), which lead to cluster formation especially with monodentate phosphines, but also with the bridging bis(phosphine) dppm (**9**). Additionally, a higher number of N-atoms results in stronger acidity of the pyridine-amine ligand and thus spontaneous deprotonation may occur (**c-(6-cu-9)**, **c-(7-Cu-9)**). While there is no major steric influence observed for cationic mononuclear complexes, the steric demand of the pyridine-amide ligands PyrBim (**5**), PyrTri (**6**) and PyrTet (**7**) determines the cluster type in polynuclear complexes, as was seen for complexes **c-(5-Cu-8-PF₆)** and **c-(6-Cu-8-PF₆)**. Some general rules for structural predictions can be derived when comparing the effects of the different phosphines: a) When using the monodentate PPh₃ (**8**), cluster-type structures are obtained with all pyridine-amide ligands, where the N^N ligand acts as bridging ligand. The extent of dissociation reactions and cluster formation is significantly larger in complexes with anionic than neutral ligands (compare Schemes 3 and 4), which can not only be a result of more coordination sites. The increased electron density of the anionic pyridine-amide ligands enables coordination to more than one copper(I) atom, which is then facilitated by the monodentate PPh₃ for cluster formation. b) The bis(phosphine) dppm (**9**) gives polynuclear complexes, where the P^P but also the N^N ligands act as bridging ligands due to more accessible coordination sites (in contrast to complex **c-(5-H-Cu-9)** based on a neutral N^N ligand). c) Only the chelating DPEPhos (**10**) ligand with a flexible bite angle as well as high steric demand favors the formation of mononuclear neutral complexes. Thus, the denticity plays a major role for the coordination chemistry of copper(I) with anionic N^N ligands, in contrast to section 3.1.2.1.

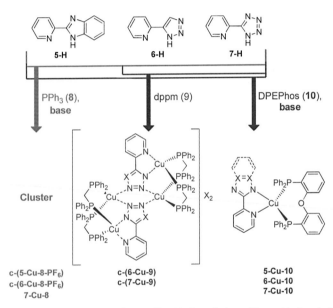

Scheme 4. The coordination chemistry of copper(I) with the anionic pyridine-amide ligands PyrBim (**5**), PyrTri (**6**), PyrTet (**7**) and monodentate or bidentate phosphines (x means C or N, while X denotes the counterions BF₄⁻ or PF₆⁻).

Influences of the stoichiometry of reactants on the coordination motifs are not tested here, but seems to be less important than observed for copper(I) halide complexes,[134,135,136,137] since the complex structures do not have to reflect the ratios of used reactants (see **c-(6-cu-9)**, **c-(7-cu-9)**) and only one species is obtained for each compound. But the stoichiometry might play a larger role for complexes based on anionic ligands and monodentate phosphine PPh₃ (**8**), which is prone to ligand exchange reactions. In general, the right combination of (chelating) ligands is crucial to obtain the desired complex species.

3.1.3 Photophysical Characterization

3.1.3.1 Copper(I) Complexes with Pyridine-amine Ligands

In Figure 24 the photoluminescence spectra of the mononuclear cationic complexes
$[(N^\wedge N)Cu(PPh_3)_2]BF_4$ **5-H-Cu-8**, **6-H-Cu-8**, and **7-H-Cu-8** with the phosphine PPh_3 (**8**) and
$[(N^\wedge N)Cu(DPEPhos)]BF_4$ **5-H-Cu-10**, **6-H-Cu-10**, and **7-H-Cu-10** based on the bis(phosphine)
DPEPhos (**10**) are shown. The dinuclear compound $[(PyrBimH)_2Cu_2(dppm)_2](BF_4)_2$ (**c-(5-H-Cu-9)**)
exhibits no photoluminescence. Since a similar complex based on a different pyridine-amine shows blue
luminescence, the absence of photoluminescence for **c-(5-H-Cu-9)** might be attributed to the ligand
PyrBimH instead of the coordination motif.[132] The emission maxima of the mononuclear
compounds **5-H-Cu-8** – **7-H-Cu-10** range from 480 to 570 nm and the spectra are broad and
unstructured as typically found for luminescent copper(I) complexes. This is ascribed to the
charge-transfer character of the emission, which is further smeared out by energetic inhomogeneity in
amorphous organic materials.[179] Photoluminescence quantum yields reach from 7% up to 46% (see
Table 9), but strongly depend on the molecular packing and environment and thus also on the
purification methods (precipitation and recrystallization).[93,180] While for compound **6-H-Cu-8** a higher
PLQY of 24% is found for the crystalline form than for the amorphous powder with 15%, for
complex **5-H-Cu-8** not only an increase of PLQY to 22% but also a strong blue-shift of the emission
maximum to 534 nm is observed for the crystalline form. This can be explained by the different extent
of interactions in the crystal compared to amorphous powder, where the molecules might not have
enough time to form a short-range order during fast precipitation. As observed in the crystal structure,
the interactions of the counterion with PyrBimH ligands of adjacent complex molecules induce a dense
and rigid packing, which reduces non-radiative processes and enhances the luminescence. This was
formerly reported for $[(N^\wedge N)Cu(P^\wedge P)]^+$ complexes based on phenanthroline ligands,[181] but is applicable
to all complexes with a ligand prone to $\pi\pi$-stacking, and is thus expected for both complexes **5-H-Cu-8**
and **5-H-Cu-10** and reported literature values confirm this statement.[182] Not only electrostatic or
$\pi\pi$-interactions, but also the counterions and embedded solvents in the amorphous powder control the
environment of the complex molecules[183] and the photophysical properties, as becomes obvious from
the comparison of the compounds **6-H-Cu-8** and **6-H-Cu-8-PF$_6$**: The larger PF_6^- counterion
in **6-H-Cu-8-PF$_6$** compared to BF_4^- enhances the photoluminescence by a factor of 2 and results in a
slight blue-shift. To compare the photoluminescence of different complexes, the aggregation and
morphology should be generally comparable, but since the effects of the molecular environment and
packing are usually small against the influences of different ligand types, some general statements from

the comparison of the compounds [(N^N)Cu(PPh₃)₂]BF₄ **5-H-Cu-8**, **6-H-Cu-8**, **7-H-Cu-8** and

[(N^N)Cu(DPEPhos)]BF₄ **5-H-Cu-10**, **6-H-Cu-10**, **7-H-Cu-10** with different pyridine-amine ligands

can be derived: For both complex series a blue-shift of the emission in the order PyrBimH, PyrTetH,

PyrTriH is observed, which will be related to theoretical calculations below (**5-H-Cu-10**: 571 nm,

7-H-Cu-10: 485 nm, **6-H-Cu-10**: 522 nm and **5-H-Cu-10**: 564 nm, **7-H-Cu-10**: 518 nm,

6-H-Cu-10: 480 nm). The trend of the PLQYs is not as clear as the trend of the emission maxima due

to a stronger control by the molecular environment, but de facto the PLQYs of complexes **5-H-Cu-8**

and **5-H-Cu-10** based on PyrBimH around 10% are much lower than of the complexes **6-H-Cu-8**,

7-H-Cu-8, **6-H-Cu-10**, **7-H-Cu-10** based on PyrTriH and PyrTetH from 21–46%, which might be a

result of JORTNER's energy gap law. This states an increase of non-radiative processes with decreasing

emission energy.[184]

a)   b)

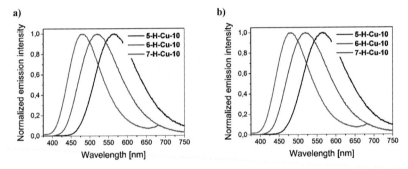

Figure 24. Photoluminescence spectra of the mononuclear cationic complexes a) [(N^N)Cu(PPh₃)₂]BF₄ **5-H-Cu-8**, **6-H-Cu-8 7-H-Cu-8** and b) [(N^N)Cu(DPEPhos)]BF₄ **5-H-Cu-10**, **5-H-Cu-10**, **7-H-Cu-10** as amorphous powders at room temperature when excited at 350 nm. The increase of PL intensity on the red edge of spectrum is assigned to the 2λ peak of the excitation source.

Table 9. Photoluminescence characteristics of the mononuclear cationic complexes [(N^N)Cu(PPh₃)₂]BF₄ **5-H-Cu-8**, **6-H-Cu-8 7-H-Cu-8** and [(N^N)Cu(DPEPhos)]BF₄ **5-H-Cu-10**, **5-H-Cu-10**, **7-H-Cu-10** as amorphous powders at room temperature if not otherwise noted.

Complex	λ_{em} [nm]	Φ_{em}	Complex	λ_{em} [nm]	Φ_{em}
5-H-Cu-8	571 (522[a])	0.07 (0.22[a])	**5-H-Cu-10**	564	0.09
6-H-Cu-8	485 (485[a])	0.15 (0.24[a])	**6-H-Cu-10**[a]	480	0.45
6-H-Cu-8-PF₆	478	0.34	**7-H-Cu-10**	518	0.21
7-H-Cu-8	522	0.46			

[a] Measured as crystalline sample.

To further determine the influence of the pyridine-amine ligands on the photophysical properties of the

corresponding complexes, absorption spectra of the mononuclear complexes are recorded. Since a

similar emission trend is observed for the complexes [(N^N)Cu(PPh₃)₂]BF₄ **5-H-Cu-8**, **6-H-Cu-8**,

7-H-Cu-8 as for the DPEPhos based complexes **5-H-Cu-10, 6-H-Cu-10, 7-H-Cu-10**, focus is laid on

the latter ones to later compare their emission properties to the respective neutral complexes **5-Cu-10**,

6-Cu-10, 7-Cu-10 (chapter 3.1.3.2). The [(N^N)Cu(DPEPhos)]BF$_4$ complexes exhibit broad and

intense absorption bands between 260–335 nm (ε = 1.8–2.3 × 10^4 M^{-1}cm^{-1}) for **5-H-Cu-10**

and 250–280 nm (ε = 1.7–2.7 × 10^4 M^{-1}cm^{-1}) for **6-H-Cu-10, 7-H-Cu-10**, which can be attributed to

ligand-centered transitions of the pyridine-amine and bis(phosphine) ligands (Figure 25). A red-shift of

the ligand-centered $\pi\pi^*$ absorption bands of the complexes compared to the free pyridine-amine ligands

is observed, which is attributed to a stabilization of the ligands' $\pi\pi^*$ states when coordinating to

copper(I). Interestingly, this stabilization is weakest for the ligand PyrTriH. Additional absorption bands

at 384 nm (ε = 3.6 × 10^3 M^{-1}cm^{-1}) for complex **5-H-Cu-10**, and at 355 nm (4.0 × 10^3 M^{-1}cm^{-1})

and 347 nm (3.0 × 10^3 M^{-1}cm^{-1}) for **6-H-Cu-10** and **7-H-Cu-10** are assigned to charge transfer

transitions from the metal to a ligand.

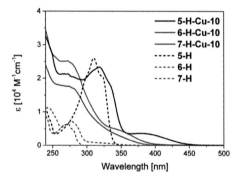

Figure 25. Absorption spectra of complexes [(N^N)Cu(DPEPhos)]BF$_4$ **5-H-Cu-10, 5-H-Cu-10, 7-H-Cu-10** and free pyridine-amine ligands. The absorption of the complexes and PyrBimH (**5-H**) are recorded in CH$_2$Cl$_2$, and of PyrTriH (**6-H**) and PyrTetH (**7-H**) in EtOH at a concentration of 10^{-5} mol/l.

The complexes **5-H-Cu-10, 6-H-Cu-10, 7-H-Cu-10** are furthermore studied by DFT calculations. For

all three complexes the HOMO is predominantly localized on the central copper(I) and the two

coordinating phosphorous atoms, while the LUMO is distributed on the pyridine-amine

ligands (Figure 26). The corresponding luminescent transition is assigned to a metal-to-ligand charge

transfer (MLCT) character d(Cu) → π^*(pyridine-amine ligand).

5-H-Cu-10	6-H-Cu-10	5-H-Cu-10
LUMO	LUMO	LUMO

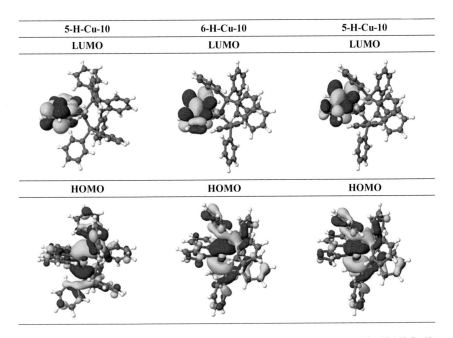

HOMO	HOMO	HOMO

Figure 26. Selected frontier orbitals of the mononuclear cationic complexes [(N^N)Cu(DPEPhos)]⁺ **5-H-Cu-10**, **5-H-Cu-10**, **7-H-Cu-10** as calculated by DFT using the B3LYP functional with def2-SV(P) basis set.[v]

Although the LUMO level of the complex is mainly controlled by the LUMO of the pyridine-amine ligand, the complex's emission color cannot be predicted from the LUMO of the free ligand as is typically found for copper(I) halide complexes.[185] While the frontier orbitals of the free ligands are stabilized in the order PyrBimH, PyrTriH, PyrTetH due to the lower electron donating ability when increasing the number of N-atoms in the ligand, the LUMO energies of the complexes **5-H-Cu-10**, **6-H-Cu-10**, **7-H-Cu-10** do not follow this trend (Table 10), and thus the coordination to copper(I) cannot be neglected for the prediction of the emission color. However, the HOMO-LUMO energy gaps give a good indication for the emission color of the complexes: The energy gaps increase in the order PyrBimH < PyrTetH < PyrTriH, which reflects the blue-shift of the complex's emission in the same order. The increasing HOMO-LUMO energy gap is also in accordance with the position of the MLCT absorption bands: A strong blue-shift of the absorption band at 384 nm of complex **5-H-Cu-10** to 355 nm for **6-H-Cu-10** and 347 nm for **7-H-Cu-10** is observed. However, it has to be taken into account that the N^N ligand in the complex **7-H-Cu-10** is deprotonated to an extent of 50% and thus

---

[v] DFT calculations were performed by Dr. Jana Friedrichs, CYNORA GmbH (see also 5.1.7).

the absorption spectra do not reflect the pure cationic species [(PyrTetH)Cu(DPEPhos)]BF$_4$. A slightly red-shifted absorption band is expected for the pure cationic compound. The very similar positions of the MLCT absorption bands of complex **6-H-Cu-10** and **7-H-Cu-10**, and also predicted HOMO-LUMO gaps and excitation energies indicate only minor electronic differences between the two complexes. Thus, the strong blue-shift of emission maxima from compound **7-H-Cu-10** to **6-H-Cu-10** is attributed to an increasing HOMO-LUMO energy gap in general and partially to packing effects, as the emission of complex **6-H-Cu-10** is measured in crystalline form.

Table 10. Calculated HOMO and LUMO energies, as well as excitation energies of the complexes [(N^N)Cu(DPEPhos)]$^+$ **5-H-Cu-10**, **5-H-Cu-10**, **7-H-Cu-10**. The frontier orbitals are obtained by DFT calculations using the B3LYP functional with def2-SV(P) basis set, and the excitation energies are calculated by TD-B3LYP.

Complex	HOMO [eV]	LUMO [eV]	$\Delta E_{HOMO\text{-}LUMO}$ [eV]	Excitation energy [eV]
5-H-Cu-10	−7.74 eV	−4.63 eV	3.11 eV	2.53
6-H-Cu-10	−7.85 eV	−4.33 eV	3.52 eV	2.91
7-H-Cu-10	−8.18 eV	−4.71 eV	3.47 eV	2.87

To conclude, the photoluminescence is mainly controlled by the pyridine-amine ligand and emission colors are in accordance with the calculated trend of HOMO-LUMO energy gaps. For complexes [(N^N)Cu(PPh$_3$)$_2$]BF$_4$ **5-H-Cu-8**, **6-H-Cu-8**, **7-H-Cu-8** based on PPh$_3$ the same concept can be applied when going from PyrBimH to PyrTriH to PyrTetH.

In addition to the strong influence of the pyridine-amine ligand on the complex's photophysical properties, also a small effect can be observed by using different phosphines, although not as significant. When comparing a complex [(N^N)Cu(PPh$_3$)$_2$]BF$_4$ with its corresponding DPEPhos analogue [(N^N)Cu(DPEPhos)]BF$_4$ a small blue-shift is observed. Other groups have reported different approaches to explain the effects of different phosphines on the complexes emission: Since DPEPhos exhibits a higher electron-donating ability than PPh$_3$, the emission of the [(N^N)Cu(DPEPhos)]BF$_4$ complex is proposed to be red-shifted due to a destabilization of the copper(I) d-orbitals.[186] Furthermore, the P-Cu-P bite angle was reported to control the extent of MLCT character and leads to a blue-shift of the emission color when it is large.[187] In the herein discussed complexes, the trend of emission color is contrary to the statements discussed in literature (**5-H-Cu-8**: 127°, 571 nm; **5-H-Cu-10**: 116°, 564 nm; **7-H-Cu-8**: 130°, 522 nm, **7-H-Cu-10**: 115°, 518 nm), and a mainly steric effect of the phosphines resulting in only a slight difference of emission wavelengths for PPh$_3$ and DPEPhos is assumed.

To sum up the photophysical behavior of the mononuclear cationic complexes **5-H-Cu-8, 6-H-Cu-8, 7-H-Cu-8** and **5-H-Cu-10, 6-H-Cu-10, 7-H-Cu-10**, an MLCT transition from metal d-orbitals with contributions from the phosphorous atoms to the pyridine-amine ligand is found, which is responsible for the luminescence. The emission color can be tuned either by using different pyridine-amine ligands or to a negligible extent by phosphines. The emission efficiency is further controlled by the molecular packing and environment, as was seen with ligands prone to ππ-stacking, different counterions and aggregation phases.

3.1.3.2 Copper(I) Complexes with Anionic Pyridine-amide Ligands

Emission spectra of the cluster-type complexes **c-(5-Cu-8-PF₆)**, **c-(6-Cu-8-PF₆)**, **c-(6-Cu-9)**, **c-(7-Cu-9)** and the mononuclear neutral complex **7-Cu-8** are shown in Figure 27 and detailed photophysical data is given in Table 11. As these compounds are mainly purified by recrystallization, they are studied as crystalline samples. Comparisons of the luminescence properties of different coordination motifs are difficult, thus the photophysical behavior of each individual complex will be discussed.

The copper(I) cluster [(PyrBim)₃Cu₄(PPh₃)₆]BF₄ (**c-(5-Cu-8-PF₆)**) shows green emission at 516 nm with a low PLQY of 3%, while the trinuclear complex [(PyrTri)₂Cu₃(PPh₃)₆]PF₆ (**c-(6-Cu-8-PF₆)**) exhibits greenish-blue emission at 496 nm with a quantum yield of 29%. Although in both compounds two types of copper(I) centers are present, either coordinated to two PPh₃ molecules and the pyridine-amide ligand in a bidentate mode, or surrounded by the pyridine-amide ligand in a monodentate mode via a third coordination site, only one emission band is observed with typical full-width-half-maximum (FWHM) values for copper(I) complexes (**c-(5-Cu-8-PF₆)**: 0.549 eV, **c-(6-Cu-8-PF₆)**: 0.527 eV). In contrast to the rather low emission efficiencies of the polynuclear complexes **c-(5-Cu-8-PF₆)** and **c-(6-Cu-8-PF₆)**, the mononuclear neutral complex [(PyrTet)Cu(PPh₃)₂] (**7-Cu-8**) shows strong, greenish-blue luminescence at 512 nm with a PLQY of 85%, which results from a high mutual steric hindrance of the two PPh₃ molecules and a strong and rigid coordination of the copper(I) atom to the PyrTet ligand preventing non-radiative pathways. The emission of **7-Cu-8** is assigned to a (ML+IL)CT transition from the metal center with contributions from the amide-moiety to the whole pyridine-amide ligand, which will be discussed in detail for the compounds **5-Cu-10 – 7-Cu-10** of the same coordination motif.

Comparing the complexes [(PyrTri)₂Cu₄(dppm)₄](X)₂ **c-(6-Cu-9)** and [(PyrTet)₂Cu₄(dppm)₄](X)₂ **c-(7-Cu-9)** based on the different pyridine-amide ligands **6** and **7**, a strong blue-shift from 505 to 463 nm is observed, which can be assigned to the different electronic effects of the N^N ligands PyrTri and PyrTet in this coordination motif. Furthermore, the influences of molecular packing effects are examined here: In both compounds **c-(6-Cu-9)** and **c-(6-Cu-9-PF₆)** which have different counterions, the complex molecules are orientated parallel to each other throughout the crystal, but in **c-(6-Cu-9)** at much further distances to each other. The counterions BF₄⁻ lie in between two complex molecules and thus separate them, while in **c-(6-Cu-9-PF₆)** the counterions fill voids of the crystal packing. These packing effects result in higher photoluminescence quantum efficiency for complex **c-(6-Cu-9)** compared to **c-(6-Cu-9-PF₆)**. In contrast, the influence of the counterions on the crystal packing is negligible in the complexes **c-(7-Cu-9)** and **c-(7-Cu-9-PF₆)**, and comparable emission efficiencies are found.

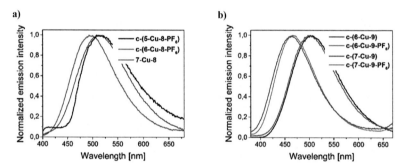

Figure 27. Photoluminescence spectra of the complexes a) [(PyrBim)₃Cu₄(PPh₃)₆]PF₆ (**c-(5-Cu-8-PF₆)**), [(PyrTri)₂Cu₃(PPh₃)₆]PF₆ (**c-(6-Cu-8-PF₆)**), [(PyrTet)Cu(PPh₃)₂] (**7-Cu-8**) and b) [(N^N)₂Cu₄(dppm)₄](X)₂ **c-(6-Cu-9)**, **c-(6-Cu-9-PF₆)**, **c-(7-Cu-9)**, **c-(7-Cu-9-PF₆)** at room temperature when excited at 350 nm. All samples are measured as crystalline samples except the mononuclear complex, which is recorded as amorphous powder. The emission band around 420 nm in the spectrum of **c-(5-Cu-8-PF₆)** arises from the filter (395 nm) used during the measurement, while the increase of PL intensity on the red edge of spectrum and is assigned to the 2λ peak of the excitation source.

Table 11. Photoluminescence characteristics of the complexes **c-(5-Cu-8-PF₆)**, **c-(6-Cu-8-PF₆)**, **7-Cu-8** and [(N^N)₂Cu₄(dppm)₄](X)₂ **c-(6-Cu-9)**, **c-(6-Cu-9-PF₆)**, **c-(7-Cu-9)**, **c-(7-Cu-9-PF₆)** as crystalline powders at room temperature if not otherwise noted.

Complex	λ_{em} [nm]	Φ_{em}	Complex	λ_{em} [nm]	Φ_{em}
c-(5-Cu-8-PF₆)	516	0.03	**c-(6-Cu-9)**	505	0.44
c-(6-Cu-8-PF₆)	496	0.26	**c-(6-Cu-9-PF₆)**	500	0.17
7-Cu-8[a]	512	0.85	**c-(7-Cu-9)**	463	0.15
			c-(7-Cu-9-PF₆)	467	0.18

[a] Measured as amorphous powder

Proceeding with the mononuclear neutral complexes [(N^N)Cu(DPEPhos)] **5-Cu-10**, **6-Cu-10** and **7-Cu-10**, emission maxima in the wavelengths region 510–551 nm with PLQYs up to 78% for compound **7-Cu-10**, but below 10% for compounds **5-Cu-10** and **6-Cu-10**, are found. The emission maxima of the complexes blue-shift from PyrBim (**5-Cu-10**: 551 nm) to PyrTri (**6-Cu-10**: 532 nm) to PyrTet (**7-Cu-10**: 510 nm), which hints at electronic effects of the pyridine-amide ligands and will be examined in the following (Figure 28).

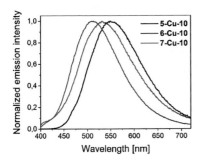

complex	λ_{em} [nm]	Φ_{em}
5-Cu-10	551	0.10
6-Cu-10	532	0.02
7-Cu-10	510	0.78

Figure 28. Photoluminescence spectra and characteristics of the mononuclear neutral complexes [(PyrBim)Cu(DPEPhos)] (**5-Cu-10**), [(PyrTri)Cu(DPEPhos)] (**6-Cu-10**), [(PyrTet)Cu(DPEPhos)] (**7-Cu-10**) as amorphous powders at room temperature when excited at 350 nm. The emission band around 420 nm arises from the filter (395 nm) used during the measurement.

In the absorption spectra of the complexes **5-Cu-10**, **6-Cu-10** and **7-Cu-10**, strong ligand-centered absorption bands between 280–343 nm (ε = 1.9–2.3 × 10^4 $M^{-1}cm^{-1}$) for **5-Cu-10** and 246–315 nm (ε = 1.4–2.2 × 10^4 $M^{-1}cm^{-1}$) for **6-Cu-10**, **7-Cu-10** are found (Figure 29). The ligand-centered $\pi\pi^*$ absorption bands of the complexes **5-Cu-10**, **6-Cu-10** are red-shifted compared to the free pyridine-amine ligands. This can be attributed to a stabilization of the ligand's frontier orbitals by coordination to copper(I) as already observed for the cationic analogues [(N^N)Cu(DPEPhos)]BF$_4$ **5-H-Cu-10**, **6-H-Cu-10** and **7-H-Cu-10**, as well as to a stronger conjugation of the ligands **5** and **6** after deprotonation, whereas the increase of conjugation for ligand **7** is minor. The charge-transfer absorption bands of the mononuclear neutral copper(I) complexes are hardly detectable and almost vanish underneath the ligand-centered bands. For compound **5-Cu-10**, the maximum of the absorption band can be estimated around 395 nm (ε = 3.4 × 10^3 $M^{-1}cm^{-1}$), while for **6-Cu-10** a charge-transfer shoulder at 356 nm (3.8 × 10^3 $M^{-1}cm^{-1}$) and for **7-Cu-10** at 341 nm (5.6 × 10^3 $M^{-1}cm^{-1}$) is determined. Also for other mononuclear neutral complexes, strong overlap of the charge-transfer absorption bands with the ligand-centered bands was observed.[101,126]

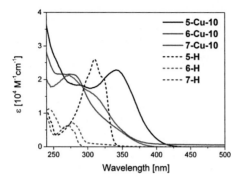

Figure 29. Absorption spectra of complexes [(N^N)Cu(DPEPhos)] **5-Cu-10**, **6-Cu-10**, **7-Cu-10** and of the free pyridine-amine ligands. The absorption of the complexes and PyrBimH (**5-H**) are recorded in CH₂Cl₂, and of PyrTriH (**6-H**) and PyrTetH (**7-H**) in EtOH at a concentration of 10^{-5} mol/l.

DFT calculations are performed for the mononuclear neutral complexes and counter-intuitively, a different localization of frontier orbitals is found for compound **5-Cu-10** compared to **6-Cu-10** and **7-Cu-10** (see Figure 30). In complex **5-Cu-10** based on the pyridine-amide ligand PyrBim (**5**), the HOMO exhibits a metal d-character with contributions of the benzimidazolate moiety, whereas the LUMO is localized on parts of the DPEPhos ligand, and the LUMO+1 lies partially on the DPEPhos and to a smaller extent on the pyridine-amide ligand. According to DFT results the transition responsible for emission is composed of HOMO → LUMO (43.1%) and HOMO → LUMO+1 (53.7%) transitions, and thus has to be described as (ML+IL)CT from the copper(I) center with contributions of the benzimidazolate ring to the DPEPhos and pyridine-amide ligand. In contrast to these observations, no participation of the bis(phosphine) ligand is found for complexes **6-Cu-10** and **7-Cu-10**. Here, the HOMO (and HOMO–1) orbitals are localized on the copper(I) atom, the triazolate or tetrazolate moiety, respectively, and on the pyridine N-atom, while the LUMO is distributed over the whole pyridine-amide ligand. The DFT calculations show, that the excitation transition is composed of HOMO–1 → LUMO (**6-Cu-10**: 60.6%, **7-Cu-10**: 61.6%) and HOMO → LUMO (**6-Cu-10**: 35.6%, **7-Cu-10**: 35.2%) transitions. The emission of the complexes **6-Cu-10** and **7-Cu-10** is characterized as (ML+IL)CT transition, where only the electronic effects of the pyridine-amide ligand influence the emission color, whereas no alteration of the complex's emission is observed with different phosphines (compare **7-Cu-8**: 512 nm and **7-Cu-10**: 510 nm). The different localizations of frontier orbitals for the complex [(PyrBim)Cu(DPEPhos)] (**5-Cu-10**) compared to [(N^N)Cu(DPEPhos)] **6-Cu-10** or **7-Cu-10**, based on PyrTri (**6**) and PyrTet (**7**), can be attributed to the high electron-donating ability of the anionic PyrBim (**5**) ligand, which shifts the LUMO of the complex to the bis(phosphine) ligand of lower energy.

5-Cu-10	6-Cu-10	7-Cu-10

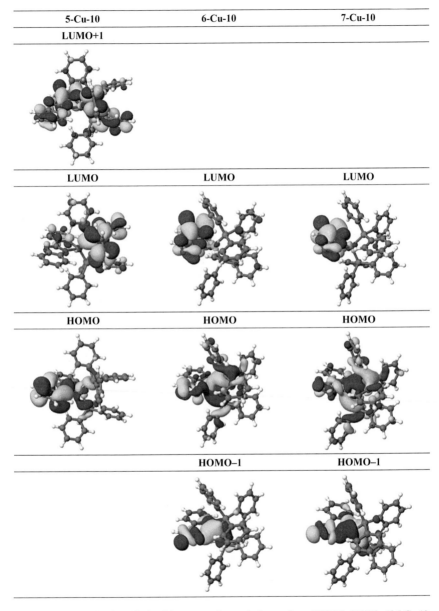

Figure 30. Selected frontier orbitals of the mononuclear cationic complexes [(N^N)Cu(DPEPhos)] 5-Cu-10, 6-Cu-10, 7-Cu-10 as calculated by DFT (B3LYP/def2-SV(P)).[V]

From the comparison of the calculated HOMO-LUMO energy gaps of the mononuclear neutral [(N^N)Cu(DPEPhos)] complexes, a redder emission is predicted for the PyrBim based complex **5-Cu-10**, whereas **6-Cu-10** and **7-Cu-10** are expected to show similar emission wavelengths (Table 12). However, the charge-transfer absorption bands of the complexes show a trend of redder absorption in the order PyrBim > PyrTri > PyrTet. The photophysical properties of the neutral complexes [(N^N)Cu(DPEPhos)] are thus better described by the excitation energies from DFT calculations, which are consistent with the trend of absorption bands, and also explain the trend of bluer emission wavelengths in the order of the anionic ligand PyrBim < PyrTri < PyrTet.

Table 12. Calculated HOMO and LUMO energies, as well as excitation energies of the complexes [(N^N)Cu(DPEPhos)] **5-Cu-10**, **6-Cu-10**, **7-Cu-10**. The frontier orbitals are obtained by DFT calculations using the B3LYP functional with def2-SV(P) basis set, and the excitation energies are calculated by TD-B3LYP.

Complex	HOMO [eV]	LUMO [eV]	$\Delta E_{HOMO\text{-}LUMO}$ [eV]	Excitation energy [eV]
5-Cu-10	−4.75 eV	−1.14 eV	3.61 eV	3.06
6-Cu-10	−5.23 eV	−1.19 eV	4.04 eV	3.23
7-Cu-10	−5.41 eV	−1.37 eV	4.04 eV	3.28

In addition to the modulation of the emission color of the complexes **5-Cu-10**, **6-Cu-10** and **7-Cu-10** by the N^N ligand, the electronic effects of the pyridine-amide ligands induce a stronger coordination of the copper(I) atom in the order PyrBim < PyrTri < PyrTet. Increasing the number of N-atoms in the pyridine-amide ligand leads to a delocalization of the anionic charge and to a stronger coordination to the copper center as proposed by Pearson's HSAB concept. This was already observed in the crystal structures of **5-Cu-10** and **7-Cu-10** and is also expected for **6-Cu-10**. The more rigid coordination in **7-Cu-10** probably results in less non-radiative processes and enhances the photoluminescence, as reflected in the photoluminescence quantum yield of 78%.

Finally, the mononuclear neutral complexes [(N^N)Cu(DPEPhos)] **5-Cu-10**, **6-Cu-10**, **7-Cu-10** are compared with the cationic complexes [(N^N)Cu(DPEPhos)]BF₄ **5-H-Cu-10**, **6-H-Cu-10**, **7-H-Cu-10**. Despite the same coordination motif and similar bond lengths and angles, the localization of frontier orbitals differs significantly for cationic versus neutral complexes: Whereas in cationic complexes a pure MLCT transition d(Cu) → π*(pyridine-amine ligand) is observed, a mixed (ML+IL)CT transition is found in the neutral analogues. Furthermore, for the neutral complexes a different charge-transfer nature is found for the pyridine-amide ligands PyrBim (**5**), PyrTri (**6**) and PyrTet (**7**), i.e. in compound **5-Cu-10** the DPEPhos ligand contributes to the lowest unoccupied molecular orbital of the complex, while in complexes **6-Cu-10** and **7-Cu-10** the LUMO is located solely on the pyridine-amide ligand. Comparing the cationic and neutral complex for each N^N ligand, a small blue-shift of the emission maxima is found for the compounds **5-Cu-10** and **7-Cu-10** based on the ligands PyrBim (**5**) and PyrTet (**7**), accompanied with an increase of the PLQY, especially in the latter case. In contrast, a red-shift and lowering of the photoluminescence efficiency is found when comparing the cationic complex **6-H-Cu-10** to the neutral one **6-Cu-10** based on the ligand PyrTri (**6**). In the UV-Vis spectra, charge transfer absorption bands are observed at 384 nm, 355 nm, 347 nm for the cationic complexes **5-H-Cu-10**, **6-H-Cu-10**, **7-H-Cu-10**, and around 395 nm, 356 nm, 341 nm for the neutral complexes **5-Cu-10**, **6-Cu-10**, **7-Cu-10**. Since the absorption bands of the neutral complexes **5-Cu-10**, **6-Cu-10**, **7-Cu-10** appear only as shoulders of strong ligand-centered bands, a direct comparison to the charge-transfer bands of the cationic complexes might be difficult. However, the blue-shift of the emission can be attributed to the additional intraligand charge-transfer character for the neutral complexes **5-Cu-10**, **6-Cu-10**, **7-Cu-10** and the consequently larger HOMO-LUMO energy gaps,[126] while higher emission efficiencies are ascribed to a stronger coordination and less vibrational quenching due to loss of the (N)*H*-atom and the counterion.[188] Complex **6-Cu-10** does not follow this concept strictly, which might be attributed as an intrinsic feature of triazolate based complexes, since a similar behavior was found for a mononuclear neutral complex based on a pyridine-1,2,4-triazolate and PPh₃ and its cationic analogue.[116] DFT calculations reveal that mainly the LUMO orbital increases the MLCT energy gap of the neutral complexes compared to their cationic analogues and results in a blue-shift of the emission (compare HOMO and LUMO energies of neutral and cationic complexes, Tables 10, 12). This can be illustrated as follows: When deprotonating the pyridine-amine ligands, the electron density is significantly increased in the benzimidazolate, triazolate or tetrazolate moieties and raises the HOMO and LUMO levels. Although the pyridine-amide ligand contributes to both frontier orbitals of the complex, the LUMO orbitals are stronger affected and thus lead to larger energy gaps for the neutral complexes **5-Cu-10**, **6-Cu-10**, **7-Cu-10**.

In this chapter, the coordination chemistry of copper(I) complexes with the neutral or anionic N^N ligands PyrBim(H), PyrTri(H), PyrTet(H) **5–7** with increasing number of N-atoms, and the phosphines PPh₃ (**8**), dppm (**9**), DPEPhos (**10**) of different denticity and bite angle was systematically studied and a set of prediction rules derived (Schemes 3 and 4). For mononuclear neutral complexes the right combination of pyridine-amide and phosphine ligand has to be chosen to obtain the desired coordination motif, although with DPEPhos (**10**) mainly mononuclear complexes are achieved. The photoluminescence behavior of each complex was investigated, and the found properties related to its molecular structure. For mononuclear neutral copper(I) complexes based on the pyridine-amide ligands **5–7** and DPEPhos (**10**), a (ML+IL)CT emission was predicted by DFT calculations. The electronic effects of the pyridine-amide ligands play a major role in the character of the luminescent transition and thus, for the complex based on the ligand PyrBim (**5**) a participation of the bis(phosphine) in the complex's LUMO was found, whereas it does not contribute to the frontier orbital in the complexes **6-Cu-10** and **7-Cu-10** based on PyrTri (**6**) and PyrTet (**7**). The complexes **7-Cu-8** and **7-Cu-10** based on pyridine-tetrazolate were found to show highest emission efficiencies compared to other pyridine-amides, and even compared to their cationic counterparts, which was ascribed to a stronger and thus more rigid coordination of the copper(I) atom and less vibrational quenching due to loss of the (N)*H*-atom and the counterion. However, the dissociation tendency of the monodentate phosphine PPh₃ (**8**) brings along synthesis and stability challenges, hence DPEPhos with its flexible bite angle is the ligand of choice for stable copper(I) complexes with superior emission properties.[123,188,165] The most promising copper(I) complex candidate [(PyrTet)Cu(DPEPhos)] (**7-Cu-10**) based on PyrTet and DPEPhos is subject to further investigations in the next chapters.

3.2 Mononuclear Neutral Copper(I) Complexes based on Tetrazolate Ligands

3.2.1 The Model Complex [(PyrTet)Cu(DPEPhos)]

In the previous chapter 3.1 a synthesis route to mononuclear neutral copper(I) complexes [(N^N)Cu(P^P)] based on the pyridine-amide ligands PyrBim (**5**), PyrTri (**6**), PyrTet (**7**) and the bidentate DPEPhos (**10**) was presented, and the photoluminescence properties of the obtained compounds investigated. Since not only luminescence, but also thermal and redox stability are important parameters for efficient emitting materials in OLEDs, the complexes **5-Cu-10**, **6-Cu-10**, **7-Cu-10** as well as modified complexes are further investigated in cyclovoltammetry, thermogravimetric analysis and temperature-dependent photoluminescence measurements.

Since the complex [(PyrTri)Cu(DPEPhos)] (**6-Cu-10**) exhibits a surprisingly low PL quantum efficiency of only 2%, especially when compared to its cationic counterpart [(PyrTriH)Cu(DPEPhos)]BF$_4$ (**6-H-Cu-10**) (compare Table 9 and Figure 28), the modified complexes **16-Cu-10** and **17-Cu-10** based on 2-(1,2,4-triazol-5-yl)pyridine ligands **16-H**, **17-H** are also studied. The low photoluminescence efficiency of **6-H-Cu-10** might be a result of molecular packing in the amorphous powder, but to rule out intrinsic effects of the 1,2,3-triazolate ligand, also the mononuclear neutral complexes **16-Cu-10**, **17-Cu-10** with 1,2,4-triazolate ligands are taken into account (Figure 31).

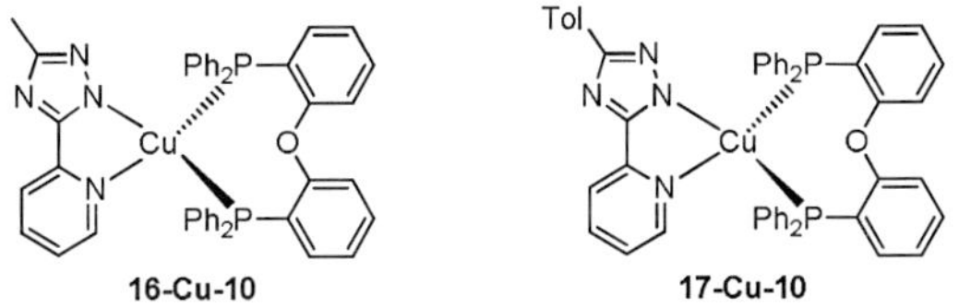

16-Cu-10   **17-Cu-10**

Figure 31. Molecular structures of the neutral copper(I) complexes [(5-(pyridin-2-yl)-3-methyl-1,2,4-triazolate)Cu(DPEPhos)] (**16-Cu-10**) and [(3-(*p*-tolyl)-5-(pyridin-2-yl)triazolate)Cu(DPEPhos)] (**17-Cu-10**) based on 1,2,4-triazolates.

The ligands 2-(3-methyl-1,2,4-triazol-5-yl)pyridine (**16-H**) and 2-(3-(*p*-tolyl)-1,2,4-triazol-5-yl)pyridine (**17-H**) are obtained either by reaction of the intermediate amidrazone **15** with *p*-toloylchloride in a basic medium or with acetic anhydride in acetic acid, and subsequent ring closure in a condensation reaction (Table 13).[189,190]

Table 13. Reaction conditions for the synthesis of (1,2,4-triazol-5-yl)pyridine-derivatives **16-H** and **17-H** via the amidrazone **15**.

	Reagent	Reaction conditions	Product	Yield
method **A**	HAc, Ac$_2$O	0 °C → rt, 4 h	**16-H**	31%
method **B**	Na$_2$CO$_3$, p-toloylchloride	DMF, 0 °C → rt, 12 h	**17-H**	48%

Mononuclear neutral complexes **16-Cu-10** and **17-Cu-10** are obtained in moderate to good yields by deprotonation of the (1,2,4-triazol-5-yl)pyridine ligands **16-H** and **17-H** with potassium hydroxide and complexation with the copper(I) salt **13** and the bis(phosphine) DPEPhos (**10**), as developed in chapter 3.1 (see Scheme 2).

As first step, the photophysical properties of the complexes **16-Cu-10** and **17-Cu-10** are compared to complex [(PyrTri)Cu(DPEPhos)] (**6-Cu-10**) to ensure a luminescent transition between same excited states for the 1,2,4-triazolate based complexes as for the complex with 1,2,3-triazolate. The similar shape of absorption spectra hints at same electronic effects of the 1,2,4-triazolate and 1,2,3-triazolate ligands, which is also confirmed by the excitation values calculated by DFT (Tables 14, 15, further details in Figure 87 in 5.3.1 and Figure 100 in 5.3.2). The frontier orbitals of both complexes depict a (ML+IL)CT transition from the copper(1) center and the triazolate moiety to the pyridine-amide ligand, while in complex **16-Cu-10** also the bis(phosphine) ligand partially contributes to the LUMO orbital.

Table 14. Calculated HOMO and LUMO energies, as well as excitation energies for the complexes [(5-(pyridin-2-yl)-3-methyl-1,2,4-triazolate)Cu(DPEPhos)] (**16-Cu-10**), in comparison to [(5-(pyridin-2-yl) 1,2,3-triazolate)Cu(DPEPhos)] (**6-Cu-10**). The frontier orbitals are obtained by DFT calculations using the B3LYP functional with def2-SV(P) basis set, and the excitation energies are calculated by TD-B3LYP.[V]

Complex	HOMO [eV]	LUMO [eV]	$\Delta E_{HOMO\text{-}LUMO}$ [eV]	Excitation energy [eV]
6-Cu-10	−5.23 eV	−1.19 eV	4.04 eV	3.23
16-Cu-10	−5.04 eV	−1.10 eV	3.94 eV	3.26

The complexes **16-Cu-10** and **17-Cu-10** exhibit yellow emission at 550 nm and 548 nm, respectively, with emission efficiencies of 22% and 21%, which are significantly higher than of the 1,2,3-triazolate based complex **6-Cu-10** (PLQY: 2%). The emission color is red-shifted compared to complex **6-Cu-10** due to the electron-donating effect of the substituents on the 1,2,4-triazolate ring, thus destabilizing the HOMO of the complex (and the LUMO to a lesser extent) and decreasing the HOMO-LUMO energy gap. Interestingly, the different substituents on the triazolate ring in 2-(3-methyl-1,2,4-triazol-5-yl)pyridine (**16-H**) and 2-(3-(*p*-tolyl)-1,2,4-triazol-5-yl)pyridine (**17-H**) have the same effect on the emission properties of the complexes. Furthermore, the photoluminescence quantum efficiencies of the complexes **16-Cu-10** and **17-Cu-10** are assumed to be affected by vibrational quenching of CH_2- or CH_3-groups from the substituents, and thus an even higher photoluminescence emission efficiency could be expected for the 1,2,3-triazolate based complex **6-Cu-10** without these substituents. Since this is not the case despite similar electronic systems of the 1,2,3-triazolate and 1,2,4-triazolate based complexes **16-Cu-10** and **6-Cu-10**, the low PL quantum efficiency of **6-Cu-10** might be attributed to molecular packing effects.

Table 15. Photoluminescence characteristics of the complexes **6-Cu-10**, **16-Cu-10**, **17-Cu-10** based on 2-(triazol-5-yl)pyridine ligands; sh denotes a shoulder in the absorption spectra.

Complex	λ_{abs} [nm] (ε [$10^4 M^{-1} cm^{-1}$])	λ_{em} [nm]	Φ_{em}
6-Cu-10	273 (2.14), 313 (1.49), 356 (sh, 0.38)	532	0.02
16-Cu-10	273 (2.80), 322 (1.64), 361 (sh, 0.44)	550	0.22
17-Cu-10	275 (2.06), 317 (1.42), 358 (sh, 0.37)	548	0.21

Now, the (electro)chemical and thermal stabilities of the complexes [(N^N)Cu(DPEPhos)] **5-Cu-10**, **16-Cu-10** and **7-Cu-10** based on the pyridine-amide ligands PyrBim (**5**), PyrTri$_{1,2,4}$ **16** and PyrTet (**7**) can be examined. To start with the electrochemical stability, which also draws conclusions for the sensitivity against oxygen, cyclovoltammetric measurements are recorded for the complexes [(PyrBim)Cu(DPEPhos)] (**5-Cu-10**), [(5-(pyridin-2-yl)-3-methyl-1,2,4-triazolate)Cu(DPEPhos)] (**16-Cu-10**), [(PyrTet)Cu(DPEPhos)] (**7-Cu-10**) (Figure 32). Irreversible oxidation peaks are found at +0.33 V, +0.33 V and +0.59 V for **5-Cu-10**, **16-Cu-10** and **7-Cu-10**, which are attributed to the oxidation of the copper(I) metal center (full circle for complex **7-Cu-10** in Figure 106 in 5.3.3). Compared to cationic complexes [(N^N)Cu(P^P)]BF$_4$, the oxidation potentials are relatively low and reflect the higher sensitivity towards oxygen, especially in solution.[101,132,178,191,192] The observed trend is in accord with the shallow HOMO energies in the order PyrBim > PyrTri > PyrTet obtained by DFT calculations, and was already experimentally observed in chapter 3.1.2.2. To conclude, the complex [(PyrTet)Cu(DPEPhos)] (**7-Cu-10**) based on the tetrazolate ligand is least sensitive towards oxygen in the series of investigated complexes.

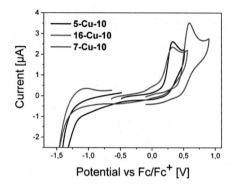

Figure 32. Cyclovoltammetric scans of the complexes [(PyrBim)Cu(DPEPhos)] (**5-Cu-10**), [(5-(pyridin-2-yl)-3-methyl-1,2,4-triazolate)Cu(DPEPhos)] (**16-Cu-10**), [(PyrTet)Cu(DPEPhos)] (**7-Cu-10**), calibrated against $FeCp_2/FeCp_2^+$.[VI]

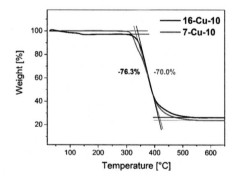

Figure 33. Thermogravimetric analysis of the complexes a) [(5-(pyridin-2-yl)-3-methyl-1,2,4-triazolate)Cu(DPEPhos)] (**16-Cu-10**) and b) [(PyrTet)Cu(DPEPhos)] (**7-Cu-10**). The negligible weight loss at 132 °C of **16-Cu-10** is attributed to evaporation of solvent embedded in the precipitated powder.[VII]

Thermal stabilities of the complexes [(5-(pyridin-2-yl)-3-methyl-1,2,4-triazolate)Cu(DPEPhos)] (**16-Cu-10**) and [(PyrTet)Cu(DPEPhos)] (**7-Cu-10**) are examined by thermogravimetric analysis (Figure 33). Both complexes are stable until above 300 °C and then decompose, probably by loss of the bis(phosphine) ligand, as this would amount to 71% weight loss for **16-Cu-10** and to 72%

[VI] Cyclovoltammetric measurements were recorded by Dr. Georgios Liaptsis and Dr. Sebastian Dück, CYNORA GmbH (see also 5.1.3).
[VII] Thermogravimetric analyses were measured by Dr. Mathias Schwotzer, Karlsruhe Institute of Technology (see also 5.1.4).

for **7-Cu-10**, resembling the recorded weight losses of 70% and 76% around 310–420 °C in the thermogravimetric measurements, although for complex **7-Cu-10** it seems to be a stepwise loss of ligand, indicating the formation of an intermediate complex species. Compared to other copper(I) complexes,[107,164,193] the herein studied compounds exhibit remarkably high thermal stabilities and loss of ligands is only observed above 300 °C.

Proceeding with the luminescence properties of the mononuclear neutral complexes [(N^N)Cu(DPEPhos)] **5-Cu-10**, **16-Cu-10**, **17-Cu-10** and **7-Cu-10** based on the pyridine-amide ligands PyrBim (**5**), 5-(pyridin-2-yl)1,2,4-triazolate **16** and **17**, and PyrTet (**7**), an increase of photoluminescence efficiency and blue-shift of the emission maxima is found in this order (Table 16), as was already described in the detailed study in chapter 3.1. The blue-shift is not as significant from complex [(PyrBim)Cu(DPEPhos)] (**5-Cu-10**) to the complexes [(5-(pyridin-2-yl)1,2,4-triazolate)Cu(DPEPhos)] **16-Cu-10** and **17-Cu-10** due to introduction of electron-donating substituents on the triazolate moiety, thus reducing the HOMO-LUMO energy gap compared to the unsubstituted complex. Even the 1,2,4-triazolate based complexes **16-Cu-10** and **17-Cu-10** do not achieve the remarkably high emission efficiencies of [(PyrTet)Cu(DPEPhos)] (**7-Cu-10**), despite similar emission characters, namely (ML+IL)CT from copper(I) d-orbitals with contributions from the amide-ring to the whole pyridine-amide ligand.

Table 16. Photoluminescence characteristics of the mononuclear neutral complexes **5-Cu-10**, **16-Cu-10**, **17-Cu-10**, **7-Cu-10** as amorphous powders when excited at 350 nm.

Complex	λ_{em} [nm]	Φ_{em}
5-Cu-10	551	0.10
16-Cu-10	550	0.22
17-Cu-10	548	0.21
7-Cu-10	510	0.78

Furthermore, the thermally activated delayed fluorescence behavior of the complexes [(5-(pyridin-2-yl)-3-methyl-1,2,4-triazolate)Cu(DPEPhos)] (**16-Cu-10**), [(3-(p-tolyl)-5-(pyridin-2-yl)-triazolate)Cu(DPEPhos)] (**17-Cu-10**) and [(PyrTet)Cu(DPEPhos)] (**7-Cu-10**) is investigated briefly. A red-shift and an increase of the emission decay times are observed for all compounds when going from 293 K to 77 K (Figure 34 and Table 17, and Figure 88 in 5.3.1). Especially the drastic elongation of emission decay times indicates emission from a different excited state at low temperature, and thus TADF behavior is assumed for all three complexes. But DFT calculations predict large energy gaps for the triazolate based (and benzimidazolate based) complexes of 0.56 eV for **16-Cu-10** (0.57 eV for **5-Cu-10**) compared to 0.20 eV for **7-Cu-10** (TD-B3LYP). Significantly longer phosphorescence lifetimes of the complexes **16-Cu-10** and **17-Cu-10** together with a small red-shifts hint at only weak TADF emission, i.e. 293 K is not sufficiently high temperature to completely repopulate the excited singlet state by thermal activation. In contrast, complex **7-Cu-10** is expected to show efficient thermally activated delayed fluorescence at ambient temperature, which will be subject to further investigation in chapter 3.3.

a)   b)

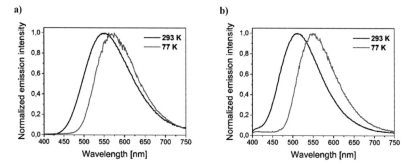

Figure 34. Emission spectra of the mononuclear neutral complexes a) [(5-(pyridin-2-yl)-3-methyl-1,2,4-triazolate)Cu(DPEPhos)] (**16-Cu-10**) and b) [(PyrTet)Cu(DPEPhos)] (**7-Cu-10**) as amorphous powder at 293 K and 77 K when excited at 350 nm.

Table 17. Photoluminescence characteristics of the mononuclear complexes **16-Cu-10**, **17-Cu-10**, **7-Cu-10** as amorphous powders at 293 K and 77 K.

	7-Cu-10	16-Cu-10	17-Cu-10
PL maximum at 293 K [nm]	510	550	548
PL maximum at 293 K [nm]	549	571	574
emission decay time at 77 K [μs]	22.0	9.9[a]	12.7
emission decay time at 77 K [μs]	111.5	124.2[a]	173.7[a]

[a] PL emission decay time is composed of two components. A weighted-average lifetime is used (τ_{ave}) and calculated by the equation $\tau_{ave} = \Sigma A_i \tau_i / \Sigma A_i$ with A_i as the pre-exponential factor for the lifetime.

Summarizing the above described experimental results, the most promising mononuclear neutral copper(I) complex class is based on 5-(pyridin-2-yl)tetrazolate ligands and the bis(phosphine) DPEPhos (**10**), as it shows superior emission properties with outstanding emission quantum yields and promising TADF behavior, and exhibits high thermal and electrochemical stability. Furthermore, the chosen ligands have proven as good combination to yield mononuclear neutral copper(I) complexes in high yields and purity. In the next sections, the model compound [(PyrTet)Cu(DPEPhos)] (**7-Cu-10**) will be modified by ligand substitution and variation to investigate the influence on the luminescence and chemical properties, and gain a deeper understanding of the photophysical processes occurring in the complexes after excitation.

3.2.2 Synthesis of Ligands and Complexes

3.2.2.1 Ligand Variation and Substitution

Starting from the model complex [(PyrTet)Cu(DPEPhos)] (**7-Cu-10**), both the 5-(pyridin-2-yl)-tetrazolate and the bis(phosphine) ligand can be varied for different emission characters, luminescence colors and efficiencies (Figure 35). When the tetrazole moiety is varied against benzimidazole, 1,2,3-triazole or 1,2,4-triazole, as discussed in chapters 3.1.3.2 and 3.2.1, the localization of frontier orbitals as well as their contributions to the luminescent transition change. While for the benzimidazolate based complex **5-Cu-10** also the bis(phosphine) participates in the LUMO and LUMO+1, the frontier orbitals are localized mainly on the copper(I) center, the phosphorous atoms and the pyridine-amide ligand for the complexes **6-Cu-10** and **16-Cu-10** with pyridine-triazolate ligands, and [(PyrTet)Cu(DPEPhos)] (**7-Cu-10**). Assuming the same emission origin in the whole class of tetrazolate complexes as for **7-Cu-10**, the modification of the pyridine moiety by ring variation or introduction of electron withdrawing or donating substituents will change the energy levels of the frontier orbitals and induce a color-shift. Taking a closer look on the localization of the frontier orbitals provided by DFT (Figure 30), this will mainly affect the energy of the LUMO orbital, as the HOMO is localized on the copper(I) center with contributions from the phosphorous atoms and the tetrazolate ring, and the LUMO is distributed over the whole pyridine-amide. On the other hand, the bis(phosphine) ligand only contributes to the luminescent transition by its phosphorous atoms and, due to comparable bite angles and donor abilities for different bis(phosphines), is expected to have only a minor impact on the complexes' luminescence. If this applies to all bis(phosphines) is examined by variation of the DPEPhos ligand by slightly modified derivatives and further bis(phosphines).

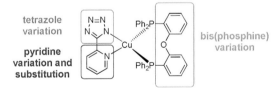

Figure 35. Variation possibilities of the model compound [(PyrTet)Cu(DPEPhos)] (**7-Cu-10**) to investigate the ligands' influence on the photophysical properties. While tetrazole variation was already discussed in detail in chapter 3.1.3.2 and 3.2.1, the bis(phosphine) and pyridine variation and substitution is addressed in the following sections.

Bis(phosphines)

Several bis(phosphines) of different steric hindrance, bite angle and flexibility are chosen to study their influence on the photophysical properties of the corresponding, mononuclear copper(I) complexes [(PyrTet)Cu(P^P)] (Figure 36). They can be mainly divided into DPEPhos derivatives **18–21** and further commonly used bis(phosphines) **22–24**, such as (–)-2,3-O-isopropylidene-2,3-dihydroxy-1,4-bis(diphenylphosphinobutane) ((–)-DIOP, **22**) or (R)-2,2'-bis(diphenylphosphino)-1,1'-binaphthyl (R-BINAP, **23**). As already mentioned above, the electronic effect of the bis(phosphine) on the complexes' emission is limited to the donor ability of its phosphorous atoms, since only these contribute to the luminescent transition. While the donor abilities of the DPEPhos derivatives **18–21** are quite similar, (-)-DIOP **(22)** represents a strong electron-donor, and R-BINAP **(23)** probably shows lower donating ability.[163,186] However, the main criterion for the chosen bis(phosphines) is their steric effect, in particular different bite angles and bulkiness. To examine the influence of different bite angles and/or Cu-P bond lengths on the complex's emission, rigid bis(phosphines) such as 4,5-bis(diphenylphosphino)-9,9-dimethylxanthene (XantPhos, **19**) and 4,6-bis(diphenylphosphino)-dibenzofuran (DBFPhos, **20**) with large bite angles, and the chiral phosphines **22**, **23** of narrow coordination angles are selected (Table 18). The ligands DPEPhos (**10**), bis(2-diphenylphosphino-p-tolyl)ether (PTEPhos, **18**) and bis(2-diphenylphosphinophenyl)thioether (DPTPhos, **21**) can adopt their coordination geometry to the metal center due to their flexible backbone, while the rigid (S)-4,12-bis(diphenylphosphino)-[2.2]-paracyclophane (S-Phanephos, **24**) with a large P-P distance induces large bite angles.[194] The variation of the P-P distances and thus the bite angle is challenging to induce a significant effect on the copper(I) coordination geometry while still forming a complex. A further aspect of the flexibility and the steric hindrance of the bis(phosphine) backbone concerns the proposed flattening distortion of mononuclear copper(I) complexes in the excited state. Steric bulkiness of the ligands is assumed to prevent this flattening and thus non-radiative processes, e.g. a copper(I) complex with the very rigid S-PhanePhos (**24**) was found to be highly luminescent in powder, film and solution despite decreasing rigidity of the environment.[145] As the DPEPhos (**10**) ligand has proven to be a good candidate for highly luminescent, mononuclear copper(I) complexes so far, the question is raised, if the oxygen atom interacts with the formal Cu(II) center in the excited state. Its sulfur analogue DPTPhos (**21**) is thus synthesized to investigate the effect of the ether-bridge atom in the copper(I) complex.

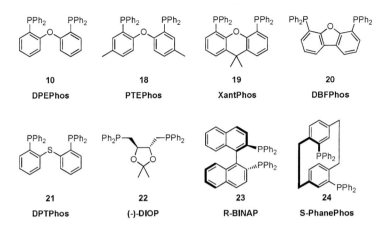

Figure 36. Selected bis(phosphines) of various steric hindrance, bite angle and flexibility for the incorporation into mononuclear neutral copper(I) complexes of the type [(PyrTet)Cu(P^P)].

Table 18. Preferred chelation angles as calculated from the steric hindrance of the ligand backbone.

Ligand	Natural bite angle	Flexibility range
DPEPhos (**10**)[172]	102°	86°–120°
PTEPhos (**18**)[195]	108°	93°–124°
XantPhos (**19**)[172]	112°	97°–135°
DBFPhos (**20**)[172]	131°	117°–147°
DIOP (**22**)[169]	102°	90°–120°
BINAP (**23**)[169,196]	92°, 85°a	

[a] Bite angle typically found in crystal structures of Pd(II) or Ni(II) complexes.

Except PTEPhos (**18**), DBFPhos (**20**) and DPTPhos (**21**), all bis(phosphines) are commercially available. The non-commercial candidates are obtained by directed ortho-lithiation of the respective diphenylether or -thioether, respectively, with n-buthyllithium and subsequent reaction with diphenylphosphinous chloride (Scheme 5).[197] Reaction yields are found to be low and could be optimized by using sec- or $tert$-butyllithium, but this is not the scope of this work.

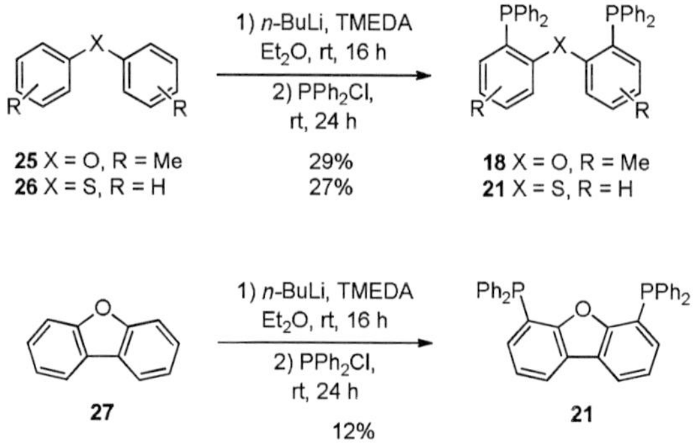

1) *n*-BuLi, TMEDA
Et$_2$O, rt, 16 h

2) PPh$_2$Cl,
rt, 24 h

25 X = O, R = Me   29%   18 X = O, R = Me
26 X = S, R = H   27%   21 X = S, R = H

1) *n*-BuLi, TMEDA
Et$_2$O, rt, 16 h

2) PPh$_2$Cl,
rt, 24 h

27   12%   21

Scheme 5. Synthesis of the bis(phosphines) PTEPhos (**18**), DBFPhos (**19**) and DPTPhos (**20**) from *p*-tolyl ether (**25**), diphenyl sulfide (**27**) or dibenzofuran (**26**).

Tetrazoles

Besides the more steric influence of the bis(phosphines) on the photoluminescence of the copper(I) complexes, the tetrazolate ligand strongly controls their emission color (and efficiency) by its electronic character, as was already discussed in chapter 3.1.3.2. By modification of the pyridine moiety of the PyrTetH (**7-H**) ligand, better understanding of the electronic system of the mononuclear neutral copper(I) complexes based on bidentate tetrazole ligands is explored. This is approached by ring variation such as enlargement of the aromatic system (**29-H**, **30-H**), substitution of the pyridine against five-membered heterocycles (**32-H–40-H**), or by introduction of electron donating substituents (**41-H–46-H**) (Figure 37).

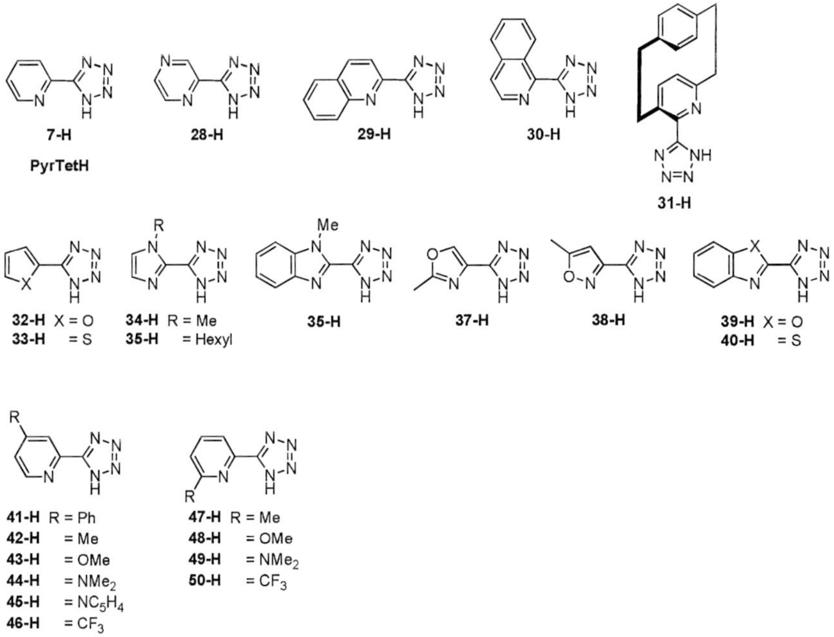

Figure 37. Selected imine-tetrazole ligands **28-H–50-H** to modify the pyridine moiety of PyrTetH (**7-H**) by ring variation or introduction of substituents.

Similar to the choice of ligands in chapter 3.1, a second – potentially coordinating – N-atom is introduced in 2-(tetrazol-5-yl)pyrazine (**28-H**) to reduce the electron donating ability of the heterocycle and thus lower the LUMO energy of the complex and red-shift the emission color. An analogical effect is achieved by enlargement of the aromatic system in 2-(tetrazol-5-yl)quinoline (**29-H**) and 1-(tetrazol-5-yl)isoquinoline (**30-H**). The position of the tetrazole towards the quinoline is varied to induce a higher steric hindrance around the copper(I) center in **29-H** and prevent geometry changes after excitation leading to non-radiative deactivation. Also with the novel pyridinophane-based ligand **31-H** the influence of a more rigid backbone is aimed to study.

Substitution of the pyridine moiety leads to 5-(furan-2-yl)tetrazole (**32-H**), 5-(thiophen-2-yl)-tetrazole (**33-H**) and 2-(tetrazol-5-yl)imidazoles **34-H–36-H** as simplest five-membered heterocycles with different coordinating atoms O, S, N to investigate their formation ability concerning mononuclear neutral copper(I) complexes and the luminescence. The electronic character of the electron rich five-membered heterocycles is changed by introduction of further heteroatoms, and their position varied in 2-methyl-4-(tetrazol-5-yl)oxazole (**37-H**) and 5-methyl-3-(tetrazol-5-yl)isoxazole (**38-H**). Thus not only the electron density of the heterocycle is reduced, but also the influence of further potentially coordinating atoms is investigated. Additionally, the aromatic system of the five-membered heterocycles is enlarged to benzimidazole (**36-H**), benzoxazole (**39-H**) and benzothiazole (**40-H**).

As further modification of the pyridine moiety of the PyrTetH (**7-H**) ligand, electron donating substituents such as $-Ph$, $-Me$, $-OMe$, and $-NC_5H_4$ groups are introduced to raise the HOMO and LUMO energies of the ligand and thus induce a blue-shift of the complexes emission. Pyridine-tetrazolate ligands substituted by a $-NMe_2$ group are synthetically not available due to low reaction yields and difficult isolation, which will be discussed later on. Also the ligands with an electron withdrawing group $-CF_3$, which is aimed to decrease the LUMO level of the respective copper(I) complex and result in an emission shift to longer wavelengths, cannot be tested for mononuclear copper(I) complexes, since the respective carbonitrile is not available (see section "Carbonitriles"). If the position of the substituent effects the electronic system of the ligand and thus of the complexes, as reported for some iridium(III) complexes,[198,199] or mainly induces different steric hindrance, is examined by substitution of the pyridine in para-(**41-H–44-H**) and ortho-position (**45-H, 46-H**).

The tetrazoles **28-H–46-H** are synthesized from their respective carbonitriles **28-CN–46-CN** in a [3+2] cycloaddition reaction with sodium azide (Table 19). For pyrazine-2-carbonitrile (**28-CN**) the SHARPLESS method is applied with zinc bromide as Lewis acid in aqueous medium,[200] but this route does not prove to be efficient as a Zn(II) complex is formed with the desired product, which makes the isolation of 2-(tetrazol-5-yl)pyrazine (**28-H**) difficult. For all other tetrazoles **29-H–46-H** the reaction is performed with sodium azide in acidic medium by three different methods **A**, **B**, **C**. Method **A** based on ammonium chloride is herein used as standard method.[174,175] Similar as method **B** with acetic acid, it is advantageous due to the isolation of the product by direct extraction from the reaction mixture into water, but uses the toxic DMF as solvent.[201] Less reactive carbonitriles can be converted to tetrazoles with pyridinium hydrochloride in method **C** at lower temperature and shorter reaction times.[202] For all tetrazoles high temperatures and long reaction times are required, and especially electron-rich or sterically hindered carbonitriles, e.g. imidazole-2-carbonitriles **34-CN** and **35-CN** or pyridinophane-carbonitrile **31-CN**, lead to low conversion. As purification of the products by column chromatography is not possible due to their polarity, the tetrazoles are usually obtained from the aqueous phase by acidification, which leads to precipitation of the pure product. Besides the lower reactivity of the methoxy-substituted pyridinecarbonitriles **43-CN** and **46-CN**, also their high polarity reduces reaction yields when precipitating from water. For the 4-dimethylamino-pyridine-2-carbonitrile (**47-CN**) and

2-dimethylamino-pyridine-6-carbonitrile (**48-CN**) only very low conversion is observed during the reaction and isolation of the very polar products from the water phase and inorganic salts is not possible in several attempts. Thus, the less polar 2-(tetrazol-5-yl)-4-piperidylpyridine (**44-H**) with a piperidyl-group is further synthesized. Except the beforehand discussed examples, all tetrazoles are obtained in high yields, and characterized by FAB mass spectrometry and elemental analysis.

Table 19. Synthesis methods, conditions and reaction yields for the reaction of carbonitriles **28-CN–46-CN** to the respective tetrazoles **28-H–46-H**.

Entry	R=	Method	Conditions	Product	Yield
28-CN	pyrazine-2-	NaN₃, ZnBr₂	100 °C, 12 h	**28-H**	10%
29-CN	quinoline-2-	A	130 °C, 48 h	**29-H**	84%
30-CN	isoquinoline-1-	A	130 °C, 48 h	**30-H**	65%
31-CN	(rac)-[2](1,4)benzo[2](2,5)pyridinophane-13-	A	150 °C, 44 h	**31-H**[a]	76%
32-CN	furan-2-	A	100 °C, 24 h	**32-H**	71%
33-CN	thiophen-2-	A	120 °C, 48 h	**33-H**	78%
34-CN	N-methylimidazole-2-	A	120 °C, 48 h	**34-H**[b]	44%
35-CN	N-hexylimidazole-2-	A	120 °C, 85 h	**35-H**[b]	51%
36-CN	N-methylbenzimidazole-2-	A	130 °C, 85 h	**36-H**[b]	79%
37-CN	2-methyloxazole-4-	C	100 °C, 40 h	**37-H**	63%
38-CN	5-methylisoxazole-3-	C	100 °C, 48 h	**38-H**	44%
39-CN	benzoxazol-2-	C	100 °C, 48 h	**39-H**	77%
40-CN	benzothiazol-2-	C	100 °C, 48 h	**40-H**	82%
41-CN	4-phenylpyridine-2-	A	120 °C, 48 h	**41-H**	92%
Entry	R=	Method	Conditions	Product	Yield
41-CN	4-phenylpyridine-2-	B	80 °C, 12 h	**41-H**	64%
42-CN	4-methylpyridine-2-	A	130 °C, 48 h	**42-H**[b]	71%
43-CN	4-methoxypyridine-2-	B	80 °C, 120 h	**43-H**	54%
44-CN	4-piperidiylpyridine-2-	A	130 °C, 48 h	**44-H**	77%
45-CN	2-methylpyridine-6-	A	120 °C, 48 h	**45-H**	79%
46-CN	2-methoxypyridine-6-	B	80 °C, 72 h	**46-H**	43%
47-CN	4-dimethylamino-pyridine-2-	A	100 °C, 15 h 120 °C, 24 h	-	-
48-CN	2-dimethylamino-pyridine-6-	A	100 °C, 15 h 120 °C, 24 h	-	-

[a] The tetrazole is provided by Dr. Joshua Kramer. [b] The tetrazoles are generously provided by Dr. Daniel Zink, CYNORA GmbH.

The pyridinophane-tetrazole **31-H** is generously provided by Joshua Kramer of this research group and the synthesis of the compound was already published in his dissertation.[203]

Carbonitriles

The carbonitrile precursors **28-CN–46-CN** for the tetrazole ligands **28-H–46-H** are either commercially available (**28-CN–30-CN, 32-CN, 33-CN, 41-CN, 42-CN, 45-CN, 46-CN**) or are synthesized by introduction of the nitrile-group or amide dehydration. In particular, the five-membered ring carbonitriles **34-CN–36-CN** are obtained in a copper(I) catalyzed cyanization reaction from the unsubstituted heterocycles *N*-methylimidazole (**34**), *N*-hexylylimidazole (**35**) and *N*-methylbenzimidazole (**36**) in reasonable yields (Table 20).[204]

Table 20. Reaction conditions and yields for the synthesis of *N*-alkylimidazole-2-carbonitrile derivatives **34-CN–36-CN**.

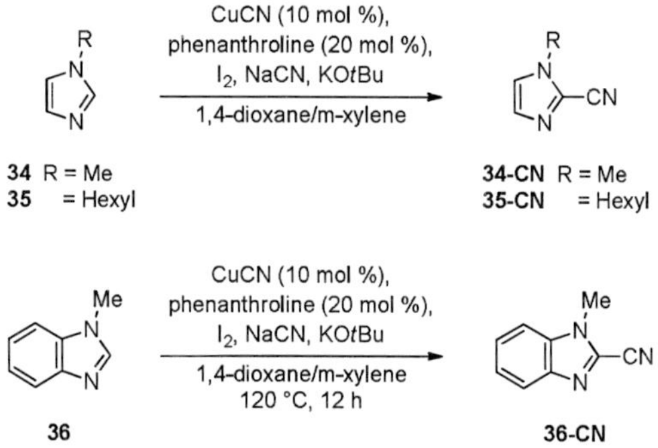

Entry	Reaction conditions	Product	Yield[a]
N-methylimidazole (**34**)	110 °C, 24 h	*N*-methylimidazole-2-carbonitrile (**34-CN**)	31%
N-hexylimidazole (**35**)	120 °C, 36 h	*N*-hexylimidazole-2-carbonitrile (**35-CN**)	16%
N-methylbenzimidazole (**36**)	120 °C, 12 h	*N*-methylbenzimidazole-2-carbonitrile (**36-CN**)	25%

[a] The reactions are performed by Dr. Daniel Zink, CYNORA GmbH.

The oxazole, isoxazole and thiazole based carbonitriles **37-CN, 38-CN, 39-CN, 40-CN** on the other hand are synthesized from the respective carboxylates **37-CO₂R″, 38-CO₂R″, 40-CO₂R″** or aldoximine **39-NOH**, as shown in Scheme 6. Conversion of the carboxylate-group to an amide by reaction with aqueous ammonium hydroxide solution, and subsequent dehydration of the amide to nitrile afforded the carbonitriles **37-CN, 38-H** and **40-CN** in good yields (Table 21),[205] while benzoxazole-2-carbonitrile (**39-H**) is obtained from synthesis of the aldoximine **39-NOH** and following reaction with thionyl chloride.[206]

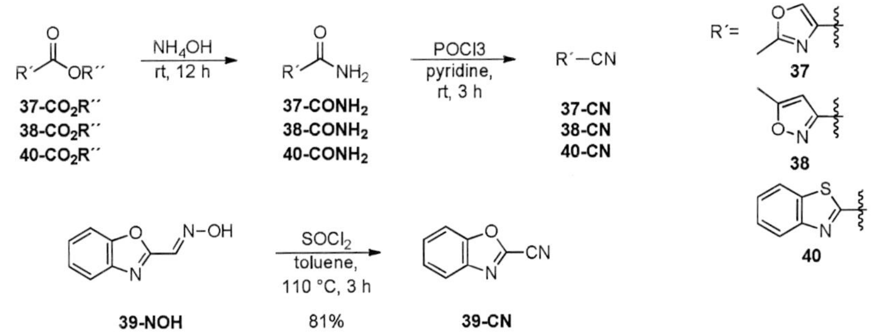

Scheme 6. Synthesis route to oxazole, isoxazole and thiazole based carbonitriles **37-CN**, **38-CN**, **40-CN** from their respective carboxylates **37-CO₂R**, **38-CO₂R**, **40-CO₂R** or the aldoximine **39-NOH** (R'' = Me, Et).

Table 21. Reaction yields for the synthesis of carbonitriles **37-CN**, **38-CN**, **40-CN** (product 2) via the respective amides **37-CONH₂**, **38-CONH₂**, **40-CONH₂** (product 1).

Entry	Product 1	Yield	Product 2	Yield
2-methyloxazole-4-(methyl)carboxylate (37)	2-methyloxazole-4-carboxamide (37-CONH₂)	81%	2-methyloxazole-4-carbonitrile (37-CN)	77%
5-methylisoxazole-3-(ethyl)carboxylate (38)	5-methylisoxazole-3-carboxamide (38-CONH₂)	63%	5-methylisoxazole-3-carbonitrile (38-CN)	66%
benzothiazole-2-(methyl)carboxylate (40)	benzothiazole-2-carboxamide (40-CONH₂)	84%	benzothiazole-2-carbonitrile (40-CN)	95%

Introduction of the nitrile-group on the pyridine-ring is achieved by the REISSERT-HENZE route via N-oxides (Scheme 7). The substituted pyridines **43**, **49**, **50–52** are treated with *m*CPBA to afford the N-oxides **43-O**, **49-O**, **50-O**, **51-O**, **52-O** in moderate to good yields,[207] and then further reacted with trimethylsilyl cyanide and dimethylcarbamoyl chloride to give the pyridine-2-carbonitriles **43-CN**, **47-CN** and **48-CN** (Table 22).[208,209] Long reaction times are required for the nucleophilic substitution reaction, and despite stirring for 10 days no conversion is observed for the trifluoromethyl-substituted pyridines **51** and **52**, but starting material is isolated.

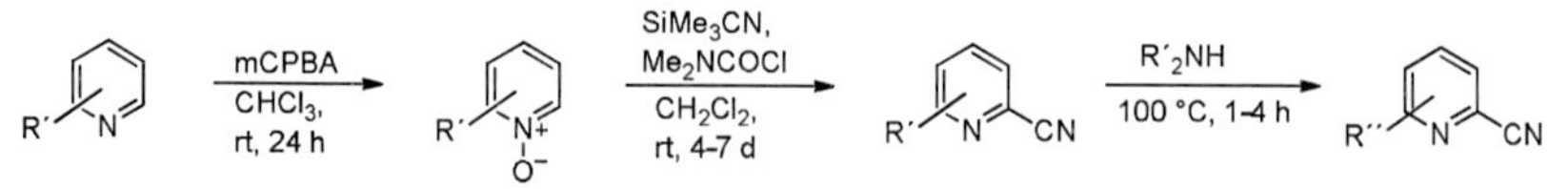

43 R' = 4-OMe   **43-O** R' = 4-OMe   **43-CN** R' = 4-OMe
49   = 4-Cl   **49-O**   = 4-Cl   **49-CN**   = 4-Cl   **44-CN** R' = 4-NC$_5$H$_4$
                                                   47-CN   = 4-NMe$_2$
50   = 6-Cl   **50-O**   = 6-Cl   **50-CN**   = 6-Cl   **48-CN**   = 6-NMe$_2$
51   = 4-CF$_3$ **51-O**   = 4-CF$_3$
52   = 6-CF$_3$ **52-O**   = 6-CF$_3$

Scheme 7. Synthesis route for the pyridine-carbonitriles **43-CN**, **44-CN**, **47-CN**, **48-CN**, **49-CN**, **50-CN** via the pyridine-N-oxides in a Reissert-Henze reaction.

Table 22. Reaction yields for the synthesis of pyridine-carbonitriles **43-CN**, **49-CN**, **50-CN**, **51-CN**, **52-CN** via the pyridine-N-oxides **43-O**, **49-O**, **50-CN**, **51-O**, **52-O**.

Entry	N-oxide	Yield	Product	Yield
4-methoxy-pyridine (43)	4-methoxypyridine-N-oxide (43-O)[a]	71%	4-methoxypyridine-2-carbonitrile (43-CN)[a]	77%
4-chloropyridine (49)	4-chloropyridine-N-oxide (49-O)	48%	4-chloro-pyridine-2-carbonitrile (49-CN)	50%
2-chloropyridine (50)	2-chloropyridine-N-oxide (50-O)	63%	2-chloropyridine-6-carbonitrile (50-CN)	46%
4-trifluoromethyl-pyridine (51)	4-trifluoromethylpyridine-N-oxide (51-O)	73%	-	-
2-trifluoromethyl-pyridine (52)	2-trifluoromethylpyridine-N-oxide (52-O)	51%	-	-

[a] The synthesis of carbonitrile **43-CN** via the N-oxide **43-O** is performed by Dr. Daniel Zink, CYNORA GmbH.

The dialkylamino-substituted pyridines **44-CN**, **47-CN**, **48-CN** are obtained by ipso-substitution of the respective chloropyridine-2-carbonitriles **49-CN**, **50-CN** with an amine in good yields (Table 23).[210]

Table 23. Reaction yields for the ipso-substitution of chloro-pyridine-2-carbonitriles **49-CN**, **50-CN** to dialkylamino-pyridine-2-carbonitriles **44-CN**, **47-CN**, **48-CN**.

Entry	Reagent	Product	Yield
4-chloro-pyridine-2-carbonitrile (49-CN)	piperidine	4-piperidilpyridine-2-carbonitrile (44-CN)[a]	65%
	Me$_2$NH (40% in water)	4-dimethylamino-pyridine-2-carbonitrile (47-CN)	78%
2-chloropyridine-6-carbonitrile (50-CN)	Me$_2$NH (40% in water)	2-dimethylamino-pyridine-6-carbonitrile (48-CN)	60%

[a] The carbonitrile is generously provided by Dr. Michael Danz, CYNORA GmbH.

To sum up, except the trifluoromethyl-substituted pyridine-carbonitriles **51-CN** and **52-CN**, all desired carbonitriles are obtained by nitrile introduction in a copper(I) catalyzed reaction for the five-membered heterocycles or by nucleophilic substitution on the pyridine, or amide dehydration.

3.2.2.2 Complex Synthesis

Mononuclear neutral copper(I) complexes [(N^N)Cu(P^P)] are obtained either by complexation of [Cu(CH₃CN)₄]BF₄ (**13**) with a bis(phosphine) and the deprotonated tetrazolate ligand in the ratio 1:1:1 (GP4, see experimental section 5.2.1 for details), or by deprotonation of the respective cationic complex [(N^N)Cu(P^P)]BF₄ with a mild base (GP3, see experimental section 5.2.1 for details). Due to the poor solubility of the tetrazole and potassium hydroxide in aprotic solvents, a solvent mixture of ethanol/dichloromethane 1:3 is applied. To remove excess ligands and inorganic salts after the reaction, the solvents of the reaction mixture are evaporated, the residue extracted with few dichloromethane, and the dichloromethane filtrate precipitated in diethyl ether affording the pure complexes as solids. All complexes are characterized by FAB mass spectrometry, which gives the complex molecule as well as higher clusters in the gas phase, and elemental analysis. Thereby, solvent residuals have to be taken into account due to their inclusion during precipitation, as already observed for the complexes in chapter 3.1. Single crystals for X-ray diffraction analysis are usually obtained by layering a dichloromethane solution of the compound with diethyl ether under nitrogen atmosphere. Detailed crystallographic data can be found in the experimental part, chapter 5.4.

Applying the different bis(phosphines) **18–24** in the complexation reaction with [Cu(CH₃CN)₄]BF₄ (**13**) and the deprotonated PyrTet ligand (**7**) gives the neutral complexes **7-Cu-18**, **7-Cu-19**, **7-Cu-21**, and **7-Cu-23** in moderate to good yields. Furthermore, the cationic complexes **7-H-Cu-18**, **7-H-Cu-19** are obtained from the complexation reactions with PyrTetH (**7-H**) without base (GP2, Table 24), to later on compare their photophysical properties to the neutral counterparts **7-Cu-18** and **7-Cu-19**. The low reaction yield for **7-Cu-23** is less attributed to an inefficient complexation reaction, than more to the low polarity of the complex leading to losses during precipitation, which also prohibits isolation and purification of the complex with (-)-DIOP (**22**). On the other hand, no complex is formed with the bis(phosphine) DBFPhos (**20**), and with S-(+)-PhanePhos (**24**) a high oxygen sensitivity of the complex is observed during purification, which results in very low yields and poor purity of the compound.

Table 24. Reaction yields for the complexation of copper(I) with PyrTet (**7**), or PyrTetH (**7-H**) respectively, and the bis(phosphines) **18–24**.

Entry	Method	Complex	Yield
PyrTetH (**7-H**)	DPEPhos (**10**)	GP3, GP4	**7-Cu-10**[a]	61–68%
		GP2	**7-H-Cu-10**[a]	57%
	PTEPhos (**18**)	GP3	**7-Cu-18**	61%
		GP2	**7-H-Cu-18**	46%
	XantPhos (**19**)	GP3, GP4	**7-Cu-19**	37–75%
		GP2	**7-H-Cu-19**	77%
	DBFPhos (**20**)	GP4	-	-
	DPTPhos (**21**)	GP4	**7-Cu-21**	56%
	(-)-DIOP (**22**)	GP4	-[b]	-
	R-(+)-BINAP (**23**)	GP4	**7-Cu-23**	28%
	S-(+)-PhanePhos (**24**)	GP4	-[c]	-

[a] Already discussed in chapter 3.1.1; [b] cannot be isolated due to high solubility; [c] obtained in very low yields and elemental analysis is not correct due to high oxygen sensitivity.

To study the electronic influence of the tetrazolate ligand on the emission properties of the corresponding complex in comparison to the model complex [(PyrTet)Cu(DPEPhos)] (**7-Cu-10**), the pyridine ring is varied in the tetrazoles **28-H–31-H** and substituted against five-membered ring heterocycles in the ligands **32-H–40-H**. Complexation of the deprotonated ligands **28–30** with [Cu(CH₃CN)₄]BF₄ (**13**) and DPEPhos as described above, yields the brightly yellow compounds **28-Cu-10–30-Cu-10**, while with the structurally similar PTEPhos (**18**) the respective complex **30-Cu-18** is obtained (Table 25). A certain solubility of the quinoline and isoquinoline based complexes **29-Cu-10**, **30-Cu-10**, **30-Cu-18** accounts for losses during precipitation and thus low yields from purification. This is even more severe for the complex **31-Cu-10** with pyridinophane-tetrazolate **31**, which precipitates only from a cold diethyl ether/n-hexane mixture. Going further to the tetrazole ligands with five-membered ring heterocycles **32-H–40-H**, mononuclear neutral complexes are only obtained with the imidazole ligands **34-H–36-H**. While with 5-(furan-2-yl)tetrazole (**32-H**) and 5-(thiophen-2-yl)tetrazole (**33-H**) no complex formation is observed, the complexes based on 2-methyl-4-(tetrazol-5-yl)oxazole (**37-H**), 5-methyl-3-(tetrazol-5-yl)isoxazole (**38-H**) are higly soluble and cannot be isolated from precipitation. The reaction of ligands **39-H** and **40-H** with [Cu(CH₃CN)₄]BF₄ (**13**) and a bis(phosphine) is accomplished by first complexation of the tetrazolate ligand with the copper(I) salt and subsequent addition of the bis(phosphine) to the reaction mixture.

With 2-(tetrazol-5-yl)benzoxazole **(39-H)** dimeric structures are obtained, either as [(N^N)Cu₂(DPEPhos)₂]BF₄ **c-(39-Cu-10)** or as [(N^N)₂Cu₂(PTEPhos)₂] **c-(39-Cu-18)**.[VIII] To prevent these dimeric structures, also the bis(phosphine) XantPhos **(19)** is tested, but does not give a complex with 2-(tetrazol-5-yl)benzoxazole **(39-H)**. Difficultly soluble coordination polymers are formed with 2-(tetrazol-5-yl)benzothiazole **(40-H)** and DPEPhos **(10)**, which can be broken apart partially by very polar solvents and thus a mononuclear neutral complex **40-Cu-10** is obtained.

Table 25. Reaction yields for the complexation of copper(I) with the tetrazolates **28-H**–**40-H** and DPEPhos **(10)**, where D = O, S, N. The bis(phosphines) PTEPhos **(18)** and XantPhos **(19)** are applied in selected cases.

Entry	Method	Complex	Yield
2-(tetrazol-5-yl)pyrazine **(28-H)**	DPEPhos **(10)**	GP4	**28-Cu-10**	85%
2-(tetrazol-5-yl)quinoline **(29-H)**		GP4	**29-Cu-10**	55%
1-(tetrazol-5-yl)isoquinoline **(30-H)**		GP4	**30-Cu-10**	23%
	PTEPhos **(18)**	GP4	**30-Cu-18**	21%
13-(tetrazol-5-yl)-[2](1,4)benzo[2](2,5)pyridinophane **(31-H)**	DPEPhos **(10)**	GP4	**31-Cu-10**	34%
5-(furan-2-yl)tetrazole **(32-H)**	DPEPhos **(10)**	GP4	-	-
5-(thiophen-2-yl)tetrazole **(33-H)**		GP4	-	-
2-(tetrazol-5-yl)-*N*-methylimidazole **(34-H)**		GP4	**34-Cu-10**	68%
2-(tetrazol-5-yl)-*N*-hexylimidazole **(35-H)**		GP4	**35-Cu-10**	49%
2-(tetrazol-5-yl)-*N*-methyl-benzimidazole **(36-H)**		GP4	**36-Cu-10**	53%
2-methyl-4-(tetrazol-5-yl)oxazole **(37-H)**		GP4	-[a]	-
	PTEPhos **(18)**	GP4	-[a]	-
5-methyl-3-(tetrazol-5-yl)isoxazole **(38-H)**	DPEPhos **(10)**	GP4	-[a]	-
2-(tetrazol-5-yl)benzoxazole **(39-H)**		see text	**c-(39-Cu-10)**[b]	-
	PTEPhos **(18)**	GP4	-[b]	-
	XantPhos **(19)**	see text	-	-
2-(tetrazol-5-yl)benzothiazole **(40-H)**	DPEPhos **(10)**	see text	**40-Cu-10**	6%

[a] Not isolated due to high solubility; [b] dimeric structure.

[VIII] The structure of [(5-(benzoxazol-2-yl)tetrazolate)₂Cu₂(PTEPhos)₂] (**c-(39-Cu-18)**) is confirmed by X-ray diffraction analysis as mere characterization method.

A further modification of the pyridine moiety in the 5-(pyridin-2-yl)tetrazolate ligand (**7**) of the model complex [(PyrTet)Cu(DPEPhos)] (**7-Cu-10**), is substitution with different groups and in different positions. From complexation of the phenyl-substituted ligand **41** with copper(I) and different bis(phosphines), the mononuclear complexes **41-Cu-10**, **41-Cu-18**, **41-Cu-19**, **41-Cu-21**, **41-Cu-23** are obtained, while no complex is formed with DBFPhos (**20**), as expected from the results with PyrTet (**7**) (Table 26). On the other hand, a complex is formed with the bis(phosphine) (-)-DIOP (**22**), but cannot be isolated even when precipitating in ice cold *n*-pentane, due to its high solubility. The influence of electron donating substituents methyl-, methoxy- and piperidyl- of the pyridine moiety and their different positions on the complexes' properties are also investigated. Mononuclear complexes **42-Cu-10**, **43-Cu-10**, **44-Cu-10**, **45-Cu-10**, and **46-Cu-10** are obtained with all substituted pyridine-tetrazolate ligands **42**, **43**, **44**, **45**, **46** in reasonable to good yields. Since the formation of the copper(I) complexes also depends on the steric hindrance determined by the ligands, different reaction yields are observed for pyridine-tetrazolate ligands substituted in para- or ortho-position (compare **46-Cu-10** and **43-Cu-10**).

Table 26. Reaction yields for the complexation of copper(I) with pyridine-tetrazoles **41-H–43-H**, **45-H**, **47-H**, **48-H**, which bear substituent groups in para- or ortho-position of the pyridine, and the bis(phosphines) **10**, **18–24**.

Entry		Method	Complex	Yield
4-phenyl-2-(tetrazol-5-yl)pyridine (**41-H**)	DPEPhos (**10**)	GP4	**41-Cu-10**	62%
	PTEPhos (**18**)	GP4	**41-Cu-18**	28%
	XantPhos (**19**)	GP3, GP4	**41-Cu-19**	66–95%
	DBFPhos (**20**)	GP4	-	-
	DPTPhos (**21**)	GP4	**41-Cu-21**	60%
	(-)-DIOP (**22**)	GP4	-[a]	-
	R-(+)-BINAP (**23**)	GP4	**41-Cu-23**	57%
4-methyl-2-(tetrazol-5-yl)pyridine (**42-H**)	DPEPhos (**10**)	GP4	**42-Cu-10**	27%[b]
4-methoxy-2-(tetrazol-5-yl)pyridine (**43-H**)		GP4	**43-Cu-10**	74%
4-piperidyl-2-(tetrazol-5-yl)pyridine (**44-H**)	DPEPhos (**10**)	GP4	**44-Cu-10**	29%
2-methyl-6-(tetrazol-5-yl)pyridine (**45-H**)		GP4	**45-Cu-10**	28%
2-methoxy-6-(tetrazol-5-yl)pyridine (**46-H**)		GP4	**46-Cu-10**	42%

[a] Cannot be isolated due to high solubility; [b] losses during reaction due to flask breakage.

In the following sections, the molecular structures and photophysical properties of the mononuclear neutral copper(I) complexes will be investigated and related to DFT calculations. First, the complexes [(PyrTet)Cu(P^P)] with different bis(phosphines) will be discussed (section 3.2.3), followed by complexes [(N^N)Cu(DPEPhos)] based on tetrazolate ligands where the pyridine moiety is either varied or substituted against five-membered ring heterocycles (section 3.2.4), and concluded with complexes based on pyridine-tetrazolate ligands with electron-donating substituents in different positions of the pyridine moiety (section 3.2.5).

3.2.3 Copper(I) Complexes [(PyrTet)Cu(P^P)] with different Bis(phosphine) Ligands

Since the bis(phosphine) ligand in mononuclear neutral copper(I) complexes [(PyrTet)Cu(P^P)] based on 5-(pyridin-2-yl)tetrazolate (**7**) contributes only with its phosphorous atoms to the emission of the complex, a minor influence of the complexes' photophysical properties with various bis(phosphines) is expected. To prove this assumption, bis(phosphines) with significantly different bite angles, steric hindrance and flexibility are chosen to aim for different coordination geometries around the copper(I) center and influences on the emission properties of the respective complexes. The bis(phosphine) DPEPhos (**10**) and the model complex [(PyrTet)Cu(DPEPhos)] (**7-Cu-10**) are taken as reference compounds (detailed discussion in section 3.1.3.2).

As observed during synthesis, the formation of mononuclear copper(I) complexes [(PyrTet)Cu(P^P)] is very sensitive to the coordination geometry determined by the bis(phosphine) ligands of different bite angles. The complexes **7-Cu-10**, **7-Cu-18**, **7-Cu-19**, **7-Cu-21** and **7-Cu-23** based on the DPEPhos derivatives PTEPhos (**18**), XantPhos (**19**), DPTPhos (**21**) with flexible bite angles, and R-(+)-BINAP (**23**) with narrow bite angle (Table 18), are easily formed and are stable as powders and in solution for several days indicating a strong coordination of the copper(I) center. In contrast, no complexation is observed for the wide bite angle DBFPhos (**20**) ligand, and with the sterically constraint S-PhanePhos (**24**) the complex is very sensitive towards oxygen in solution, which hints at a weak coordination. These empirical results suggest a stable complex motif with bis(phosphines) of bite angles smaller than 120°, and especially the DPEPhos derivatives **10**, **18**, **21** with a flexible backbone, (and partially **19**) can adapt to the given steric hindrance around the copper(I) center. When comparing the actual bis(phosphine) bite angles in the crystal structures of complexes **7-Cu-10**, **7-Cu-18**, **7-Cu-19** and **7-Cu-23**, similar values around 115° are found with the DPEPhos-analogous ligands **10**, **18**, **19**, while a much smaller bite angle of 100° is found with R-(+)-BINAP (23) resulting from the short P-P distance (Table 27). All other bond lengths and angles are comparable for the four copper(I) complexes.

Table 27. Selected bond lengths (Å) and angles (°) of the mononuclear complexes [(PyrTet)Cu(P^P)] 7-Cu-10, 7-Cu-18, 7-Cu-19, 7-Cu-23.[IV]

	7-Cu-10	7-Cu-18	7-Cu-19	7-Cu-23[a]
Cu1-N1	2.0400 (15)[b]	2.0614 (14)	2.0206 (12)	2.023 (3)
Cu1-N11	2.0818 (16)[c]	2.1024 (14)	2.1138 (12)	2.136 (4)
Cu1-P1	2.2493 (5)	2.1024 (14)	2.2510 (4)	2.2376 (13)
Cu1-P2	2.2262 (5)	2.2436 (5)	2.2405 (4)	2.2596 (13)
P1-P2	3.762	3.815	3.812	3.436
N1-Cu-N11	80.525 (6)[d]	80.26 (5)	81.08 (5)	80.02 (16)
P1-Cu-P2	114.42 (2)	114.812 (17)	116.125 (15)	99.62 (5)
φ[e]	84.306	87.91	79.11	82.52
N1-C5-C6-N11	8.290 (3)[f]	10.2 (2)	6.52 (19)	1.1 (6)

[a] Two independent molecules are found in the crystal structure. For the second molecule the PyrTet ligand is slightly disordered which cannot be resolved. [b] Cu1-N47; [c] Cu1-N37; [d] N47-Cu-N37; [e] dihedral angle between N–Cu–N and P–Cu–P planes; [f] N37-C42-C43-N47.

The photoluminescence properties of the complexes 7-Cu-10, 7-Cu-18, 7-Cu-19, 7-Cu-21 and 7-Cu-23 with different bis(phosphines) as amorphous powders show surprising results: While the complexes 7-Cu-10, 7-Cu-18, 7-Cu-19 with DPEPhos (10), PTEPhos (18), XantPhos (19) exhibit strong emission in the blueish-green and yellow region of the visible spectrum with high quantum yields, no emission is observed for the copper(I) complexes 7-Cu-21 and 7-Cu-23 based on DPTPhos (21) and R-(+)-BINAP (23), which will be investigated in detail in the following sections.

3.2.3.1 The Influence of Molecular Packing[IX]

When comparing the photophysical characteristics of the three mononuclear complexes **7-Cu-10**, **7-Cu-18** and **7-Cu-19** based on the bis(phosphines) DPEPhos (**10**), PTEPhos (**18**) and XantPhos (**19**) (Figure 38), similarly high photoluminescence quantum yields are found, while a strong emission red-shift for **7-Cu-19** is noticeable (Figure 39a, Table 28). This emission shift can either be attributed to steric effects of the less flexible XantPhos compared to DPEPhos and PTEPhos ligands, or results from different electronic effects of the bis(phosphines).

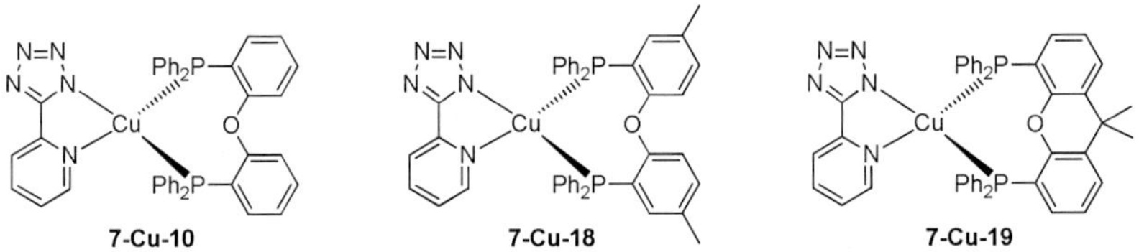

Figure 38. Molecular structures of the mononuclear neutral copper(I) complexes based on DPEPhos (**10**), PTEPhos (**18**) and XantPhos (**19**).

Therefore, UV-Vis measurements of the complexes are recorded and the position of the (ML+IL)CT absorption bands compared. This absorption band corresponds to the energy gap between the ground and excited state during excitation (vertical transition), and can be approximated as HOMO-LUMO gap. As already mentioned in chapter 3.1.3.2, these bands are hardly detectable for mononuclear neutral copper(I) complexes due to their high intramolecular CT character. Nevertheless, the absorption bands of all three complexes **7-Cu-10**, **7-Cu-18**, **7-Cu-19** are found at the same position (Figure 39b), indicating similar HOMO and LUMO levels. This is further confirmed by DFT calculations, which show only minor differences of the energy gaps between HOMO and LUMO orbitals for the different complexes (Table 29). Also the phosphorescence energies (same trend expected for excited singlet state energy), which take geometry changes of the excited states into account (adiabatic transitions), exhibit only small differences.

[IX] Parts of the chapter „The Influence of the Molecular Packing" have already been published.[167]

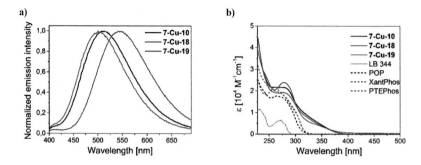

Figure 39. a) Emission spectra of the mononuclear complexes **7-Cu-10, 7-Cu-18, 7-Cu-19** as amorphous powder when excited at 350 nm, and b) absorption spectra in CH_2Cl_2 at a concentration of 10^{-5} mol/l.

Table 28. Photophysical characterization of the mononuclear, neutral complexes [(PyrTet)Cu(P^P)] with different bis(phosphine) ligands as powders at room temperature. The absorbance of the complexes is recorded in dichloromethane solutions.

Complex	λ_{abs} [nm] (ε [$10^4\,M^{-1}\,cm^{-1}$])	λ_{em} [nm]	ϕ_{em}	τ [µs]	k_r [s^{-1}]	k_{nr} [s^{-1}]
7-Cu-10	341 (0.56)	510	0.78	22.0	3.5×10^4	1.0×10^4
7-Cu-18	340 (0.50)	502	0.89	21.1	4.2×10^4	5.2×10^3
7-Cu-19	341 (0.44)	545	0.76	27.2	2.8×10^4	8.8×10^3

Table 29. Calculated HOMO, LUMO, as well as phosphorescence energies as obtained by DFT. The crystal structures are used as initial geometries and ground states and lowest triplet states optimized using the functional BP86. The phosphorescence energies are calculated as energy differences between the closed-shell singlet ground-state and the lowest (unrestricted) triplet state.[V]

Complex	HOMO [eV]	LUMO [eV]	$\Delta E_{HOMO-LUMO}$ [eV]	Phosphorescence energy [eV]
7-Cu-10	–4.58	–2.23	2.35	1.93
7-Cu-18	–4.54	–2.20	2.34	1.85
7-Cu-19	–4.54	–2.28	2.27	1.92

Hence, the emission shift of complex **7-Cu-19** has to be attributed to steric effects of the XantPhos ligand. Taking a closer look at the coordination geometry around the copper(I) center in the crystal structures of the complexes (Table 27), similar bond lengths and angles are found for **7-Cu-10, 7-Cu-18, 7-Cu-19** and despite a slightly stronger distortion of the tetrahedral coordination in **7-Cu-19** (reflected in φ) no further abnormalities are observed, and an influence of bite angles or steric constraints of the bis(phosphine) backbone on the complexes' emission can be ruled out. However, the rigid xanthene backbone in XantPhos exhibits less flexibility range than DPEPhos and enforces a predefined orientation of the phosphorous atoms around the copper(I) center (usually P-P distance around 4 Å),[169,211] thus facilitating a particular packing in the solid state, as already observed for complex **5-H-Cu-8** in chapter 3.1.3.1. These intermolecular packing effects strongly control the emission properties of the complexes in semicrystalline powders, but can be broken apart when diluting the complexes in films of an optically non-active polymer matrix (without aromatic character). Film samples of the complexes **7-Cu-10**, **7-Cu-18**, **7-Cu-19** are prepared by spincoating a dichloromethane solution of a mixture of the complex and PMMA in the ratio 20:80. Thereby, the individual complex molecules are isolated from each other and intermolecular interactions can be neglected. The photoluminescence quantum yields of the film samples are recorded under nitrogen to prevent oxygen quenching, which is found to influence the emission efficiency of the films but not of the powders, while the emission spectra are measured in air. From comparison of the photoluminescence of the film samples, the emission maxima are very similar for all three complexes and are in accordance with the results seen in the absorption spectra and DFT calculations, as expected (Table 30). To sum up the experimental observations, the emission red-shift for **7-Cu-19**, when compared to **7-Cu-10**, **7-Cu-18**, in powder is attributed to intermolecular packing effects of the XantPhos ligand with its rigid backbone, which can be prevented in diluted films of a polymer matrix.

Table 30. Photoluminescence characteristics of the mononuclear, neutral complexes [(PyrTet)Cu(P^P)] with different bis(phosphine) ligands in PMMA films (20:80). The photoluminescence is recorded at room temperature with an excitation wavelength of 350 nm. Photoluminescence quantum yields are measured under nitrogen atmosphere to rule out quenching effects by oxygen.

Complex	λ_{em} [nm]	ϕ_{em}
7-Cu-10	527	0.46
7-Cu-18	528	0.45
7-Cu-19	531	0.49

A further aspect of the different bis(phosphine) ligands is the proposed flattening distortion after excitation,[118,119,152] which leads to non-radiative deactivation and is especially pronounced in less rigid environments, such as solution but also matrix films. The planar xanthene backbone in XantPhos seems to be particularly prone to flattening distortions to a more square-planar coordination, but on the other hand its lower flexibility compared to DPEPhos and PTEPhos hinders small molecular movements. Interestingly, the photoluminescence quantum yields of the three complexes **7-Cu-10**, **7-Cu-18** and **7-Cu-19** in PMMA films, for which more possibilities of molecular movements are expected, do not show large differences. And even the non-radiative rates, or to be more precise the ratio of radiative to non-radiative rates, of the compounds in powder show only minor differences for the DPEPhos and XantPhos complexes **Cu-10** and **7-Cu-19**, and slightly less emission quenching for **7-Cu-18** (Table 28). Thus it is concluded, that the flattening distortion of the three complexes is ineffective in the solid state (which is important for the application in OLED devices) or exhibits the same extent for the different DPEPhos derivatives. The red-shift of the emission color, except for complex **7-Cu-19** due to molecular packing effects, and the reduction of photoluminescence quantum yield from powder to film are attributed to a combination of ground state distortion in the film and more flexibility for molecular movements after excitation.[179,212]

The different DPEPhos derivatives imply less an electronic effect on the complexes emission by their electron donating ability, than more a steric effect by molecular packing in semicrystalline samples and suppression of non-radiative processes, as observed from the very high photoluminescence quantum yields. If the molecular packing effects result in an emission color shift or just an increased photoluminescence quantum yield, is difficult to predict (compare **5-H-Cu-8**, **6-H-Cu-8** and **c-(6-Cu-9)** chapter 3.1.3), but obviously already minor changes of the crystal packing, e.g. by different counter ions in cationic complexes (compare **6-H-Cu-8** and **6-H-Cu-8-PF₆**),[183] affect the photoluminescence properties. Thus, when comparing the photoluminescence of different complexes in (semi-)crystalline phase, the influence of the molecular environment, induced interactions or distortions have to be taken into account.

Similar as in chapter 3.1.3, the photoluminescence properties of the complexes **7-Cu-10**, **7-Cu-18**, **7-Cu-19** are briefly compared to their cationic analogues [(PyrTetH)Cu(P^P)]BF₄ **7-H-Cu-10**, **7-H-Cu-18**, **7-H-Cu-19**. Since the same molecular packing effects are assumed for the neutral and cationic complex with a given bis(phosphine), photophysical properties are compared in powder. The yellow compounds **7-H-Cu-10**, **7-H-Cu-18** and **7-H-Cu-19** exhibit strongly red-shifted emission maxima and absorption bands, and significantly lower quantum yields (Table 31, Figure 89 in 5.3.1). This is in accordance with an emission of pure MLCT nature from the metal center (with contributions from the phosphorous atoms) to the pyridine-amine ligand $d(Cu) \rightarrow \pi^*$(pyridine-amine), in contrast to the additional intraligand CT character of the neutral complexes. Hence, the loss of the proton in neutral complexes [(PyrTet)Cu(P^P)] **7-Cu-10**, **7-Cu-18**, **7-Cu-19** results not only in a different emission character, but also in outstanding emission efficiencies due to less vibrational quenching and stronger coordination (see also chapter 3.1.3).

Table 31. Photophysical characterization of the mononuclear, cationic complexes [(PyrTet)Cu(P^P)]BF₄ with different bis(phosphine) ligands.

Complex	λ_{abs} [nm] (ε [$10^4 M^{-1} cm^{-1}$])	λ_{em} [nm]	ϕ_{em}
7-H-Cu-10	347 (0.30)	518	0.21
7-H-Cu-18	362 (0.25)	569	0.04
7-H-Cu-19	353 (0.23)	559	0.06

3.2.3.2 The Influence of the bridging Atom in DPEPhos

The bis(phosphine) DPEPhos (10) and its derivatives have proven to imply good luminophore properties to copper(I) complexes and ensure efficient emission, both in the beforehand studied complexes [(PyrTet)Cu(P^P)], as well as in many other mononuclear copper(I) complexes, e.g. the quantum efficiencies for cationic complexes based on phenanthroline ligands could be increased from below 1% to around 60% when exchanging one of the phenanthrolines by the DPEPhos ligand.[161] As there are many other bis(phosphine) ligands available, ligand design rules for an ideal coordination around copper(I) are desired. As was already discussed for the synthesis of the [(PyrTet)Cu(P^P)] complexes with different bis(phosphines), a bite angle of the P^P ligand below 120° is advantageous, and a good combination of flexibility and rigidity of the backbone are required to both adapt to the coordination sphere of the copper(I) atom and prevent non-radiative rates. On the other hand, the ether-bridge atom might also control the coordination chemistry of the copper(I) complexes, particularly in the excited states. Originally, this ether bridge was introduced in bis(phosphines) to prevent metalation, when the ligand was applied in catalytic materials with transition metals such as Ni(II) and Pd(II).[211] Since the copper atom in luminescent complexes formally adopts a +II oxidation state after excitation, which favors a square-planar coordination geometry and thus enables nucleophilic attack for a fifth (and sixth) coordination site, an interaction of the ether-bridge atom of DPEPhos further stabilizing the excited state and preventing non-radiative processes, could be assumed. To investigate this statement, the [(PyrTet)Cu(P^P)] 7-Cu-21 complex with the sulfur analogue DPTPhos (21) is synthesized and its photoluminescence recorded (Figure 40).

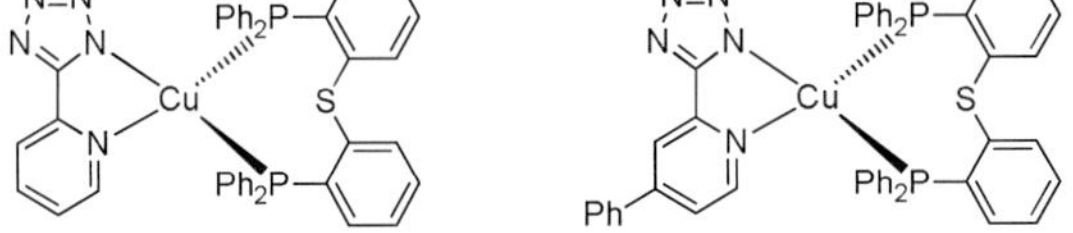

7-Cu-21   **41-Cu-21**

Figure 40. Molecular structures of the mononuclear neutral copper(I) complexes **7-Cu-21** and **41-Cu-21** based on DPTPhos (**21**).

Despite a strong charge-transfer absorption band at 346 nm ($\varepsilon = 6.2 \times 10^3$ M^{-1}cm^{-1}) in the UV-Vis spectrum, the complex shows no luminescence (Figure 41). Also its corresponding complex **41-Cu-21** based on 5-(4-phenylpyridin-2-yl)tetrazolate (**41**) does not exhibit any photoluminescence ((ML+IL)CT absorption band at 357 nm, $\varepsilon = 5.9 \times 10^3$ M^{-1}cm^{-1}), and is discussed with **7-Cu-21** in the following.

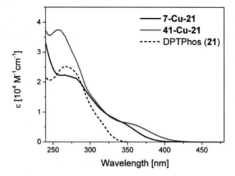

Figure 41. Absorption spectra of complexes [(N^N)Cu(DPTPhos)] **7-Cu-21** and **41-Cu-21** and of the free ligand in CH₂Cl₂ at a concentration of 10^{-5} mol/l.

The ground state of the complex **41-Cu-21** is given by its crystal structure (no single crystals are obtained for complex the **7-Cu-21**), which exhibits a similar coordination geometry as the model complex [(PyrTet)Cu(DPEPhos)] (**7-Cu-10**) and even further, a large distance of the copper(I) center and the sulfur-bridge atom of DPTPhos (**21**) is observed (Figure 42).

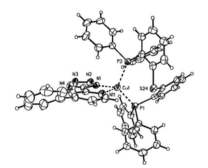

	41-Cu-21
Cu1-N1	2.050 (5)
Cu1-N11	2.104 (5)
Cu1-P1	2.2252 (16)
Cu1-P2	2.2663 (16)
P1-P2	3.737
Cu1-S24	3.390
N1-Cu-N11	80.88 (18)
P1-Cu-P2	112.61 (6)
φ^d	85.84
N1-C5-C6-N11	4.7 (7)

Figure 42. Crystal structure of [(5-(4-phenylpyridin-2-yl)tetrazolate)Cu(DPTPhos)] (**41-Cu-21**) and selected bond lengths [Å] and angles [°].[IV]

The excited state is then simulated by DFT calculations with the BP 86 functional, starting from the crystal structure as initial geometry. Strikingly, a pentacoordinated copper(I) center with a weak coordination to the sulfur atom of the DPTPhos (**21**) ligand is found for the lowest triplet state, which is not observed in the model complex [(PyrTet)Cu(DPEPhos)] (**7-Cu-10**) (Figure 43). As many complexes of both Cu(I) as well as Cu(II) with sulfur ligands are known,[213-218] the coordination of the S-atom to copper is not surprising. However, the ground state of complex **41-Cu-21** favors a tetrahedral coordination to N- and P-donors over sulfur coordination.[136,213]

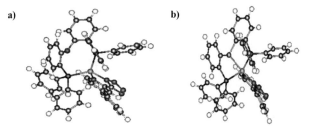

Figure 43. Calculated triplet state geometries of the complexes a) [(PyrTet)Cu(DPEPhos)] (**7-Cu-10**) and b) [(5-(4-phenylpyridin-2-yl)tetrazolate)Cu(DPTPhos)] (**41-Cu-21**).[V]

The formation of the pentacoordinated complex species can even be observed in the DFT simulation, when calculating the spin densities of the complex after excitation (ground state geometry) and in the triplet geometry (Figure 44): After relaxation into the geometry of the triplet state, the spin density distribution changed significantly and the sulfur atom moved towards the copper atom, so that a copper sulfur bond is formed (P-S distance: 2.458 Å).

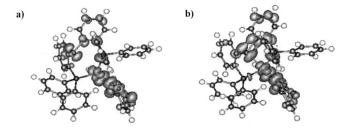

Figure 44. Spin densities for complex [(5-(4-phenylpyridin-2-yl)tetrazolate)Cu(DPTPhos)] (**41-Cu-21**) in the a) ground and b) triplet state geometry. The spin density changes significantly after relaxation to the optimized geometry, which results in a Cu-S bonding and a stronger Cu(II) character. In contrast, the change of spin density from ground to triplet state geometry is minor for the model complex **7-Cu-10**, which indicates a small Cu(II) character for the excited state.[V]

Since other copper(I) complexes containing sulfur ligands are reported to show luminescence,[213-215,219] the mere coordination of Cu(I) to sulfur in the ground state is likely not responsible for the emission quenching. But the non-radiative deactivation of complex **41-Cu-21** is attributed to the formation of a pentacoordinated copper species in the excited state, which then relaxes back to the ground state by energy dissipation through molecular movements. Such five-coordinated species have also been proposed as the reason for emission quenching of luminescent copper(I) complexes in solution.[220]

To sum up, despite a similar ground state coordination of the complexes [(N^N)Cu(DPTPhos)] **7-Cu-21**, **41-Cu-21** with PyrTet (**7**) or 5-(4-phenylpyridin-2-yl)tetrazolate (**41**) compared to the model complex [(PyrTet)Cu(DPEPhos)] (**7-Cu-10**), the exchange of the ether-bridge atom against sulfur significantly changes the coordination situation in the excited state, namely a five-coordinated copper species is observed in DFT modelling, which results in the quenching of emission.

3.2.3.3 The Electronic Effect of BINAP

The mononuclear neutral complex [(PyrTet)Cu(BINAP)] (**7-Cu-23**), as depicted in Figure 45, exhibits a different coordination geometry around the central copper(I) atom than the above discussed complexes **7-Cu-10**, **7-Cu-18** and **7-Cu-19** with DPEPhos derivatives as bis(phosphine) ligands: Due to the small distance between the two phosphorous atoms in the BINAP ligand, the bite angle to copper(I) amounts only to 100°, whereas typical bite angles of the DPEPhos derivatives lie around 115° (see crystal structures in 3.2.3). A different overlap of the bis(phosphine)'s molecular orbitals with the copper(I) d-orbitals and thus altered emission properties might be assumed for the complex **7-Cu-23**.

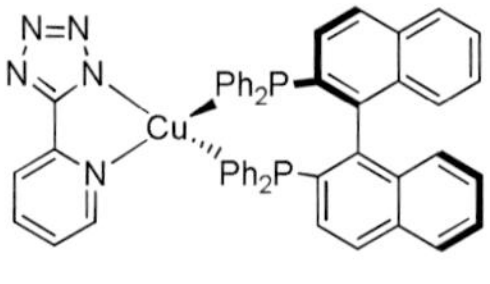

7-Cu-23

Figure 45. Molecular structure of the mononuclear neutral complex [(PyrTet)Cu(BINAP)] (**7-Cu-23**).

The photoluminescence properties indeed differ significantly for the complex **7-Cu-23** with a broad and red-shifted, structured emission compared to the model complex **7-Cu-10** (Figure 46, Table 32). Also the UV-Vis spectrum of **7-Cu-23** shows a red-shifted charge-transfer absorption band at 376 nm, which fits well to the statement by DAVID R. MCMILLIN that the MLCT character is controlled by the P-Cu-P bite angle, and a reduced energy gap between excited and ground state is proposed when it is small.[187] However, this does not explain the drastically reduced photoluminescence quantum yield from of 2% and thus further effects that influence the emission properties are suggested.

a)

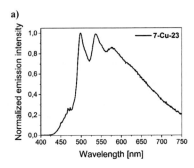

b)

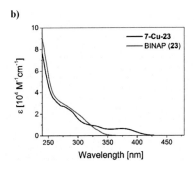

Figure 46. a) Emission spectrum of the mononuclear complex [(PyrTet)Cu(BINAP)] (**7-Cu-23**) as powder, and b) its absorption spectrum with the free ligand in CH_2Cl_2 at a concentration of 10^{-5} mol/l. The low resolution of the emission spectrum from 600 nm on is attributed to very weak emission intensity together with lower detector sensitivity at longer wavelengths.

Table 32. Photophysical characterization of the mononuclear, neutral complexes [(PyrTet)Cu(BINAP)] (**7-Cu-23**) as powders at room temperature, in comparison to the model complex [(PyrTet)Cu(DPEPhos)] (**7-Cu-10**).

Complex	λ_{abs} [nm] (ε [10^4 M^{-1} cm^{-1}])	λ_{em} [nm]	ϕ_{em}
7-Cu-10	341 (0.56)	510	0.78
7-Cu-23	376 (0.64)	499, 537, 578[a]	0.02

[a] Structured emission with three emission bands.

When a different molecular orbital overlap between the phosphorous atoms of the BINAP and the copper(I) atom due to a smaller bite angle is assumed, the localization and/or energy levels of the frontier orbitals are expected to be altered, since the bis(phosphine) contributes to the HOMO orbital of the complex with its phosphorous atoms. Interestingly, the distribution of the HOMO orbital does not change from the model complex **7-Cu-10** to the complex **7-Cu-23** based on BINAP, but the LUMO orbital is localized on the whole bis(phosphine) instead of the PyrTet (**7**) ligand (Figure 47, compare Figure 30). Thus, the lower electron donating ability of BINAP induces lower HOMO and LUMO energies and results in a shift of the complex's lowest unoccupied orbital from the N^N to the bis(phosphine) ligand, while the HOMO orbital is only marginally affected. The LUMO energy levels of the bis(phosphine) and the PyrTet ligands in this complex class must lie very close to each other, so that just minor changes of the electronic properties of the bis(phosphine) lead to strong changes of the complex's LUMO orbital. The predicted energy levels of the frontier orbitals in **7-Cu-23** based on BINAP show a strong lowering of the LUMO, and the HOMO to a much smaller extent, which confirms the observed emission and absorption shift to longer wavelengths (Table 33). The low electron donating ability of the BINAP ligand prevails the effects of a small P-Cu-P bite angle on the emission properties of the complex **7-Cu-23** compared to the model complex. Despite the different localizations of frontier orbitals of the complexes [(PyrTet)Cu(P^P)] **7-Cu-10**, **7-Cu-18**, **7-Cu-19** and **7-Cu-23**, the electron donating abilities of the different bis(phosphines) can be estimated from the energy levels of the complexes' HOMO orbital: A strong electron donating ability leads to a destabilization of the copper(I) d-orbitals, which is observed in the order **7-Cu-23** < **7-Cu-10** < **7-Cu-19** ≤ **7-Cu-18**, or expressed as electron donating property: BINAP < DPEPhos < XantPhos ≤ PTEPhos, resembling the chemical intuition.[163,186]

a)   b)

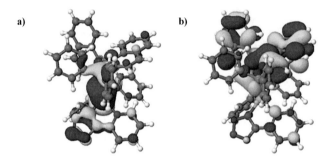

Figure 47. Frontier orbitals a) HOMO and b) LUMO of the mononuclear cationic complexes [(PyrTet)Cu(BINAP)] (**7-Cu-23**) as calculated by DFT (BP86/def2-SV(P)).[V]

Table 33. Calculated HOMO and LUMO energies of the complex [(PyrTet)Cu(BINAP)] (**7-Cu-23**) as obtained by DFT using the BP86 functional starting from the crytal structure as initial geometry. The energy gaps between the excited singlet and triplet states ΔE_{ST} are calculated by time-dependent DFT with the B3LYP functional.

Complex	HOMO [eV]	LUMO [eV]	$\Delta E_{HOMO\text{-}LUMO}$ [eV]	ΔE_{ST} [eV]
7-Cu-10	–4.58	–2.23	2.35	0.20
7-Cu-23	–4.62	–2.60	2.02	0.49

The DFT calculations further show a very large energy splitting between the excited singlet and triplet states when compared to the model complex **7-Cu-10**, which indicates only very weak thermally activated delayed fluorescence for this system and was also reported for other mononuclear copper(I) complexes based on BINAP.[221] The weak photoluminescence observed at room temperature is then assigned to phosphorescence, which is assumed to be quenched according to the energy gap law (red-shift) and due to additional CH-vibrations of the BINAP backbone. The strong participation of the BINAP ligand in the emission process is also observed in the structured emission spectrum. For the weakly luminescent copper(I) complex [(5-(4-phenylpyridin-2-yl)tetrazolate)Cu(BINAP)] (**41-Cu-23**) with a photoluminescence quantum yield of 1% the same explanation can be applied (absorption spectrum in Figure 90 in 5.3.1).

To sum up the results on the neutral copper(I) complexes [(PyrTet)Cu(P^P)] **7-Cu-10**, **7-Cu-18**, **7-Cu-19**, **7-Cu-21** and **7-Cu-23**, it is found that the different bis(phosphines) control the coordination geometry around the central copper(I) atom by their bite angles and if applying ligands of too large distances between their phosphorous atoms, even the formation of the complexes. Despite different sterical constraints of the backbone, similar geometries are found for the complexes with the DPEPhos derivatives **10**, PTEPhos (**18**) and XantPhos (**19**). The influence of these type of bis(phosphines) is mainly a sterical one, which is also observed for the photoluminescence of the copper(I) complexes. Intermolecular packing effects control the emission color despite theoretically same energy levels of the frontier orbitals. No significant differences of photoluminescence quantum yields are observed for the various phosphines (in powder and film), indicating a rigid molecular structure in solid state, but fluorescence spectroscopy in solution could give more evidences. Also the absence of luminescence for complexes **7-Cu-21** and **41-Cu-21** based on the bis(phosphine) DPTPhos (**21**) with a thioether-bridge, is more related to coordinational effects than to an electronic influence of the P^P ligand. But when applying the ligand R-(+)-BINAP (**23**) in the complexation reaction, a different electronic system is found for the complex due to lower frontier orbital energy levels of the bis(phosphine) ligand than of the N^N ligand in the complex. Thus, for the similar class of DPEPhos derivatives, the various bis(phosphines) **10**, **18**, **19** only act as steric ligand since they exhibit similar electronic properties, i.e. donating abilities, whereas the less electron donating ligand BINAP (**23**) affects the electronic and thus the emission properties of the corresponding complex.

3.2.4 Copper(I) Complexes [(D-Tet)Cu(DPEPhos)] with different Tetrazolate Ligands

The 5-(pyridin-2-yl)tetrazolate (**7**) ligand in the model complex [(PyrTet)Cu(DPEPhos)] (**7-Cu-10**) significantly controls the emission properties of the complex by both participating in the HOMO and the LUMO orbitals (Figure 30). While the energy level of the HOMO is mainly controlled by the tetrazolate moiety of the N^N ligand, the pyridine ring contributes to the LUMO orbital (together with the tetrazolate moiety). Thus, an increase of the LUMO energy level is expected when using N^N ligands with an electron rich pyridine ring, while lowering of the LUMO level is expected for electron deficient pyridines or heterocycles. The limits of the simplified electronic model visualized by frontier orbitals were already tested in chapter 3.2.3.3, when the ligand BINAP with significantly different energy levels than DPEPhos was applied in the complex **7-Cu-23**. Herein, the N^N ligand will be varied and thereby examined, if the same localization of frontier orbitals and a (ML+IL)CT character of the emission can be assumed for all complexes. This chapter will deal with the modification of the pyridine moiety as well as the substitution against five-membered ring heterocycles, while in the next chapter (3.2.5) the introduction of various substituents in different positions on the pyridine ring and their influence on the complexes properties will be investigated.

3.2.4.1 Modification of the Pyridine Moiety in the PyrTet Ligand and the Energy Gap Law

In the complexes **28-Cu-10**, **29-Cu-10**, **30-Cu-10** and **30-Cu-18** the electron donating ability of the pyridine moiety is lowered by introduction of a further N-atom in the ligand 2-(tetrazol-5-yl)-pyrazine (**28**), or by enlargement of the aromatic system to quinoline structures in the ligands 2-(tetrazol-5-yl)quinoline (**29**) and 1-(tetrazol-5-yl)isoquinoline (**30**) (Figure 48). This modification is expected to lower the energy of the LUMO orbital and thus result in a red-shift of the complex's emission. On the other hand, the different quinoline ligands **29** and **30** are applied to study if the position of the tetrazolate unit on the quinoline affects the steric hindrance around the copper(I) center or shows an electronic influence.

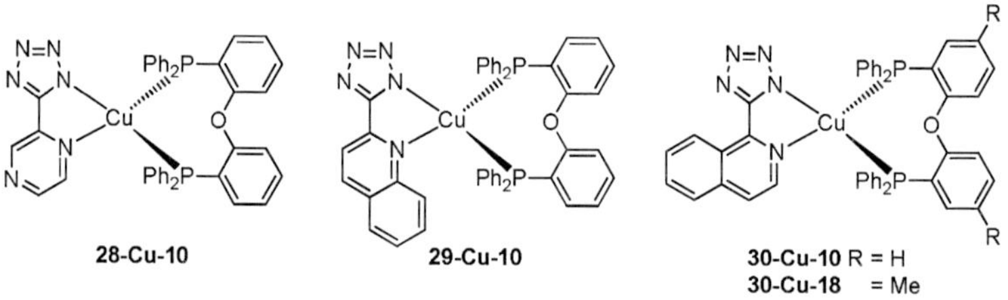

28-Cu-10   **29-Cu-10**   **30-Cu-10** R = H
                                              30-Cu-18   = Me

Figure 48. Molecular structures of the neutral copper(I) complexes based on 2-(tetrazol-5-yl)pyrazine (**28**), 2-(tetrazol-5-yl)quinoline (**29**) and 1-(tetrazol-5-yl)isoquinoline (**30**).

All complexes are obtained as yellow powders and also their dichloromethane solutions are yellow, which already hints at absorption in the blue region of the visible spectrum. This is confirmed by UV-Vis measurements with distinct charge-transfer absorption bands around 380 nm, which are not observed for the free N^N ligands (Figures 49, Figure 91 in 5.3.1). The absorption spectrum of complex **30-Cu-18** resembles that of the similar complex **30-Cu-10** and is shown in Figure 91 in the experimental part. In comparison to the model complex **7-Cu-10**, the (ML+IL)CT absorption bands of the complexes **28-Cu-10**, **29-Cu-10**, **30-Cu-10** and **30-Cu-18** are red-shifted by about 50 nm due to a smaller energy gap between the frontier orbitals HOMO and LUMO. If the different electronic effects of the N^N ligands only affect the LUMO energy levels of the complex, will be discussed in the following.

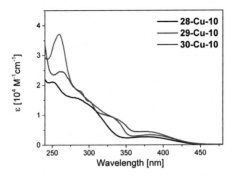

Figure 49. Absorption spectra of the complexes [(N^N)Cu(DPEPhos)] **28-Cu-10, 29-Cu-10, 30-Cu-10** in CH₂Cl₂ at a concentration of 10⁻⁵ mol/l.

A moderate luminescence at 590 nm in the orange yellow region of the visible spectrum is found for complex **28-Cu-10** based on 5-(pyrazin-2-yl)tetrazolate (**28**) (Table 35, Figure 92 in 5.3.1). Thus its emission color is red-shifted compared to the model complex with an emission maximum at 510 nm, as expected, due to the lowering of the LUMO level by the electron-deficient pyridine moiety. Also DFT calculations show a distinct decrease of the energy level of the LUMO by 0.49 eV when exchanging the PyrTet (**7**) with the pyrazine ligand **28**, but at the same time the HOMO level is slightly lowered by 0.18 eV, resulting in a smaller effect on the emission color than expected (Table 34). Taking a closer look on the localization of the frontier orbitals, also a small contribution of the coordinating N-atom of the pyridine ring to the HOMO is observed both for **28-Cu-10** as well as for the model complex **7-Cu-10** (compare Figures 30 and 101 in 5.3.2). That implies that with various N^N ligands not only the LUMO orbital energies of the complexes are varied, but also the HOMO levels are affected, even though to a lesser extent. Significant red- or blue-shifts of the emission color of the complex are thus difficult to achieve by changing the electron donating ability of the pyridine moiety.

Table 34. Calculated HOMO and LUMO energies, as well as excitation and phosphorescence energies of the complex [(5-(pyrazin-2-yl)tetrazolate)Cu(DPEPhos)] (**28-Cu-10**) in comparison to the model complex **7-Cu-10**, using the BP86 functional. The phosphorescence energies are calculated as energy differences between the closed-shell singlet ground-state and the lowest (unrestricted) triplet state.[V]

Complex	HOMO [eV]	LUMO [eV]	$\Delta E_{HOMO-LUMO}$ [eV]	excitation energy [eV]	phosphorescence energy [eV]
7-Cu-10	−4.58	−2.23	2.35	2.92	1.93
28-Cu-10	−4.76	−2.72	2.04	2.53	1.77

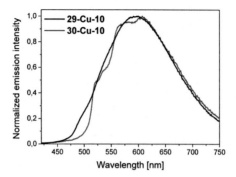

Figure 50. Emission spectra of the complexes [(N^N)Cu(DPEPhos)] **29-Cu-10** and **30-Cu-10** based on two different quinoline-tetrazolate ligands, as powder when excited at 350 nm. A vibrational structure is found for the complex **30-Cu-10**.

Table 35. Photoluminescence characteristics of the complexes [(N^N)Cu(DPEPhos)] **28-Cu-10**, **29-Cu-10**, **30-Cu-10** with different tetrazolate ligands, and the complex **30-Cu-18** as powders at room temperature, in comparison to the model complex [(PyrTet)Cu(DPEPhos)] (**7-Cu-10**). The absorbance of the complexes is recorded in dichloromethane solutions.

Complex	λ_{abs} [nm] (ε [$10^4 M^{-1} cm^{-1}$])	λ_{em} [nm]	ϕ_{em}	τ_{ave}^a [µs]	k_r [s^{-1}]	k_{nr} [s^{-1}]
7-Cu-10	341 (0.56)	510	0.78	22.0	3.5×10^4	1.0×10^4
28-Cu-10	379 (0.27)	590	0.16	4.3	3.7×10^4	1.9×10^5
29-Cu-10	384 (0.34)	596	0.13	15.2	8.5×10^3	5.7×10^4
30-Cu-10	378 (0.44)	523, 575, 607^b	0.05	40.0	1.2×10^3	2.4×10^4
30-Cu-18	380 (0.50)	527, 570, 610^b	0.23	88.6	2.6×10^3	8.7×10^3

^{a} PL emission decay time is composed of two components. A weighted-average lifetime is used (τ_{ave}) and calculated by the equation $\tau_{ave} = \Sigma A_i \tau_i / \Sigma A_i$ with A_i as the pre-exponential factor for the lifetime. ^{b} Structured emission with three emission bands.

The enlargement of the aromatic system of the ligands in the complexes **29-Cu-10**, **30-Cu-10** and **30-Cu-18** also leads to a red-shift, and orange yellow photoluminescence is recorded for the compounds based on the quinoline moiety (Table 35). However, the emission efficiencies are only moderate between 5–23%. Even the pyrazine based complex **28-Cu-10**, where only a C-atom is exchanged against nitrogen and thus no further potentially quenching C-H bonds are introduced,[42] exhibits only a PLQY of 16%. From comparison with the model complex (**7-Cu-10**, Table 28), it gets obvious that the radiative rates are similar, whereas the non-radiative rate of complex **28-Cu-10** increases by an order of magnitude (for a red-shift by 0.33 eV). This effect of lower emission efficiency with decreasing emission energy is described by the energy gap law.[184]

On the other hand, for the complexes **29-Cu-10**, **30-Cu-10** and **30-Cu-18** not only the non-radiative rate increases, but especially for the complexes **30-Cu-10**, **30-Cu-18** based on the isoquinoline ligand **30** the radiative rates of the complexes are by an order of magnitude lower than for the model complex **7-Cu-10**. In addition, the emission decay times composed of two time constants are drastically elongated up to 88.6 µs for **30-Cu-18**. Since the energy gap law would result in pronounced non-radiative processes and shorter emission decay times, the lower radiative rates of **30-Cu-10** and **30-Cu-18** might be attributed to a higher ILCT character with the isoquinoline ligand **30**. Even a vibrational structure is found in the emission spectra of **30-Cu-10** with three emission bands around 525 nm, 570 nm and 610 nm, which is not observed for the very similar complex **29-Cu-10** (Figure 50).

To examine the vibrationally structured emission of the complexes **30-Cu-10** and **30-Cu-18**, low temperature photoluminescence is recorded thus reducing molecular movements, which smear out the vibrational bands. At 77 K, the three emission bands resolve into doublets, and especially for **30-Cu-18** distinct peaks are observed (Figure 51). These are slightly shifted to shorter wavelengths compared to the room temperature emission. The compounds **30-Cu-10** and **30-Cu-18** are the first reported copper(I) complexes with such strong vibrational structures,[222] which is typically found for organic molecules, and hints at a strong contribution of the N^N ligand to the emission via the intraligand character of the charge-transfer transition. The less molecular movements at low temperature are also reflected in the lower non-radiative rate of $2.5 \times 10^3 \ s^{-1}$ versus room temperature (Table 36). Interestingly, the photoluminescence quantum yields decrease at 77 K, which results from a radiative rate constant of one order of magnitude lower and thus hints at emission from a different excited state, i.e. the triplet. Despite only a minor blue-shift of the complex's emission from 293 K to 77 K, thermally activated delayed fluorescence can be assumed for these complexes as well.

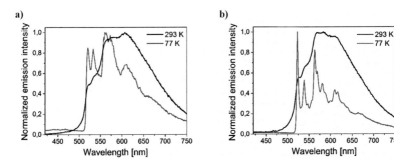

Figure 51. Emission spectra of the complexes a) **30-Cu-10** and b) **30-Cu-18** based on 5-(isoquinolin-1-yl)-tetrazolate with different bis(phosphine) ligands at room temperature and 77 K, as powder when excited at 350 nm.

Table 36. Photoluminescence characteristics of the complex [(5-(isoquinolin-1-yl)tetrazolate)Cu(P^P)] **30-Cu-18** at 77 K compared to 293 K.

Temperature [K]	λ_{em} [nm]	ϕ_{em}	$\tau_{ave}{}^a$ [µs]	k_r [s^{-1}]	k_{nr} [s^{-1}]
293	527, 570, 610	0.23	88.6	2.6×10^3	8.7×10^3
77	523 (539), 563 (581), 610 (617)^b	0.13^c	346.0	3.8×10^2	2.5×10^3

^{a} PL emission decay time is composed of two or three components. A weighted-average lifetime is used (τ_{ave}) and calculated by the equation $\tau_{ave} = \Sigma A_i \tau_i / \Sigma A_i$ with A_i as the pre-exponential factor for the lifetime. ^{b} The emission spectrum shows three emission bands (weaker second peak of resolved doublet at 77 K). ^{c} Measured in the group of Christian Strassert, University Münster, with an integrating sphere equipped with a Dewar.

Summing up the photophysical behavior of the complexes **28-Cu-10**, **29-Cu-10**, **30-Cu-10** and **30-Cu-18**, a red-shifted emission color is achieved by reducing the electron density in the pyridine moiety or enlarging the aromatic system. However, a significant color-shift is difficult, as both the LUMO and the HOMO orbitals of the complex are affected, even though the energy level of the HOMO to a lesser extent. Furthermore, lowering the LUMO energies to induce an emission color shift to longer wavelengths brings along more pronounced non-radiative processes, thus reducing the emission efficiencies. Additionally, for the complexes **30-Cu-10** and **30-Cu-18** based on the 5-(isoquinolin-1-yl)tetrazolate (**30**) ligand a structured emission is found and resolved into distinct vibrational bands at low temperature. Thus, the position of the tetrazolate unit on the quinoline does not affect the steric hindrance around the copper(I) center to reduce non-radiative processes, but changes the electronic system of the complex.

3.2.4.2 The Pyridinophane-tetrazole as New Ligand[X]

In cooperation with Dr. Joshua Kramer of this research group, a novel copper(I) complex based on the pyridinophane scaffold is synthesized and its photophysical properties investigated to learn about the electronic and steric influence of the 5-([2](1,4)benzo[2](2,5)pyridinophan-13-yl)-tetrazolate (**31**) ligand (Figure 52).

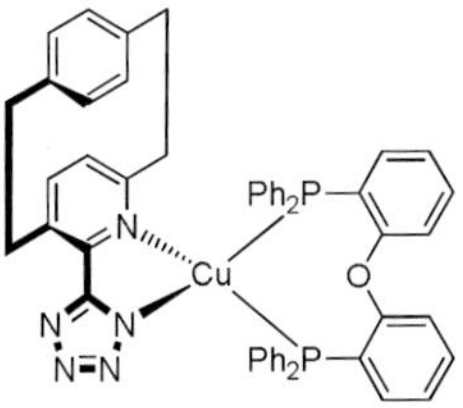

31-Cu-10

Figure 52. Molecular structure of the complex **31-Cu-10** with a pyridinophane-backbone.

The crystal structures of complex **31-Cu-10** exhibits a distorted tetrahedral coordination of the copper(I) center, where the tetrazolate moiety is coplanar to the pyridinophane-backbone (Figure 53). When comparing the coordination geometry of **31-Cu-10** with the model complex **7-Cu-10**, no significant differences are obvious. Due to the larger steric hindrance of the pyridinophane-scaffold, the copper(I) center moves slightly further away from the pyridine to the tetrazolate moiety, and the Cu-P bonds are elongated, which results in a minor narrowing of the bis(phosphine) bite angle. Thus, the tetrahedral geometry is more distorted in contrast to complex **7-Cu-10** (see also Table 8). Furthermore, in the ^{31}P NMR spectrum two signals at -12.7 and -14.6 ppm are found, indicating that the two phosphorous atoms of the DPEPhos ligand are magnetically not equivalent (**7-Cu-10**: -12.8 ppm). The overall similarities of the coordination geometry around the copper(I) center suggest a mainly electronic effect of the pyridionphane-tetrazolate **31** on the complex's emission if the same localization of frontier orbitals can be applied here as in the model complex **7-Cu-10**.

[X] The results of this collaboration, in which Dr. Joshua Kramer provided the ligand, whereas synthesis of the complex and its characterization was conducted as part of this thesis, were already published in the dissertation of Dr. Joshua Kramer.[203]

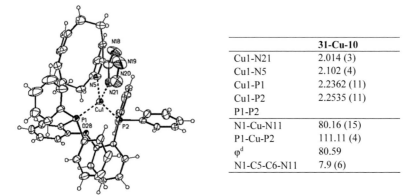

	31-Cu-10
Cu1-N21	2.014 (3)
Cu1-N5	2.102 (4)
Cu1-P1	2.2362 (11)
Cu1-P2	2.2535 (11)
P1-P2	
N1-Cu-N11	80.16 (15)
P1-Cu-P2	111.11 (4)
φ^d	80.59
N1-C5-C6-N11	7.9 (6)

Figure 53. Crystal structure of the complex **31-Cu-10**, and selected bond lengths [Å] and angles [°].[IV]

The UV-Vis spectrum of the complex **31-Cu-10** in dichloromethane shows an absorption shoulder at 354 nm, which is attributed to a charge-transfer transition from the copper(I) center to a ligand (Figure 55). The analysis of frontier orbitals confirms this assumption: While the copper(I) atom, the two phosphorous atoms of the bis(phosphine) and to a smaller extent the tetrazolate moiety contribute to the HOMO orbital, the LUMO is localized on the whole N^N ligand including the pyridinophane-scaffold, and on one of the bis(phosphine) phenyl rings (Figure 54). Under the assumption, that the HOMO → LUMO transition is responsible for the emission of the complex, the electronic picture resembles that one of the model complex **7-Cu-10**, namely (ML+IL)CT. However, the exact influence of the pyridonphane-backbone in this system is not clear yet and thus, the photoluminescence of **31-Cu-10** is studied in the following.

a)   b)

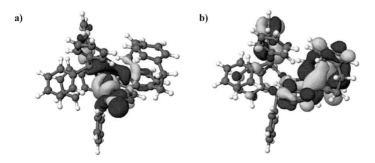

Figure 54. Frontier orbitals a) HOMO and b) LUMO of the complex [(N^N)Cu(DPEPhos)] **31-Cu-10**, as calculated by DFT using the BP86 functional with def2-SV(P) basis set. The crystal structure is used as initial geometry, and is optimized in the ground and triplet state.[V]

The complex **31-Cu-10** shows yellow emission at 549 nm of low photoluminescence quantum yield, which is surprising, when taking the steric rigidity of the ligand's backbone (to suppress molecular movements and thus quenching processes) into account (Table 37). At low temperature, the emission shifts to shorter wavelengths is contradictory to the red-shift proposed for TADF materials, which show longer wavelength phosphorescence when the thermal back-population of the excited singlet state is disabled. The emission decay time of the complex is composed of two significantly different components (20.1 µs and 107.3 µs, see also Figure 93 in 5.3.1), and increases drastically when going to low temperature (three components: 215.3 µs, 2.2 ms and 9.3 ms).

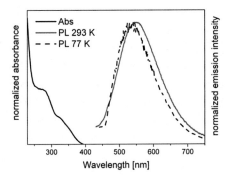

Figure 55. Absorption and emission spectra of complex **31-Cu-10**. The UV-Vis measurements are recorded in CH₂Cl₂ at a concentration of 10^{-5} mol/l. The emission spectra of the amorphous powder are recorded at room temperature and 77 K at an excitation wavelength of 350 nm. The offset at 450 nm in the emission spectrum of **31-Cu-10** arises from emission of the filter (395 nm) used during the measurement.

Table 37. Photoluminescence characteristics of complex **31-Cu-10** based on the pyridinophane-tetrazolate **31**, in comparison to the model complex **7-Cu-10**.

Complex	Temperature	λ_{abs} [nm]	λ_{em} [nm]	ϕ_{em}[a]	τ_{ave} [µs]	k_r [s⁻¹]	k_{nr} [s⁻¹]
7-Cu-10	293 K	341	510	0.68	22.0	3.1×10^4	1.5×10^4
	77 K		549	0.82	111.5	7.4×10^3	1.6×10^3
31-Cu-10	293 K	354	549	0.09	35.2[b]	2.6×10^3	2.6×10^4
	77 K		535	0.29	1546.3[b]	1.9×10^2	4.6×10^2

[a] Measured by Alexander Schinabeck, group of Hartmut Yersin, University of Regensburg, with an integrating sphere equipped with a Dewar. [b] PL emission decay time is composed of two or three components. A weighted-average lifetime is used (τ_{ave}) and calculated by the equation $\tau_{ave} = \Sigma A_i\tau_i/\Sigma A_i$ with A_i as the pre-exponential factor for the lifetime.

These experimental results indicate the following: The emission efficiency of complex **31-Cu-10** is lowered only partially by vibrational quenching of the CH₂-groups of the ethyl-bridges,[42] as becomes obvious from the non-radiative rate of 2.6×10^4 in a similar range as for the model complex **7-Cu-10**. However, the radiative rate is decreased by an order of magnitude to 2.6×10^3, as was already observed for the complexes **30-Cu-10** and **30-Cu-18** based on the 5-(isoquinolin-1-yl)tetrazolate (**30**). At low temperatures, further molecular movements are frozen out and the photoluminescence quantum yield increases to 29% due to a drastically lowered non-radiative rate of 4.6×10^2. Additionally, a blue-shift of the emission is observed, which is commonly known as rigidochromism – the suppression of structural changes leading to similar geometries of the excited and ground states[223] – and prevails the red-shift expected for TADF materials. If the complex **31-Cu-10** exhibits TADF behavior is not clearly proven, but the elongation of emission decay times at low temperatures and especially the lowering of the radiative rate by an order of magnitude hints at emission from a different excited state, namely the triplet. Since the emission decay times both at room and at low temperature are composed of significantly different components, two radiative processes or two species of different molecular environment in powder could be assumed.[179] But further experiments have to be performed to prove this.

The electronic influence of the pyridonphane-backbone seems to lower the energy level of the LUMO orbital, as the emission of the complex is red-shifted. But DFT calculations reveal that both the HOMO and LUMO orbitals of the complexes are hardly affected by the exchange of PyrTet (**7**) against pyridionphane-tetrazolate **31** (Table 38). In contrast, the excitation energies ($S_0 \rightarrow T_1$) show a smaller energy gap between ground and excited states for the pyrdinophane complex **31-Cu-10**, which is also reflected in the absorption spectrum of **31-Cu-10**. However, the phosphorescence energies ($T_1 \rightarrow S_0$) predict a bluer emission wavelength for the pyridinophane complex **31-Cu-10** compared to the model complex **7-Cu-10**, which is obviously not the case. The DFT calculations using the BP86 functional seem to give a too simple picture of a more complex electronic system, as was already observed for the neutral complexes **5-Cu-10**, **6-Cu-10**, **7-Cu-10** in chapter 3.1.3.2. Other functionals (e.g. B3LYP) have to be applied to this system to understand the electronic transitions and the red-shifted emission color compared to the model complex.

Table 38. Calculated HOMO and LUMO energies, as well as excitation and phosphorescence energies of complex **31-Cu-10** in comparison to the model complex **7-Cu-10**, using the BP86 functional. The phosphorescence energies are calculated as energy differences between the closed-shell singlet ground-state and the lowest (unrestricted) triplet state.[V]

Complex	HOMO [eV]	LUMO [eV]	$\Delta E_{HOMO\text{-}LUMO}$ [eV]	excitation energy [eV]	phosphorescence energy [eV]
7-Cu-10	–4.58	–2.23	2.35	2.92	1.93
31-Cu-10	–4.52	–2.14	2.38	2.76	2.11

To conclude, a similar coordination geometry around the copper(I) center is found for complex **31-Cu-10** based on the pyridine-tetrazolate **31** compared to the model complex **7-Cu-10**. The steric hindrance of the pyridinophane-backbone slightly distorts the tetrahedral coordination, but significant differences are not observed. The DFT calculations predict a HOMO-LUMO transition from the metal center with contributions from the phosphorous atoms and the tetrazolate moiety to the whole N^N ligand, as in the model complex. From these predictions, the pyridinophane-backbone does not change the energy level of the LUMO orbital. But the experimental absorption and emission color of the complex **31-Cu-10** cannot be described by the theoretical calculations and they give a too simplified picture of a more complex electronic system. Despite its rigid backbone, complex **31-Cu-10** exhibits only moderate emission efficiency, which is partially attributed to vibrational quenching of the ethyl-bridges, but further processes are assumed due to unusual emission decays.

3.2.4.3 Substitution of the Pyridine Moiety against Five-membered Ring Heterocycles

The aim of substitution of the pyridine moiety in 2-(tetrazol-5-yl)pyridine (**7-H**) against five-membered ring heterocycles is on the one hand to investigate the coordination chemistry of copper(I) with these ligands and on the other hand to test, if the same electronic picture that was gained of the complex class in the last sections, can be applied here.

During synthesis it is observed, that the simplest of the five-membered ring heterocycles, namely 5-(furan-2-yl)tetrazole (**32-H**) and 5-(thiophen-2-yl)tetrazole (**33-H**), do not form complexes with copper(I) and the bis(phosphine) DPEPhos (**10**). This is not surprising since the steric constraint of the coordination (bite angle around 80° for the model complex) might be to high for the weakly coordinating sulfur,[213] and even more for the oxygen atom. Only very few copper(I) complexes with ether ligands coordinating via the oxygen atom are reported,[224] due to its hard Lewis base nature in contrast to the very soft copper(I) atom. However, when the pyridine moiety is exchanged against imidazole rings, such as in 2-(tetrazol-5-yl)-N-methylimidazole (**34-H**), 2-(tetrazol-5-yl)-N-hexylimidazole (**35-H**) or 2-(tetrazol-5-yl)-N-methyl-benzimidazole (**36-H**), mononuclear neutral copper(I) complexes **34-Cu-10**, **35-Cu-10** and **36-Cu-10** are formed, based on the strong coordination of copper(I) to imines (Figure 56).[220] Substitution of the second N-atom of the imidazole against oxygen or sulfur in the oxazole, isoxazole or thiazole ligands **37-H**, **38-H**, **39-H** and **40-H** affords copper(I) complexes, but not in all cases mononuclear species are obtained. The weakly coordinating sulfur-atom in 2-(tetrazol-5-yl)benzothiazole (**40-H**) facilitates the formation of coordination polymers,[213,217] which can only be broken apart by very polar solvents to afford the mononuclear complex **40-Cu-10** (Figure 56). Interestingly, mononuclear complexes with the 2-(tetrazol-5-yl)benzoxazole (**39-H**) ligand seem unfavored and dimeric structures are formed. For the cluster compound **c-(39-Cu-10)** with DPEPhos (**10**) the composition [(N^N)Cu$_2$(DPEPhos)$_2$]BF$_4$ is suggested from elemental analysis, but unfortunately no single crystals are obtained to investigate the particular coordination chemistry. With PTEPhos (**18**) single crystals of the composition [(N^N)$_2$Cu$_2$(PTEPhos)$_2$] **c-(39-Cu-18)** are obtained, where the copper(I) atoms only coordinate to the tetrazolate moiety of two N^N ligands and one bis(phosphine) ligand, but coordination to oxygen is not observed. Why the ligand **39-H** forms dimeric structures instead of mononuclear neutral species is difficult to determine, but might also be a result of solution equilibria between different species and crystallization of the most polar one.[131-133] Unfortunately, the complex species based on 2-methyl-4-(tetrazol-5-yl)oxazole (**37-H**) and 5-methyl-3-(tetrazol-5-yl)isoxazole (**38-H**) are highly soluble (the complex formation is proven by mass spectrometry), as is also observed for the imidazole based complexes, and cannot be isolated. From the results above, mononuclear neutral complexes with the copper(I) center coordinating to the N-atom of the five-membered ring heterocycle is expected, but other complex species cannot be ruled out.

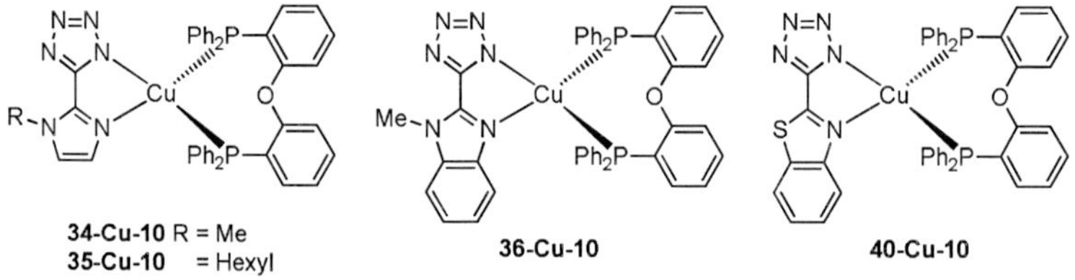

34-Cu-10 R = Me
35-Cu-10   = Hexyl   **36-Cu-10**   **40-Cu-10**

Figure 56. Molecular structures of the neutral copper(I) complexes **34-Cu-10**, **35-Cu-10**, **36-Cu-10** and **40-Cu-10** based on 2-(tetrazol-5-yl)imidazoles **34-H** and **35-H**, 2-(tetrazol-5-yl)-*N*-methyl-benzimidazole (**36-H**), and 2-(tetrazol-5-yl)benzothiazole (**40-H**).

The crystal structures of the mononuclear complexes **34-Cu-10**, **35-Cu-10** and **40-Cu-10** show a tetrahedral coordination of the central copper(I) atom (Figure 57). While the copper(I) atom is marginally disordered in **34-Cu-10** (95:5), two molecules with different orientations of the hexyl group are found in the crystal structure in the ratio 1:1 for complex **35-Cu-10**. They are denoted as **35-Cu-10a** and **35-Cu-10b** in Table 39. In the case of **35-Cu-10a**, the hexyl group on the imidazole moiety is oriented towards the bis(phosphine), whereas in **35-Cu-10b** it points away from the molecule center. The packing and orientation of the backbone of the ligands affects the geometry at the coordination centers, as gets obvious from the comparison of Cu-N and Cu-P bond lengths of the two molecules **35-Cu-10a** and **35-Cu-10b**. This also shows, that mainly steric bulkiness and molecular packing effects instead of the electronic density in the individual ligand moieties control the exact bond lengths of copper(I) to the N^N ligand, and thus makes comparison difficult. Nevertheless, similar coordination geometries are found for the complexes **34-Cu-10**, **35-Cu-10** and **40-Cu-10** based on tetrazole ligands with five-membered ring heterocycles as for the model complex **7-Cu-10**. The imidazole-tetrazolate ligands coordinate the central copper(I) atom with bite angles of around 80°, which resembles that one of PyrTet (**7**). Thus, similar steric constraints and shielding of the copper(I) atom are assumed for the N^N ligands with five-membered ring heterocycles instead of the pyridine moiety.

Minor differences arise from the distortion of the tetrahedral coordination: While in the complexes **34-Cu-10** and **35-Cu-10** based on 2-(tetrazol-5-yl)-*N*-alkyl-imidazole **34-H** and **35-H** the N-Cu-N and P-Cu-P planes lie close to perpendicular to each other, the tetrahedron in the model compound **7-Cu-10** is slightly distorted. In the complex **40-Cu-10** the orientation of the bis(phosphine) versus the N^N ligand reflect the ideal tetrahedral coordination. It is noted, that the thiazole moiety of the 5-(benzothiazol-2-yl)-tetrazolate ligand in this complex is strongly distorted due to the large sulfur atom.

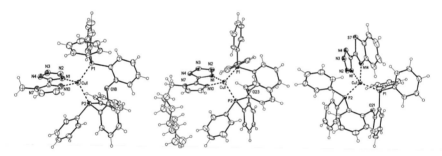

Figure 57. Crystal structures of the neutral copper(I) complexes **34-Cu-10**, **35-Cu-10**, **40-Cu-10** from left to right. Here the molecule **35-Cu-10a** is shown, in which the hexyl group is disordered.[IV]

Table 39. Selected bond lengths (Å) and angles (°) of the complexes [(D-Tet)Cu(DPEPhos)] **34-Cu-10**, **35-Cu-10**, **40-Cu-10**, in comparison to the model compound [(PyrTet)Cu(DPEPhos)] (**7-Cu-10**).

	7-Cu-10	34-Cu-10	35-Cu-10a	35-Cu-10b	40-Cu-10
Cu1-N1	2.0400 (15)[a]	2.113 (2)	2.094 (6)	2.048	2.0890 (14)
Cu1-N10	2.0818 (16)[b]	2.078 (2)	2.084 (6)	2.083	2.0808 (14)[c]
Cu1-P1	2.2493 (5)	2.2259 (7)	2.253 (2)	2.217	2.2347 (5)
Cu1-P2	2.2262 (5)	2.2474 (7)	2.282 (2)	2.272	2.2602 (5)
P1-P2	3.762	3.838	3.754	3.768	3.873
N1-Cu-N10	80.525 (6)[d]	81.21 (9)	80.3 (2)	81.18	80.48 (5)[e]
P1-Cu-P2	114.42 (2)	118.20 (3)	121.13 (17)	114.13	119.019 (17)
φ[f]	84.306	87.94	86.38	89.97	89.94
N1-C5-C6-N10	8.290 (3)[g]	6.3 (4)	1.3 (9)	1.86	2.0 (2)[h]

[a] Cu1-N47; [b] Cu1-N37; [c] Cu1-N14; [d] N47-Cu-N37; [e] N1-Cu-N14; [f] dihedral angle between N–Cu–N and P–Cu–P planes; [g] N37-C42-C43-N47; [h] N1-C5-C6-N14.

Despite their similar coordination geometries as in the model complex **7-Cu-10**, the five-membered ring heterocycles are expected to induce different electronic effects on the complexes, e.g. the imidazole-tetrazolates **34**, **35** and **36** are more electron rich compared to pyridine,[225] resulting in a raise of the LUMO level and a larger HOMO-LUMO energy gap. This is observed as a blue-shift in the photoluminescence spectra (Figure 58, Figure 94 in 5.3.1), when compared to the model complex **7-Cu-10**. The complexes **34-Cu-10** and **35-Cu-10** based on *N*-alkyl-imidazoles show similar emission maxima, which indicate a minor electronic effect of the alkyl-group on the complexes emission. As already observed for the complexes with 1-(tetrazol-5-yl)isoquinoline (**30**) in chapter 3.2.4.1, the emission spectrum of **36-Cu-10** with the benzimidiazole-tetrazolate ligand is structured and three emission bands at 416 nm, 439 nm and 467 nm are resolved. The UV-Vis measurements give only little information on the excitation energy of the charge transfer transition, as the respective absorption bands are shifted to short wavelengths and thus vanish underneath the ligand-centered absorption bands (Figure 59). Nevertheless, this supports the above statement of a larger HOMO-LUMO energy gap for complexes with the imidazole-tetrazolate ligands **34**, **35**, **36** compared to PyrTet (**7**). Strikingly, the emission efficiencies of the complexes are very low around 1–7% and the question arises, if this is an intrinsic feature of the imidazole ligand class or might be attributed to other reasons.

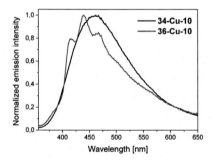

Complex	λ_{em} [nm]	ϕ_{em}
34-Cu-10	462	0.07
35-Cu-10	463	0.01
36-Cu-10	416, 439, 467	0.05

Figure 58. Photoluminescence characteristics of complexes **34-Cu-10**, **35-Cu-10**, **36-Cu-10** with imidazole-tetrazolate ligands as powders at room temperature when excited at 350 nm. The emission spectrum of complex **35-Cu-10** is not shown due to a low signal-to-noise ratio of the weak luminescence. For complex **36-Cu-10** a structured emission with three emission bands is observed.

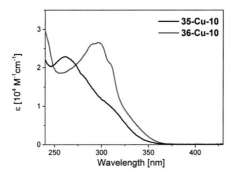

Figure 59. Absorption spectra of complexes **35-Cu-10** and **36-Cu-10** in CH₂Cl₂ at a concentration of 10⁻⁵ mol/l.

DFT calculations are performed for the complexes **34-Cu-10**, **35-Cu-10** and **36-Cu-10** and a different localization of the frontier orbitals is revealed (Figure 60). While the copper(I) atom, and to a lesser extent the tetrazolate moiety and the two phosphorous atoms contribute to the HOMO orbital, the LUMO is localized on half of the bis(phosphine) ligand for complex **34-Cu-10**. The frontier orbitals of the complexes **35-Cu-10** and **36-Cu-10** show the same behavior, whereas in the model complex **7-Cu-10** the LUMO was found on the whole PyrTet ligand without participation of the bis(phosphine) (Figure 30). Thus, the LUMO energies of the bis(phosphine) and the imidazole-tetrazolate ligands must be similar. If then the LUMO level of the N^N ligand is raised to higher energies than of the bis(phosphine), e.g. by increasing the electron density in imidazoles compared to pyridine, the localization of the frontier orbital in the complex shifts to the bis(phosphine). That in turn implies an electronic effect of the N^N ligand only on the complex's HOMO energy, while the LUMO level is expected to be the same for the complexes **34-Cu-10**, **35-Cu-10** and **36-Cu-10** with different imidazole ligands. The comparison of calculated HOMO and LUMO energies confirms this assumption, and further support the strong blue-shift of the complexes' emission maxima in contrast to **7-Cu-10** (Table 40). As observed experimentally, the alkyl-group in 2-(tetrazol-5-yl)-*N*-methylimidazole (**34-H**) or 2-(tetrazol-5-yl)-*N*-hexylimidazole (**35-H**) does not affect the electronic system, while the enlargement of the aromatic system in 2-(tetrazol-5-yl)-*N*-methyl-benzimidazole (**36-H**) results in a slightly larger HOMO-LUMO energy gap and shifts the emission to shorter wavelengths.

.

a)   b)

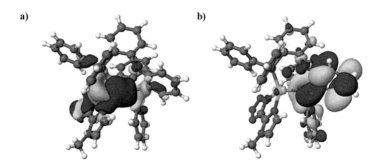

Figure 60. Frontier orbitals a) HOMO and b) LUMO of complex **34-Cu-10**, as calculated by DFT using the BP86 functional with def2-SV(P) basis set. The crystal structure is used as initial geometry, and is optimized in the ground and triplet state.[V]

Table 40. Calculated HOMO and LUMO energies, as well as excitation and phosphorescence energies of complexes **34-Cu-10**, **36-Cu-10**, **36-Cu-10** in comparison to the model complex **7-Cu-10**, using the BP86 functional. The phosphorescence energies are calculated as energy differences between the closed-shell singlet ground-state and the lowest (unrestricted) triplet state.

Complex	HOMO [eV]	LUMO [eV]	$\Delta E_{HOMO\text{-}LUMO}$ [eV]	excitation energy [eV]	phosphorescence energy [eV]
7-Cu-10	–4.58	–2.23	2.35	2.92	1.93
34-Cu-10	–4.50	–1.98	2.52	3.16	2.22
35-Cu-10	–4.48	–1.98	2.51	3.15	2.22
36-Cu-10	–4.58	–2.02	2.56	2.98	2.40

The question remains, why the photoluminescence quantum yields are only very low. One reason could be found in geometry changes after excitation, in particular the flattening distortion was proposed to be a major energy loss channel in mononuclear copper(I) complexes.[118,119,152] Thus, the calculated dihedral angles as parameter for the tetrahedron flattening are compared for ground state and excited state (triplet). Although the theoretical model assumes isolated molecules in the gas phase, the values give a hint on the extent of flattening distortion after excitation and subsequent relaxation into the triplet state geometry. For the model compound **7-Cu-10**, a dihedral angle between the N-Cu-N and P-Cu-P planes of 84° is found in the crystal structure, which is predicted to flatten to 61° in the triplet state. For complex **36-Cu-10**, a dihedral angle of 88° is calculated for the ground state geometry and fits well to the experimental values of **34-Cu-10** and **35-Cu-10**. For the triplet state of **36-Cu-10** the calculated dihedral angle amounts to 75° and thus proposes only a small flattening distortion. This would imply less non-radiative processes of the excited molecule and higher emission efficiencies. However, the photoluminescence quantum yields follow the opposite trend and are drastically lower than for the model complex. The examination of the complex **40-Cu-10** based on 5-(benzothiazol-2-yl)tetrazolate **40** might shed light on the origin of the non-radiative processes.

By exchange of the second N-atom in the 2-(tetrazol-5-yl)-*N*-alkylimidazole ligands **34-H**, **35-H** and **36-H** and enlargement of the aromatic system by benzannulation to 2-(tetrazol-5-yl)-benzothiazole (**40-H**), the electron density is significantly reduced so that the LUMO level of the N^N ligand is aimed to lie below that one of the bis(phosphine) resulting in the usual localization of frontier orbitals as in the model compound **7-Cu-10** (compare Figure 102 in 5.3.2). The calculated energies of the frontier orbitals further show both a decrease of the HOMO and LUMO levels, which indicates an even more electron deficient system in the complex with a benzothiazole than pyridine moiety (Table 41). As expected, the emission spectra of complex **40-Cu-10** shows photoluminescence less in the blue region than the complexes **34-Cu-10**, **35-Cu-10** and **36-Cu-10**, but is even further red-shifted than for the model compound **7-Cu-10** despite similar HOMO LUMO energy gaps (Figure 61). Also a vibrational structure is detectable in the emission spectrum of **40-Cu-10** and only very low photoluminescence quantum yields of 10% are recorded. As already discussed above, this cannot be attributed to a flattening distortion after excitation since DFT calculations predict hardly a geometry change of the tetrahedral coordination from the ground state to the triplet state geometries. Furthermore, the different localization of frontier orbitals do not influence the emission efficiencies, since both the complexes **34-Cu-10**, **35-Cu-10** and **36-Cu-10** based on 2-(tetrazol-5-yl)-*N*-alkylimidazole ligands **34–36**, and the complex **40-Cu-10** based on 2-(tetrazol-5-yl)benzothiazole **40**, exhibit very low PLQYs. This in turn suggests, that the bis(phosphine) ligand adopts a rigid orientation in all complexes and molecular movements are only small, otherwise the complex **40-Cu-10** would exhibit a significantly higher quantum yields than the complexes **34-Cu-10**, **35-Cu-10**, **36-Cu-10**, where the bis(phosphine) is strongly involved in the frontier orbitals. Thus, the low emission efficiencies must be an intrinsic feature of the tetrazolate ligands with five-membered ring heterocycles.

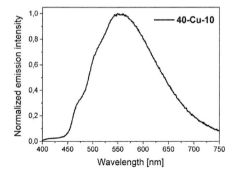

Complex	40-Cu-10
λ_{em} [nm]	555
ϕ_{em}	0.10
τ [µs]	19.6
k_r [s^{-1}]	5.1×10^3
k_{nr} [s^{-1}]	4.6×10^4

Figure 61. Photoluminescence characteristics of complex [(5-(benzothiazol-2-yl)tetrazolate)Cu(DPEPhos)] (**40-Cu-10**) as powder at room temperature when excited at 350 nm.

Table 41. Calculated HOMO and LUMO energies, as well as dihedral angles between the N-Cu-N and P-Cu-P planes in the ground and triplet state geometry of complex [(5-(benzothiazol-2-yl)tetrazolate)Cu(DPEPhos)] (**40-Cu-10**) in comparison to the model complex **7-Cu-10**, using the BP86 functional. The phosphorescence energies are calculated as energy differences between the closed-shell singlet ground-state and the lowest (unrestricted) triplet state.

Complex	HOMO [eV]	LUMO [eV]	$\Delta E_{HOMO-LUMO}$ [eV]	φ_{exp} [°]	φ_{GS} [°]	φ_{TS} [°]
7-Cu-10	–4.58	–2.23	2.35	84	85	61
40-Cu-10	–4.74	–2.42	2.32	90	88	84

When comparing all of the so far discussed tetrazolate complexes in chapter 3.2, a trend of low photoluminescence quantum yields for complexes based on 1-(tetrazol-5-yl)isoquinoline (**30**) (3.2.4.1), 5-([2](1,4)benzo[2](2,5)pyridinophan-13-yl)-tetrazolate **(31)** (3.2.4.2) and tetrazoles with five-membered ring heterocycles (3.2.4.3) is observed. Common to all these complexes are the low radiative rates, while the non-radiative rates lie in the same range or are even lower compared to the model complex (e.g. **30-Cu-18**). Additionally, only minor changes of the emission spectra or blue-shifts are observed when cooling down to 77 K (see also Figure 95 in 5.3.1). In ligands with an enlarged aromatic system, the resolution of a vibrational structure in the emission spectrum is further observed at low temperatures, which is smeared out by molecular movements at ambient temperature. Thus, the low emission efficiencies of these complexes is not attributed to molecular movements and quenching processes than to the low radiative rates induced by the respective N^N ligands. Since a general prediction of the complex's emission color by the electron donating ability of the tetrazolate ligand is still valid, it can be assumed that the energy level scheme is the same for the different ligands, but transition probabilities vary. Since no significant differences for the excitation coefficients of the charge-transfer transition is found in the absorption spectra, which correspond to the vertical transition during excitation, the transition probabilities of different tetrazolate ligands must vary for the relaxed excited state geometry (adiabatic transition). This would also explain the photophysical properties of the complexes **29-Cu-10** and **30-Cu-18** based on quinoline ligands, where the tetrazolate unit is located in a different position on the quinoline: While both complexes exhibit similar absorption bands, the emission spectrum is structured for complex **30-Cu-10**, and the emission intensity graph of **29-Cu-10** just enfolds these vibronic features (Figure 50). This proves the same overall electronic effects of the two electronically equivalent ligands on the complexes' emission. But the emission decay times are drastically longer for complex **30-Cu-10** (and **30-Cu-18**) due to different steric effects of the 1-(tetrazol-5-yl)isoquinoline (**30**) on the excited state geometry and thus the transition probabilities to the ground state.

To conclude the chapter on mononuclear neutral copper(I) complexes based on tetrazolate ligands with five-membered ring heterocycles, only the complexes **34-Cu-10**, **35-Cu-10**, **36-Cu-10** and **40-Cu-10** based on imidazole- and benzothiazole-tetrazolate ligands are obtained and investigated by absorption and emission spectroscopy, and the results related to DFT calculations. The coordination geometries of the complexes resemble the model compound **7-Cu-10**, but the photochemistry differs significantly. While with the electron-rich imidazole ligands **34-H**, **35-H** and **36-H** the LUMO levels of the complexes are raised to such high energies that the LUMO localization shifts to the bis(phosphine), by benzannulation and introduction of a sulfur atom in the five-membered ring the electron donating ability is decreased and the typical localization of frontier orbitals is achieved as in **7-Cu-10**. That also means that shifting the emission color of the complexes to shorter wavelengths is only possible to a certain extent, as tetrazolate ligands with very high LUMO energies will not contribute to the LUMO orbital of the complex. All four complexes **34-Cu-10**, **35-Cu-10**, **36-Cu-10** and **40-Cu-10** exhibit only low photoluminescence quantum yields, which cannot be attributed to a flattening distortion after excitation of pronounced molecular movements, but to the low radiative rates of the complexes with five-membered ring heterocycles. It is suggested, that the low emission efficiencies originate from small transition probabilities with particular N^N ligands in the excited state geometry. The comparison of all complexes with varying tetrazolate ligands in chapter 3.2.4 shows that particular changes of the size or steric hindrance of the pyridine moiety in PyrTet (**7**), such as benzannulation to isoquinoline structures, attachment of a pyridinophane backbone or even substitution by five-membered ring heterocycles are critical for the photophysical properties of the complexes. So far the pyridine moiety has proven as best counterpart to the tetrazole to achieve highly luminescent copper(I) complexes. In the next chapter, the effects of different substituents, which will be introduced on the pyridine, are discussed.

3.2.5 Copper(I) Complexes [(R-PyrTet)Cu(P^P)] with different Substituents on the Pyridine-Tetrazolate Ligand

In the previous chapter it was established that particular modifications of the pyridine moiety lead to a decrease of the radiative rate in the luminescent complexes [(N^N)Cu(DPEPhos)], whereas the introduction of (small) substituent groups on the pyridine ring is expected to maintain the high photoluminescence quantum yields found for the model complex **7-Cu-10**. Various pyridine-tetrazolate ligands with electron donating substituents in the para- and ortho-position are applied in the complexation with copper(I) and a bis(phosphine) ligand to shift the emission color of the model complex **7-Cu-10** further into the blue region of the visible spectrum. Since the introduction of a phenyl-group on the pyridine moiety leads to different photophysical properties of the complexes than purely electron donating substituents, the complexes [(5-(4-phenylpyridin-2-yl)tetrazolate)Cu(P^P)] with different bis(phosphine) ligands will be discussed first and related to the results from chapter 3.2.3, followed by the photophysical investigation of the complexes [(R-PyrTet)Cu(DPEPhos)], where R denotes alkyl-, alkoxy-, and dialkylamino-substituents.

3.2.5.1 The Influence of different Energetic Sites on the Emission Color

Since the influence of different bis(phosphines) on the emission properties of [(PyrTet)Cu(P^P)] complexes was already investigated in chapter 3.2.3.1, the complexes [(5-(4-phenylpyridin-2-yl)-tetrazolate)Cu(P^P)] based on the phenyl-substituted PyrTet ligand **41** will just be discussed briefly in this section.

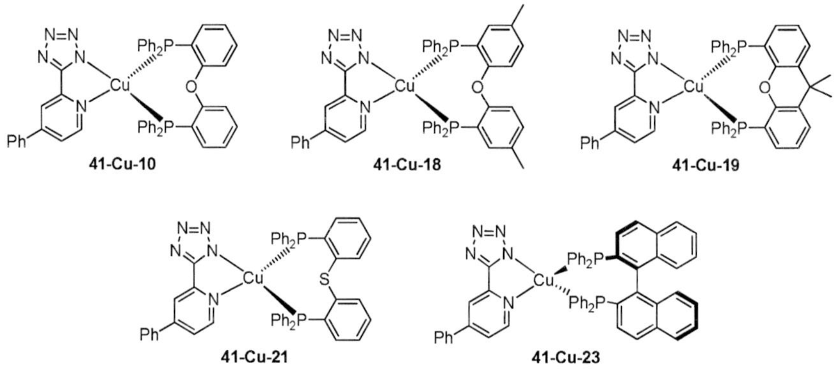

Figure 62. Molecular structures of the complexes [(5-(4-phenylpyridin-2-yl)tetrazolate)Cu(P^P)] **41-Cu-10**, **41-Cu-18**, **41-Cu-19**, **41-Cu-21** and **41-Cu-23** with different bis(phosphine) ligands.

Complex formation of copper(I) and 5-(4-phenylpyridin-2-yl)tetrazolate (**41**) is accomplished with bis(phosphines) of bite angles smaller than 120°, such as DPEPhos (**10**), PTEPhos (**18**), XantPhos (**19**), DPTPhos (**21**), and R-(+)-BINAP (**23**), to give the mononuclear neutral complexes **41-Cu-10**, **41-Cu-18**, **41-Cu-19**, **41-Cu-21** and **41-Cu-23** (Figure 62). The wide bite angle DBFPhos (**20**) does not coordinate the copper(I) atom due to its large P-P distance and the different orientation of phosphorous lone pairs, which disfavors monometallic coordination. [226] The coordination geometries of the complexes **41-Cu-10**, **41-Cu-19**, **41-Cu-21**, for which crystal structures are obtained, are consistent with the model complex **7-Cu-10**. Interestingly, the phenyl ring in para-position of the pyridine does not lie perpendicular to the pyridine moiety, but is only slightly distorted by an angle of around 30° (torsion angle C7-C8-C12-C13 in **41-Cu-10**: 28.8°, **41-Cu-19**: 24.8°, **41-Cu-21**: 35.0°).

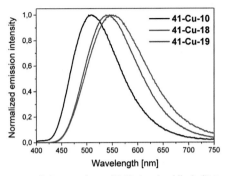

Figure 63. Emission spectra of the complexes [(5-(4-phenylpyridin-2-yl)tetrazolate)Cu(P^P)] **41-Cu-10**, **41-Cu-18**, **41-Cu-19** with different bis(phosphine) ligands as powders at room temperature when excited at 350 nm.

Table 42. Photophysical characterization of the complexes [(5-(4-phenylpyridin-2-yl)tetrazolate)Cu(P^P)] **41-Cu-10**, **41-Cu-18**, **41-Cu-19** with different bis(phosphine) ligands at room temperature.

Complex	λ_{em} [nm]	ϕ_{em}	τ_{ave} [µs]	k_r [s^{-1}]	k_{nr} [s^{-1}]
41-Cu-10	510	0.74	18.6	4.0×10^4	1.4×10^3
41-Cu-10 crystal	556	0.53	15.9[a]	3.3×10^4	3.0×10^4
41-Cu-18	551	0.52	8.5[a]	6.1×10^4	5.6×10^4
41-Cu-19	541	0.64	12.7	5.1×10^4	2.8×10^4

[a] PL emission decay time is composed of two components. A weighted-average lifetime is used (τ_{ave}) and calculated by the equation $\tau_{ave} = \Sigma A_i \tau_i / \Sigma A_i$ with A_i as the pre-exponential factor for the lifetime.

The non-luminescent copper(I) complexes **41-Cu-21** and **41-Cu-23** were already discussed together with their unsubstituted analogues **7-Cu-21** and **7-Cu-23** in chapters 3.2.3.2 and 3.2.3.3. In contrast, the complexes **41-Cu-10**, **41-Cu-18**, and **41-Cu-19** based on DPEPhos (**10**) and its derivatives PTEPhos (**18**) and XantPhos (**19**) exhibit bright photoluminescence in the green to yellow region of the visible spectrum. The results of chapter 3.2.3.1 suggest a mainly steric influence of the different bis(phosphine) ligands due to their similar electronic properties, and thus same emission colors are expected for the complexes **41-Cu-10**, **41-Cu-18** and **41-Cu-19**. On the other hand, it was observed that molecular packing effects in the semicrystalline powders can influence the emission color and only in diluted films with separated complex molecules, the predicted trend is followed. Also in the series of phenyl-substituted complexes **41-Cu-10**, **41-Cu-18** and **41-Cu-19** molecular packing effects lead to different emission colors although similar HOMO-LUMO energy gaps are assumed (Figure 63). While in chapter 3.2.3.1 the complex based on XantPhos (**19**) was found to show an abnormal photophysical behavior, in this complex series of the type [(5-(4-phenylpyridin-2-yl)tetrazolate)Cu(P^P)], the DPEPhos (**10**) ligand induces a blue-shift compared to **41-Cu-18** and **41-Cu-19**. Interestingly, the complex exhibits yellow emission in the crystal similar to its PTEPhos and XantPhos analogues (Table 42).

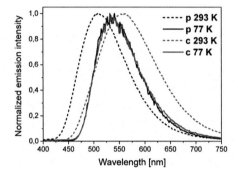

Figure 64. Emission spectra of the complex [(5-(4-phenylpyridin-2-yl)tetrazolate)Cu(DPEPhos))] **41-Cu-10** as powder (p) and crystalline (c) sample at 293 K and 77 K when excited at 350 nm.

When cooling down to 77 K, both the crystalline sample and the powder exhibit the same emission spectra with a maximum at 536 nm (Figure 64). That implies that the energy level scheme of both aggregation phases is in principle the same, but the molecular environment leads to different energetic sites in the solid state, which are in thermal equilibrium at ambient temperature. Electronic sites can either be molecules of slightly different geometry induced by the molecular environment or of different sterical constraints (and thus potential for molecular movements after excitation). Few studies on the luminescence of mononuclear phenanthroline copper(I) complexes of the type $[(N^\wedge N)_2 Cu]^+$ or $[(N^\wedge N)Cu(P^\wedge P)]^+$ in solid state have been reported. It was found that molecules of different environment in the crystal show significantly different emission spectra ($\Delta\lambda_{max}$ up to 0.25 eV) with the molecule of less steric constraint emitting at shorter wavelengths.[179,227,228] For the complex **41-Cu-10** as powder sample, the range of energetic sites is very large and, together with different transition probabilities for each site, results in a blue-shifted emission spectrum compared to the crystal. At 77 K the complex emits from the triplet state and the thermal population of higher lying energetic sites is disabled. Furthermore, molecular movements are frozen out and thus the range of different sites reduced, leading to a consistent emission spectrum at low temperature for both the crystal and powder sample. However, the prediction of such strong molecular packing effects on the emission behavior of the complex for different bis(phosphines) is difficult, since it is not an intrinsic feature of the $P^\wedge P$ ligand, but more attributed to the particular combination of $P^\wedge P$ and $N^\wedge N$ ligands in the complex. From the experimental results so far, the effect of different emission colors in powder and crystal (or film) is mainly observed for ligands prone to ππ-stacking or which exhibit large planar systems, such as the benzimidazolate ligand **5** in **5-Cu-10** (3.1.3.2), XantPhos (**19**) in **7-Cu-19** (3.2.3.1) or the 5-(4-phenylpyridin-2-yl)tetrazolate (**41**) in **41-Cu-10**. Although no particular ππ-interactions are observed in the crystal structures, these ligands facilitate the formation of molecules with different environments and particularly different emission properties.

In general, all complexes **41-Cu-10**, **41-Cu-18** and **41-Cu-19** exhibit red-shifted emission compared to the model compound **7-Cu-10** (510 nm), which can be attributed to the "enlargement" of the aromatic system by the phenyl-substituent, since the conjugation between the pyridine and the phenyl ring is partially maintained due to almost coplanar orientation. Consistent with the shift of the emission color to longer wavelengths, the charge transfer absorption band of **41-Cu-19** at 351 nm ($\varepsilon = 0.98 \times 10^4$ M^{-1} cm^{-1}) is red-shifted compared the unsubstituted complex **7-Cu-19** (341 nm, $\varepsilon = 0.44 \times 10^4$ M^{-1} cm^{-1}), and of unexpectedly high intensity (Figure 96 in 5.3.1). To confirm that the altered photophysical properties are due to an enlargement of the aromatic system of the pyridine-tetrazolate ligand, cyclovoltammetric measurements are performed for the complexes **41-Cu-19** and **7-Cu-19**. Using the oxidation potential determined from the cyclovoltammetric scans, it is possible to estimate the HOMO energy. As expected, the oxidation potentials found for **41-Cu-19** at +0.57 V and **7-Cu-19** at +0.53 V are similar (Figure 107 in 5.3.3). The phenyl-substituent on the pyridine moiety of the N^N ligand influences only the LUMO level of the complex, which is further confirmed by DFT calculations (Table 43). While the HOMO energies of the complex **41-Cu-10** are not affected by the substituent, the enlargement of the aromatic system leads to a lowering of the LUMO level and thus an emission shift to longer wavelengths.

Table 43. Calculated HOMO and LUMO energies, and excitation and phosphorescence energies of the complex [(5-(4-phenylpyridin-2-yl)tetrazolate)Cu(DPEPhos)] (**41-Cu-10**) in comparison to the model complex **7-Cu-10**, using the BP86 functional. The phosphorescence energies are calculated as energy differences between the closed-shell singlet ground-state and the lowest (unrestricted) triplet state.[V]

Complex	HOMO [eV]	LUMO [eV]	$\Delta E_{HOMO-LUMO}$ [eV]	excitation energy [eV]	phosphorescence energy [eV]
7-Cu-10	−4.58	−2.23	2.35	2.92	1.93
41-Cu-10	−4.56	−2.52	2.03	2.76	1.80

In this chapter, the influence of molecular packing effects on the emission behavior of a complex is further investigated. The molecular environment in the solid state induces different steric constraints and thus controls the energy levels and transition probabilities of the excited states for each molecule. A particular combination of N^N and P^P ligands in the mononuclear neutral copper(I) complexes facilitate a broader range of energetic sites and/or significantly different transition probabilities of these sites, which leads to emission color abnormalities or different properties of the powder and the crystal.

3.2.5.2 The influence of the Substituent's Position

The investigation of different modifications of the tetrazolate ligand in copper(I) complexes [(N^N)Cu(DPEPhos)] has shown that the variation of the complex's emission color into the red or blue region of the visible spectrum is limited, since altering the electronic properties of the pyridine or other heterocycles both shifts the LUMO and the HOMO energies of the complex. When applying very electron rich N^N ligands, the lowest unoccupied molecular orbital shifts to the bis(phosphine) instead of the tetrazolate ligand, and thus cannot be electronically varied by the N^N ligand. Furthermore, especially sterically crowded tetrazolate ligands, such as 5-([2](1,4)benzo[2](2,5)pyridinophan-13-yl)-tetrazolate (31) but also the five-membered ring heterocycles, influence the transition probabilities of the relaxed excited state and lead to low photoluminescence quantum yields. It is assumed, that by introduction of (small) substituents on the pyridine moiety the emission color of the complexes can be varied while still maintaining high emission efficiencies. In the following it will be examined, if the electronic alteration of PyrTet (7) by electron-donating substituents, such as alkyl-, alkoxy-, and dialkylamino-groups, fits into the picture that was gained of the complex's electronic system so far. In particular, the influence of a substituent in different positions on the pyridine moiety will be examined regarding its steric and electronic effects.

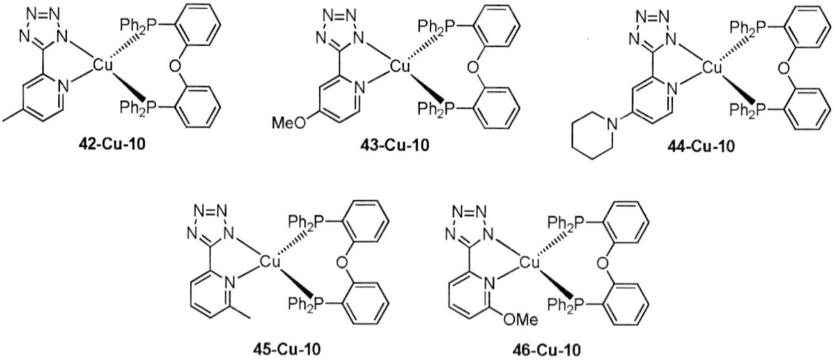

Figure 65. Molecular structures of the neutral copper(I) complexes [(R-PyrTet)Cu(DPEPhos)] **42-Cu-10**, **43-Cu-10**, **44-Cu-10**, **45-Cu-10**, **46-Cu-10** with different substituents in para- and ortho-position.

With all substituted pyridine-tetrazolates, namely 5-(4-methylpyridin-2-yl)tetrazolate (**42**),

5-(4-methoxypyridin-2-yl)tetrazolate (**43**), 5-(4-piperidylpyridin-2-yl)tetrazolate (**44**),

5-(2-methylpyridin-6-yl)tetrazolate (**45**) and 5-(2-methoxypyridin-6-yl)tetrazolate (**46**), mononuclear

neutral copper(I) complexes are obtained. Their molecular structures are depicted in Figure 65.

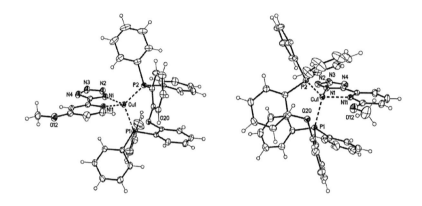

Figure 66. Crystal structures of copper(I) complexes [(R-PyrTet)Cu(DPEPhos)] **43-Cu-10** and **46-Cu-10** with a methoxy-substituent in para- or ortho-position on the pyridine moiety.[IV]

Table 44. Selected bond lengths (Å) and angles (°) of the complexes [(R-PyrTet)Cu(DPEPhos)] **43-Cu-10**, **46-Cu-10** with a methoxy-substituent in para- or ortho-position on the pyridine moiety.

	43-Cu-10	**46-Cu-10**
Cu1-N1	2.0236 (13)	2.0512 (14)
Cu1-N11	2.1194 (13)	2.0763 (14)
Cu1-P1	2.2410 (4)	2.2694 (5)
Cu1-P2	2.2138 (4)	2.2042 (5)
P1-P2	3.724	3.771
N1-Cu-N11	80.53 (5)	80.14 (5)
P1-Cu-P2	113.442 (16)	114.913 (19)
φ^a	89.76	89.68
N1-C5-C6-N11	3.2 (2)	1.3 (2)

[a] Dihedral angle between N–Cu–N and P–Cu–P planes.

Their crystal structures exhibit a tetrahedral coordination of the copper(I) atom by the

pyridine-tetrazolate and the bis(phosphine) ligand with overall comparable bond lengths and angles,

here exemplarily shown for the complexes **43-Cu-10** and **46-Cu-10** (Figure 66). However, slight

differences are observed for the two complexes with a substituent in para- or ortho-position of the

pyridine moiety: While in complex **43-Cu-10** the N^N ligand coordinates the copper(I) atom in an

asymmetric mode with the tetrazolate slightly closer than the pyridine moiety, similar Cu-N bond lengths are found for complex **46-Cu-10** despite a higher steric hindrance of the methoxy-group in ortho-position (Table 44). Since the complexes **44-Cu-10** and **45-Cu-10** with piperidyl- or methyl-substituents in para- and ortho-position show the same behavior, it cannot be attributed to electrostatic interactions or molecular packing effects, but counter-intuitively to different electronic properties. Taking into account the results from chapter 3.1.2.2 on the coordination chemistry of copper(I) with different pyridine-amide ligands, a stronger and thus shorter bonds is expected with less electron donating ligands, i.e. the 5-(2-methoxypyridin-6-yl)tetrazolate (**46**) seems to be less electron-rich than its isomer **43** with the methoxy-substituent in para-position. The sterically crowded complex center in **46-Cu-10** leads to an outwards orientation of the methoxy-group, which is also the reason for slightly different Cu-P bond lengths. In the isomeric complex **43-Cu-10** the methoxy-substituent points away from the complex center, while the piperidyl-group in **44-Cu-10** adopts a chair conformation but is only twisted by 5.5° against the pyridine moiety (torsion angle C7-C8-N12-C17).

To begin the photophysical investigation of the complexes, UV-Vis measurements are recorded in dichloromethane. Surprisingly, the absorption spectra for the complexes **43-Cu-10**, **44-Cu-10** exhibit different shapes than for **45-Cu-10**, **46-Cu-10**, while complex **42-Cu-10** resembles features of both cases (Figure 67). For the complexes **45-Cu-10** and **46-Cu-10** with substituents in ortho-position on the pyridine moiety, charge transfer absorption bands around 335 nm are observed as shoulders ($\varepsilon = 6.3 \times 10^3\,\mathrm{M^{-1}\,cm^{-1}}$), which reach into the red region up to 390 nm. For the para-substituted complexes **43-Cu-10** and **44-Cu-10** the CT bands vanish underneath the strong and red-shifted ligand-centered bands between 305–350 nm. Since the absorption edges only reach up to 375 nm, bluer charge-transfer bands compared to the complexes **45-Cu-10** and **46-Cu-10** and thus larger HOMO LUMO energy gaps can be assumed. For complex **42-Cu-10** the charge-transfer band is also not detectable underneath the strong ligand-centered bands, but its absorption edge reaches up to 395 nm. These experimental results imply different electronic effects of the 5-(2-methoxypyridin-6-yl)-tetrazolate (**46**) ligand and its isomer **43** with the methoxy-substituent in para-position, as was already observed in the crystal structure. In contrast, the charge transfer bands of complex **42-Cu-10** and its isomer **45-Cu-10** are assumed to be at a similar position estimated from their absorption edges, which indicates comparable electronic influences of the isomeric methyl-substituted pyridine-tetrazolate ligands **42** and **45** on the complex.

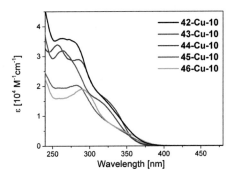

Figure 67. Absorption spectra of the complexes [(R-PyrTet)Cu(DPEPhos)] **42-Cu-10**, **43-Cu-10**, **44-Cu-10**, **45-Cu-10**, **46-Cu-10** in CH_2Cl_2 at a concentration of 10^{-5} mol/l.

All complexes **42-Cu-10**, **43-Cu-10**, **45-Cu-10** and **46-Cu-10** exhibit bright, blueish-green photoluminescence with quantum yields up to 76%, except the piperidyl-substituted complex **44-Cu-10**, which shows only a low emission quantum yield of 5% at an emission maximum of 493 nm (Table 45). This is the result of both a lower radiative and a drastically increased non-radiative rate compared to the other complexes in this series, due to introduction of a large substituent with CH_2-groups, which can potentially quench the emission by vibrational movements.[42] The emission maxima of the complexes **42-Cu-10**, **43-Cu-10**, **44-Cu-10**, **45-Cu-10** and **46-Cu-10** are blue-shifted by maximally 25 nm (0.13 eV) compared to the model complex **7-Cu-10** due to the higher electron density of the pyridine-tetrazolate ligand with electron donating substituents. When taking a closer look at the emission spectra, a stronger shift to short wavelengths is observed for the complexes **45-Cu-10** and **46-Cu-10** with a substituent in ortho-position (λ_{em} = 485 nm), while the complexes **42-Cu-10**, **43-Cu-10** and **45-Cu-10** exhibit very broad photoluminescence spectra with maxima around 495 nm (Figure 68). As already observed in absorption spectroscopy, the electronic system of the two complex isomers **43-Cu-10** and **46-Cu-10** as well as for **42-Cu-10** and **45-Cu-10** differ and lead to a blue-shift for the ortho-substituted compared to the para-substituted complexes, i.e. the sterically stronger hindered complexes **45-Cu-10** and **46-Cu-10** with a substituent in ortho-position of the coordinating pyridine moiety exhibit a smaller Stokes-shift.

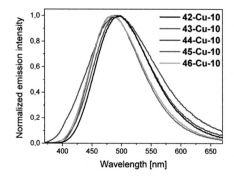

Figure 68. Emission spectra of the complexes [(R-PyrTet)Cu(DPEPhos)] **42-Cu-10, 43-Cu-10, 44-Cu-10, 45-Cu-10, 46-Cu-10** as powders at room temperature when excited at 350 nm.

Table 45. Photophysical characterization of the complexes [(R-PyrTet)Cu(DPEPhos)] **42-Cu-10, 43-Cu-10, 44-Cu-10, 45-Cu-10, 46-Cu-10** in comparison to the model complex **7-Cu-10**, as powders at room temperature.

Complex	λ_{em} [nm]	ϕ_{em}	τ_{ave} [μs]	k_r [s^{-1}]	k_{nr} [s^{-1}]
7-Cu-10	510	0.78	19.9	3.9×10^4	1.1×10^4
42-Cu-10	495	0.78	17.4	4.5×10^4	1.3×10^4
43-Cu-10	494	0.52	7.8^a	6.7×10^4	6.2×10^4
44-Cu-10	493	0.05	5.8^a	8.6×10^3	1.6×10^5
45-Cu-10	485	0.71	14.5	4.9×10^4	2.0×10^4
46-Cu-10	485	0.76	14.3	5.3×10^4	1.7×10^4

^{a} PL emission decay time is composed of two components. A weighted-average lifetime is used (τ_{ave}) and calculated by the equation $\tau_{ave} = \Sigma A_i \tau_i / \Sigma A_i$ with A_i as the pre-exponential factor for the lifetime.

To understand the different electronic effects of the substituents on the complexes' emission and especially the unusual behavior of the isomeric complexes **43-Cu-10** and **46-Cu-10** based on the methoxy-substituted ligands **43** and **46**, a systematic study of pyridine-tetrazolate ligands with methyl, methoxy- and dimethylamino-substituents in ortho-, meta- and para-positions is performed by DFT calculations (Table 46). The unsubstituted PyrTet (**7**) and the corresponding model complex **7-Cu-10** are taken as reference.

Table 46. LUMO energies of pyridine-tetrazolate ligands with different substituents in ortho-, meta- and para-position on the pyridine moiety, and of the corresponding [(N^N)Cu(DPEPhos)] complexes as obtained by DFT calculations (BP86). The localization of the LUMO orbital on the pyridine-tetrazolate or bis(phosphine) ligand is denoted as N^N or P^P.[V]

Ligand	Position of substituent	LUMO of ligand [eV]	LUMO of complex [eV]	localization of LUMO
PyrTet	-	-1.96	-2.23	N^N
(Me-substituted pyridine-tetrazolate)	-ortho	-2.70	-2.15	N^N
	-meta	-2.67	-2.18	N^N
	-para	-2.74	-2.15	N^N
(MeO-substituted pyridine-tetrazolate)	-ortho	-2.68	-2.13	N^N
	-meta	-2.44	-2.09	N^N
	-para	-2.66	-2.02	P^P
(Me₂N-substituted pyridine-tetrazolate)	-ortho	-2.44	-2.09	P^P
	-meta	-2.14	-2.01	P^P
	-para	-2.33	-1.95	P^P

From the comparison of the LUMO energies of the different N^N ligands and of the corresponding mononuclear neutral complexes [(N^N)Cu(DPEPhos)] in this study, several statements can be made: The extent of the electronic effect induced by a electron-donating substituent varies for the different positions (ortho-, meta-, para-) on the pyridine moiety of PyrTet, and is usually strongest for meta-position, which also resembles the rules of directing substituents in aromatic substitution reactions.[229] The LUMO energies differ most for the ortho-, meta- and para-positions with the strongly electron donating dimethylamino-group, and least with the weakly activating methyl-substituent. This explains the different photophysical properties of the isomeric complexes **43-Cu-10** and **46-Cu-10** substituted by a methoxy-group compared to the similar absorption behavior of the methyl-substituted complexes **42-Cu-10** and **45-Cu-10**. Since the LUMO energies of the N^N ligands are obtained for its neutral form, and furthermore the complex coordination cannot be neglected, the trend of the ligands' LUMO energies is usually not followed for the complexes' LUMO levels. But when the LUMO energy of the pyridine-tetrazolate ligand exceeds that one of the bis(phosphine) in the complex, the LUMO orbital shifts to the bis(phosphine), as already observed in chapter 3.2.3.3 and 3.2.4.3. Thus, for the isomeric complexes **42-Cu-10** and **45-Cu-10** the usual localization of frontier orbitals with a participation of the N^N ligand in the LUMO is found, while the LUMO shifts to the bis(phosphine) when going from complex **43-Cu-10** to **46-Cu-10**.

The shift of the lowest unoccupied molecular orbital from the N^N ligand to the bis(phosphine) in the mononuclear copper(I) complexes is illustrated in Figure 69: While the LUMO of the complex is located on the N^N ligand for all (methylpyridinyl)tetrazolate isomers, it is distributed on the (methoxypyridinyl)tetrazolate ligand for the ortho- and meta-substituted isomers, but shifts completely to the bis(phosphine), when the methoxy-group is in para-position of the pyridine ring (see Figures 103 and 104 in 5.3.2). For complexes with a (dimethylaminopyridinyl)tetrazolate substituent, the LUMO is solely located on the P^P ligand for all substituent-positions, and the frontier orbitals of the complex **44-Cu-10** substituted with a piperidyl-group are consistent with these results (Figure 105 in 5.3.2). To conclude, the LUMO level of the complex can only be controlled by the methyl-substituted pyridine-tetrazolate ligands **42**, **45**, and the 5-(2-methoxypyridin-6-yl)tetrazolate (**46**). These results explain the different shape of absorption spectra of the complexes **43-Cu-10** and **44-Cu-10** compared to **45-Cu-10** and **46-Cu-10** and the intermediate role of complex **42-Cu-10**.

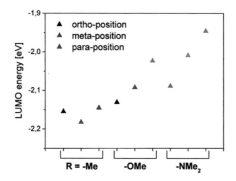

Figure 69. Calculated LUMO energies of the complexes [(N^N)Cu(DPEPhos)] based on pyridine-tetrazolate ligands with different substituents in ortho-, meta- and para-position of the pyridine moiety. Dark color symbols depict a LUMO localization on the N^N ligand, while for light color symbols the LUMO shifts completely to the bis(phosphine) ligand.[V]

A further tool to investigate isomeric complexes are photoluminescence measurements at low temperature: If two isomeric pyridine-tetrazolate ligands, such as 5-(4-methoxypyridin-2-yl)-tetrazolate (**43**) and 5-(2-methoxypyridin-6-yl)tetrazolate (**46**), induce different electronic effects on the corresponding complexes, the energy level schemes differ for the two complexes, and different emission spectra at ambient temperature, but even more at low temperature are expected. On the other hand, the complexes' emission spectra at low (and room) temperature should be consistent for isomeric pyridine-tetrazolate ligands with similar electronic effects, such as in **42-Cu-10** and **45-Cu-10**. Since at ambient temperature molecular packing effects might shift the emission spectra, as discussed in the previous chapter 3.2.5.1, low temperature photoluminescence is more reliable. From comparison of the 77 K emission maxima of the complexes **43-Cu-10** and **46-Cu-10**, a strong red-shift is observed by 30 nm, while the spectra of **42-Cu-10** and **45-Cu-10** shift by the same extent at low temperature (Figure 70). The minor electronic differences between the para- and ortho-substituted methylpyridine-tetrazolates **42** and **45** do not change the photophysical properties of the complexes significantly, while only minor differences of the methoxypyridine-tetrazolates result in different localizations of frontier orbitals.

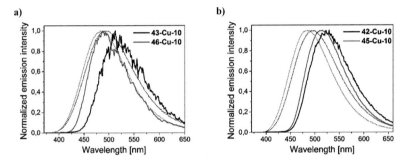

Figure 70. Emission spectra of the complexes [(R-PyrTet)Cu(PTEPhos)] a) **43-Cu-10** and **46-Cu-10** and b) **42-Cu-10** and **45-Cu-10** at 77 K (solid line) and 293 K (dashed line), as amorphous powder when excited at 350 nm.

The above results determine that by introduction of different substituents on the pyridine moiety of the N^N ligand, the emission color of the corresponding complex can be shifted to shorter wavelengths only to a certain limit. Furthermore, also the HOMO level is slightly affected by the electron-donating substituent resulting in only small alteration of the energy gap when compared to the model complex **7-Cu-10** (Table 58 in 5.3.2). Using cyclic voltammetry, the effects of the various ligands on the oxidation potential is examined, which might further be influenced by different steric effects besides the thoroughly investigated electronic effects of the ligands. During the cyclovoltammetric cycle, the copper(I) atom in the complexes is oxidized to a formal Cu(II) species, which favors a square-planar geometry. When substituents in ortho-position of the pyridine-tetrazolate ligand sterically hinder the flattening from the tetrahedral to square-planar geometry, the oxidation is less favored resulting in higher oxidation potentials.[230] The cyclovoltammetric scan of the complexes based on methoxy-substituted pyridine-tetrazolate shows a lower oxidation potential for complex **43-Cu-10** at +0.43 V than for **46-Cu-10** at +0.48 V, i.e. the ligand 5-(2-methoxypyridin-6-yl)tetrazolate (**46**) sterically hinders flattening of the complex (Figure 71). Also a lower oxidation potential at +0.47 V for the complex **42-Cu-10** compared to +0.57 V (first peak) for **45-Cu-10** is found. It is valid to attribute the different oxidation potentials to steric effects in this case, as the ligands **43** and **46**, or **42** and **45** induce similar effects on the HOMO orbital of the complex, and same energy values are found from the DFT calculations (Table 58 in 5.3.2). The higher PLQY of **46-Cu-10** might be attributed to the lower flattening extent in this complex,[146] whereas the complex **43-Cu-10** exhibits a higher non-radiative rate and lower emission efficiency (Table 45).

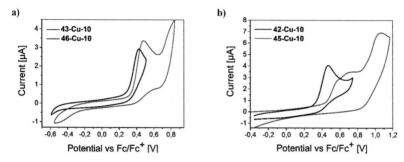

Figure 71. Cyclovoltammetric spectra of the complexes [(R-PyrTet)Cu(DPEPhos)] a) **43-Cu-10** and **46-Cu-10**, and b) **42-Cu-10** and **45-Cu-**10, calibrated against $FeCp_2/FeCp_2^+$.[VI]

In this chapter, the copper(I) complexes [(R-PyrTet)Cu(DPEPhos)] with methyl-, methoxy- and piperidyl-substituents in ortho- and para-position on the pyridine moiety of the N^N ligand are examined in detail. It is found that very electron-rich pyridine-tetrazolate ligands lead to a different localization of the frontier orbitals in the complex and shift the LUMO orbital from the N^N ligand to the bis(phosphine), which further alters the absorption spectra and emission properties of the complexes. A systematic study of different substituents and in different positions on the pyridine moiety is conducted by DFT calculations. The position of the substituent group in the pyridine tetrazolate ligand causes differently strong electronic effects in the complexes, e.g. for the methoxy-substituted, isomeric complexes **43-Cu-10** and **46-Cu-10** different coordination geometries, absorption and emission properties are found. In addition to the electronic effects, the substituents in ortho- and para-positions also induce different sterical hindrance, which can be followed by cylovoltammetric measurements.

3.3 Investigation of the Thermally Activated Delayed Fluorescence by Time-resolved Spectroscopy[XI]

Besides the development and photophysical characterization of various luminescent, mononuclear neutral copper complexes [(N^N)Cu(P^P)] based on tetrazolate ligands to study the electronic system controlling the emission properties (chapter 3.2), the thermally activated delayed fluorescence of this complex class is the focus of this research work. The TADF behavior of copper(I) complexes enables high emission efficiencies, short decay times and the use of all excitons for the emission of light in an OLED, despite their low spin-orbit coupling when compared to phosphorescent materials based on iridium(III) and platinum(II) (ξ_{Cu}=857 cm^{-1} versus ξ_{Pt}= 4481 cm^{-1}, ξ_{Ir}=3909 cm^{-1})[98] (see also introductory chapter 1.2.2). The narrow energy gap between the excited singlet S$_1$ and triplet T$_1$ states of 1000 cm^{-1} or lower and a stable, but relatively long-lived triplet state enables the back transfer of the excitons from the triplet reservoir to the excited singlet state at ambient temperatures and finally deactivation by delayed fluorescence.

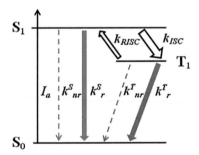

Scheme 8. Schematic representation of the interplay of radiative (r, solid arrows) and non-radiative (nr, dashed arrows) transitions in a TADF material. After excitation to the excited singlet state (I_a), intersystem crossing to the triplet state occurs (k_{ISC}) and at sufficient temperature, the excitons are repopulated to the excited singlet state (k_{RISC}). From there thermally activated delayed fluorescence occurs. k$_S$ and k$_T$ depict rate constants from the excited singlet or triplet state, respectively.

A current challenge in the study of TADF systems is unravelling the various excited state processes that occur, including radiative and non-radiative decay rates from the excited singlet, the rate of intersystem crossing from excited singlet to triplet, the rates of radiative and non-radiative losses in the triplet, and rates of reverse intersystem crossing from the triplet to singlet (Scheme 8).

[XI] Parts of the chapter "Investigation of the Thermally Activated Delayed Fluorescence by Time-resolved Spectroscopy" have published previously.[138]

Determining these rates can be challenging, especially since no spontaneous fluorescence is observed in copper(I) complexes, and thus clear unambiguous observation of any of the key rates is valuable in TADF materials. In this chapter, the radiative and non-radiative transitions occurring in a copper(I) complex after excitation are investigated by time-resolved emission (and absorption) spectroscopy, and better understanding of the TADF mechanism and its limiting factors are explored.

3.3.1 Intersystem Crossing in the Copper(I) Complex [(PyrTet)Cu(DPEPhos)]

The non-radiative transitions intersystem crossing and reverse intersystem crossing are important parameters in the TADF mechanism, which can be indirectly determined for organic TADF materials by analysis of the spontaneous fluorescence,[31,32] but are difficult to examine for copper(I) complexes due to the apparent lack of this spontaneous fluorescence in a system with fast intersystem crossing rates (see also 1.1.3). The non-luminescent copper(I) bis(imine) complexes have been studied in detail by time-resolved emission and absorption spectroscopy due to their interesting dynamics in solution, and ISC rates have been determined to 7–15 ps.[122,151-154] However, these complexes exhibit very inefficient TADF with large energy gaps[139] and their photophysical properties are strongly controlled by a flattening distortion after excitation.[118] And since the higher degree of freedom in solution might not represent the processes in the solid state of an OLED, the intersystem crossing rate of the luminescent model complex [(PyrTet)Cu(DPEPhos)] (**7-Cu-10**) is aimed to be studied in the solid state, in particular neat film to rule out molecular packing effects. [XII] But the presence of long-lived delayed fluorescence and back population of the excited singlet state S_1 by RISC challenges the experimental methods used to determine ISC and the assignment of the observed dynamics. Thus, time-resolved measurements are performed on a microsecond time-range to obtain the overall lifetime of the thermally activated delayed fluorescence, and on a picosecond time-range to monitor the initial photophysical processes after excitation. Time-resolved photoluminescence spectra support the interpretation of kinetics and especially the assignment of the intersystem crossing rate, which is further studied in transient absorption measurements.

[XII] Neat film samples are prepared by spincoating a dichloromethane solution of the complex **7-Cu-10**.

The complex **7-Cu-10** exhibits green emission with a peak at 535 nm and a photoluminescence quantum yield of 37% in neat film (Figure 72). The photoluminescence maximum at 512 nm in the amorphous powder is red-shifted to 535 nm in the neat film, which is due to a combination of more flexibility in the amorphous film compared to the semi-crystalline powder and thus easier distortion of the excited states with a distortion of the ground state geometry.[212]

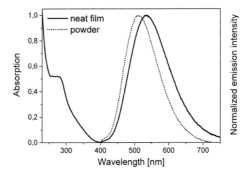

Figure 72. Absorption spectrum of [(PyrTet)Cu(DPEPhos)] (**7-Cu-10**) in neat film (solid line), as well as emission spectra of the neat film (solid line) and amorphous powder sample (dashed line) at room temperature when excited at 350 nm.

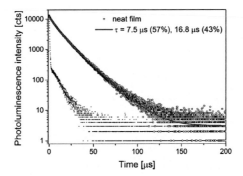

Figure 73. Microsecond luminescence decay dynamics of [(PyrTet)Cu(DPEPhos)] (**7-Cu-10**) in neat film detected at 535 nm (PL maximum) in nitrogen atmosphere. The red solid line represents a fit to a biexponential decay with a time constant of 7.5 µs (0.57) and a component of 16.8 µs (0.43), which gives a weighted-average lifetime of 11.5 µs by the equation $\tau_{ave} = \Sigma A_i \tau_i / \Sigma A_i$ with A_i as the pre-exponential factors for the lifetime. The instrument-response function is depicted as dotted line.

Time-resolved photoluminescence is initially measured on a microsecond time-range, and the luminescence decay dynamics at the emission maximum show long-lived emission with two lifetime components (and pre-exponential components) of 7.5 µs (0.57) and 16.8 µs (0.43) (Figure 73), which may correspond to different coordination environments in the film.[179,227,228] A weight-averaged lifetime of 11.5 µs is obtained for the thermally activated delayed fluorescence. In this measurement, fast (picosecond) dynamics are not observed owing to the long temporal response of the time correlated single photon counting method (TCSPC).

Luminescence decay dynamics on the picosecond time-range of complex **7-Cu-10** as neat film are recorded with a streak camera (2 ps temporal response) with a 375 nm laser with a repetition rate of 20 kHz as an excitation source. Since the emission is very weak, the signals had to be averaged from several scans over the wavelength region 540–600 nm, which is around the emission maximum and on the red-side photoluminescence. The observed decay can be fitted to a single exponential with a time constant of 27 ps when convolved with the instrument response function (Figure 74). This decay sits on a very strong background of long-lived emission longer than 12 ns (the repetition rate of the laser that triggers the camera being 80 MHz), making observation of the 27 ps decay challenging.

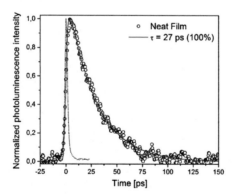

Figure 74. Picosecond luminescence decay dynamics of [(PyrTet)Cu(DPEPhos)] (**7-Cu-10**) in neat film at 540–600 nm (PL maximum and red-side) when excited at 375 nm. The solid line represents a fit to a monoexponential decay with a time constant of 27 ps. The instrument-response function is depicted as dotted line with 2.5 ps full-width half-maximum.

To confirm that this decay does not embody a spectral shift, time-resolved photoluminescence spectra are recorded (Figure 75). Emission spectra in the time windows of 0–20 ps and 40–60 ps, and all emission longer than 12 ns, show a fast decrease of emission intensity, reflecting the picosecond luminescence decay of 27 ps (Figure 75a), while normalized emission spectra are essentially the same, and agree with the PL spectrum for the long-lived (> 12 ns) background that is representative of TADF emission (Figure 75b). To make sure that the delayed fluorescence background does not alter the shapes of the picosecond spectra, the long-lived spectrum is subtracted.

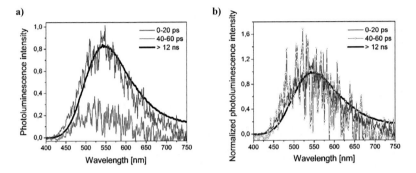

Figure 75. a) Time-resolved photoluminescence spectra at different delay times after excitation with a laser pulse. The accumulated background of the long-lived (> 12 ns) fluorescence is subtracted from the spectra before analysis. As prompt fluorescence in copper(I) complexes is very weak, signals had to be averaged over the time windows 0–20 ps (red line), 40–60 ps (green line with symbols), and long-lived PL for times longer than 12 ns (black line). A fast intensity decrease can be observed in the time-window of 150 ps after excitation. b) Normalized time-resolved photoluminescence spectra. The spectra do not change within the time-window of 150 ps.

To understand the excited state processes in the copper(I) complex [(PyrTet)Cu(DPEPhos)] (**7-Cu-10**), the photoluminescence on picosecond, nanosecond and microsecond timescales is examined to aid in the assignments of the intersystem crossing rate. The emission band at 535 nm in the steady-state emission spectrum represents the $S_0 \leftarrow S_1$ (^{1}MLCT) transition. This transition either occurs directly after population of the S_1 state upon excitation, or subsequently upon back transfer of the excited state from the triplet state T_1 (^{3}MLCT). Long-lived emission on the microsecond timescale characterizes the delayed fluorescence component, consistent with the $S_0 \leftarrow S_1$ transition after back transfer from the triplet to singlet excited states. Since this process is controlled by the triplet lifetime and the reverse intersystem crossing (RISC) rate, an overall lifetime of 11.5 µs is obtained. From photoluminescence dynamics on the picosecond timescale a single fast decay of 27 ps is observed. Since the excitons in the triplet state do not emit much (or more likely, any) light in the observed 150 ps time window (emission decay time of triplet state is several µs), the rapid picosecond decay of the photoluminescence in the wavelength region 540–600 nm is attributed to emission from the excited singlet state S_1. The two main processes that can occur after photoexcitation are relaxation to lowest excited states or lowest energetic sites due to environmental inhomogeneity,[231 , 232] and intersystem crossing to the triplet state.

Time-resolved spectra in the 0–20 ps and 40–60 ps windows are essentially the same, indicating that this decay is not linked to any spectral relaxation in the singlet state. To further confirm that hopping of the exciton to lowest energy sites is not occurring on this timescale, the picosecond luminescence of complex **7-Cu-10** is also recorded in THF solution, and although the emission is weak, an almost identical decay time of 26 ps is found (Figure 76).

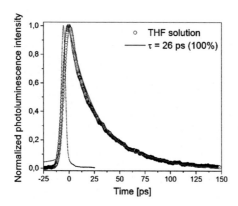

Figure 76. Picosecond luminescence decay dynamics of [(DPEPhos)Cu(PyrTet)] (**7-Cu-10**) in THF solution at 470–600 nm (PL maximum and red-side) when excited at 400 nm. The solid line represents a fit to a monoexponential decay with a time constant of 26 ps. The instrument-response function is depicted as dotted line with 2.5 ps full-width half-maximum.

Substantial structural changes of the molecule such as the flattening distortion have been reported for copper(I) bis(imine) complexes,[119,124,220] however this apparently does not occur in neat film in the observed picosecond time-window, as this would result in a spectral red-shift with time. Consequently, the 27 ps process is attributed to intersystem crossing, consistent with the observation that after this process no further emission is observed on a timescale shorter than 12 ns (the triggering frequency of the camera), i.e. all subsequent emission is delayed fluorescence. After the excitons have established in the triplet, reverse intersystem crossing back to the excited singlet and subsequently TADF emission occurs. Measurement of the RISC rate will be extremely challenging as only a small fraction of the triplets will ever cross back to the singlet at any one time (controlled by the energy splitting of excited singlet and triplet states), making detection either by transient emission or absorption very difficult. No rise in emission within the 2 nanosecond time window is observed, indicating that RISC is on a timescale > 2 ns. To conclude, the time constant of 27 ps found for the luminescence decay dynamics from 540–600 nm can be attributed to intersystem crossing $S_1 \rightarrow T_1$. This gives a rate of ISC of 3.7×10^{10} s^{-1} ($= \tau_{ISC}^{-1}$).

Transient absorption measurements on a broadband setup qualitatively complement the assignment of the observed kinetics to intersystem crossing. Thereby a sample is excited by a short laser pulse of 100 fs (400 nm) and subsequently the time-resolved absorbance of the excited material recorded, thus the temporal dynamics of the absorption bands corresponding to $S_1 \rightarrow S_n$ or $T_1 \rightarrow T_n$ can be observed. The picosecond transient absorption dynamics for [(PyrTet)Cu(DPEPhos)] (**7-Cu-10**) in neat film are measured for probe wavelengths in the range 400–800 nm (Figure 97 in 5.3.1). The strongly emissive nature of the material make the transient absorption difficult to measure, as large amounts of photoluminescence (535 nm) are detected which overlap with the continuum probe, and thus the negative signal of excited state absorption is very noisy despite subtraction of the background fluorescence before analysis. The transient absorption measurements are not very sensitive to intersystem crossing and thus comparison of absolute numbers has to be taken with care. However, a decay of an excited state absorption is found at a wavelength region 530–700 nm, which is believed to be the loss of the excited singlet state by ISC to the triplet, and supports the time-scale of the intersystem crossing to tens of picoseconds with similar dynamics.

The small spin-orbit coupling constant of copper ($\xi_{Cu}=857$ cm^{-1}) is reflected in the rather slow intersystem crossing of 27 ps compared to iridium(III) complexes ($\xi_{Ir}=3909$ cm^{-1}) with ISC time constants of 100 fs.[18] When compared to reported intersystem crossing constants of copper bis(imine) complexes in solution (7–15 ps),[124,151-154] the time constant of 27 ps in the studied TADF [(PyrTet)Cu(DPEPhos)] (**7-Cu-10**) in the solid state appears to be slower than one would expect. This indicates a weak *effective* spin-orbit coupling in the molecular orbitals involved in the emission, which is intrinsically controlled by ligands and/or halides coordinated to the copper(I) center and the specific coordination geometry.[124] Yersin reported a strong effective SOC for small energy gaps between energy states of different d-orbital character, i.e. singlet and triplet (sub-)states have to stem from different d-orbitals of the metal atom to experience strong spin-orbit coupling and thus small intersystem crossing rates.[47,144] Since these energy differences depend on the coordination geometry around the copper(I) center, the environment and morphology has a strong influence on the ISC rate by changing the molecular ground or excited state geometry. For copper(I) bis(phenanthroline) complexes it was further found that the ISC rate is controlled by the extent of flattening distortion in solution,[118] and the distortion to a square planar geometry in solution changes the energetic level scheme in such a way that other channels are opened up for intersystem crossing.[124] The herein studied complex

[(PyrTet)Cu(DPEPhos)] (**7-Cu-10**) exhibits a fundamentally different emission character than the complexes based on phenanthroline ligands, which makes direct comparisons difficult. However, DFT calculations confirmed an (ML+IL)CT transition with small percentage of copper(I) interactions (weak internal heavy atom effect),[101] which accounts for the relatively slow intersystem crossing. The intersystem crossing rate can be understood as measure of spin-orbit coupling, which ensures higher transition probabilities for the phosphorescence, and thus indirectly shortens the overall TADF lifetime. The shorter emission decay times of 1–4 µs of dinuclear copper(I) halide complexes $[Cu_2X_2(P^\wedge N)(P)_2]$[94] can be easily explained by higher SOC induced by the halides and thus shorter ISC rates are expected. The determination of intersystem crossing rates in different classes of copper(I) complexes could give further information about the influence of the coordination motif and the ligands on the strength of spin-orbit coupling.

Interestingly, the intersystem crossing rates, obtained from fitting of the emission decays, do not differ significantly in neat film and solution. No spectral shift over time is observed for the neat film in the time window of 150 ps, leading to the conclusion that geometry distortions can occur only on a shorter femtosecond, or longer nanosecond timescale when the excited state resides in the triplet. Luminescence decay dynamics on the femto- to picosecond time-range of complex **7-Cu-10** in THF solution are recorded by fluorescence upconversion with an optical gating setup (FOG). The solution sample in a rotating cell (to minimize degradation) is excited by a 400 nm laser and the kinetics are recorded at 550 nm, 600 nm and 650 nm, which corresponds to the blue- and red-side photoluminescence and the PL peak. Since the emission is very weak, the signals had to be averaged from numerous scans. Although much weaker than in neat film, long-lived emission > 12 ns is upconverted as constant background with the kinetics of interest on top of it. The observed decay at 550 nm can be fitted to a biexponential with two time components (and pre-exponential components) of 26 ps (0.51) and 1.5 ps (0.49) (Figure 77). The short components is less pronounced in the kinetics at 600 nm and not observed at 650 nm (Figure 98 in 5.3.1), while the beforehand assigned intersystem crossing of 26 ps is observed in all luminescence decays.

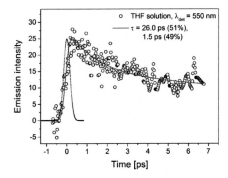

Figure 77. Picosecond luminescence decay dynamics of [(DPEPhos)Cu(PyrTet)] (**7-Cu-10**) in THF solution at 550 nm (PL blue-side) when excited at 400 nm. The solid line represents a fit to a biexponential decay with a time constant of 26 ps (0.51) and 1.5 ps (0.49). The instrument-response function is depicted as dotted line with 360 fs full-width half-maximum.

From the observed kinetics in the 0.4–7 ps window, the intersystem crossing rate is further confirmed and an additional short component of 1.5–2 ps observed. Geometric changes mainly affecting the blue-side of the photoluminescence would be consistent with this decay component, and similar dynamics with a time constant of 660 fs were reported for copper(I) bis(phenanthroline) complexes.[151] However, it is difficult to draw conclusions based on this data and since the short component is more pronounced in longer runs of excitation, it could be assigned to degradation of the concentrated solution, despite minimization of such effects through solution rotation. Different excited state processes, such as a flattening distortion, were proposed to be the less probable, i.e. their time constant would be the longer, the more sterically hindered the complex molecule,[155] and reaches up to 920 fs with bulky substituents on the phenanthroline.[233] The assignment of the short component in the luminescence decays of [(PyrTet)Cu(DPEPhos)] (**7-Cu-10**) in solution is only possible with further investigations such as time-resolved emission spectra and varying excitation powers. Nevertheless, similar decay dynamics are observed for neat film and solution in the 150 ps window, indicating that either the coordination geometry around the copper(I) center in both aggregation states is similar (and thus the 1.5 ps component is attributed to degradation) or that different geometries (flattened pseudotetrahedral in solution) affect the intersystem crossing only marginally.

To conclude, the picosecond dynamics of the luminescent TADF copper(I) complex [(PyrTet)Cu(DPEPhos)] (**7-Cu-10**) have been measured in the solid state by means of time-resolved emission and transient absorption spectroscopy. In time-resolved photoluminescence measurements a 27 ps decay is observed representing intersystem crossing from the excited singlet state S_1 to the triplet state T_1 with a rate constant of 3.7×10^{10} s^{-1}. This is the first time that intersystem crossing in the solid state of a TADF copper(I) complex is recorded, and contributes to the ability to understand the numerous excited state processes in this system. After the excited state has crossed to the triplet, reverse intersystem crossing back to the singlet occurs on a nanosecond timescale (> 2 ns) by thermal activation, giving rise to an overall average photoluminescence lifetime of 11.5 µs. The determination of the intersystem crossing rate in such highly emissive compounds is challenging, but very important for the understanding of the excited state mechanisms in TADF materials that can be used in OLEDs.

3.3.2 The TADF Mechanism in Mononuclear Neutral Copper(I) Complexes based on Tetrazolate Ligands

The thermally activated delayed fluorescence of luminescent copper(I) complexes is commonly investigated by temperature dependent time-resolved emission spectroscopy based on the two-state model in thermal equilibrium as proposed by DAVID R. MCMILLIN (chapter 1.2.3).[139] While the material emits delayed fluorescence from the excited singlet state at ambient temperature, the back-transfer from the triplet to the higher lying excited singlet is disabled at low temperature and the excitons are deactivated by long-lived phosphorescence. Usually 77 K is sufficiently low temperature to disable reverse intersystem crossing in materials with singlet-triplet splitting of 1000 cm^{-1}, and thus pure phosphorescence can be assumed at 77 K, but TADF prevails at ambient temperature. Since the radiative rates of the phosphorescence and the overall TADF emission differ, a first statement on the TADF behavior of a material can already be made from its emission properties at 293 K and 77 K: When cooling down to 77 K, an increase of the emission decay times of the complex [(PyrTet)Cu(DPEPhos)] (**7-Cu-10**) from 24.9 to 107.1 µs is observed and the radiative rates are lowered by a factor of four, due to the lower transition probability for phosphorescence $T_1 \rightarrow S_0$ than for the delayed fluorescence $S_1 \rightarrow S_0$ (Table 47). In addition to the deactivation via phosphorescence with increased lifetimes at 77 K, a red-shift is commonly observed for TADF materials which results from the naturally lower energy gap between excited triplet and singlet ground state compared to the $S_0 \rightarrow S_1$ gap. However, in some cases even a shift to shorter wavelengths is observed from superposition of a rigidochromic blue-shift with the red-shift expected for TADF materials.[117] This, together with color shifts from molecular packing effects, as was observed for many herein developed complexes, make an attribution to TADF, but even more the determination of the singlet-triplet splitting from the emission maxima at 293 K and 77 K difficult. In the model complex **7-Cu-10** a strong red-shift of about 40 nm accompanied with a slight narrowing of the emission band and an increase of the quantum yield is found, indicating less vibrational or rotational quenching processes at 77 K (Figure 78a).

Table 47. Photophysical characterization of the model complex [(PyrTet)Cu(DPEPhos)] (**7-Cu-10**) at different temperatures as powder sample.

Temperature	λ_{em} [nm]	ϕ_{em}	τ_{ave} [μs]	k_r [s^{-1}]	k_{nr} [s^{-1}]
293 K	513	0.83	24.9	3.3×10^4	6.8×10^3
77 K	550	0.92[a]	107.1	8.6×10^3	7.5×10^2

[a] Obtained by the relative method with the same sample at 293 K as reference.

The emission band is continuously shifted to longer wavelengths with decreasing temperature, as depicted in Figure 78a for the temperatures 295 K, 230 K, 155 K, and 77 K exemplarily. Due to the very fast intersystem crossing, no spontaneous fluorescence is observed when measuring with TCSPC, and the temporal resolution of fluorescence and phosphorescence is lost. Instead, the thermally activated delayed fluorescence overlaps with the phosphorescence and, depending on their ratio to each other at different temperatures, the emission band red-shifts. Along with the emission color shift, the emission decay times steadily increase with decreasing temperature (Figure 78b). The recorded luminescence dynamics can be fitted monoexponentially by convolving with the instrument response function and give the decay times τ(295 K) = 24.9 μs, τ(230 K) = 53.7 μs, τ(155 K) = 93.1 μs, and τ(77 K) = 107.1 μs. Further emission decay times at intermediate temperatures are shown in Figure 79 below.

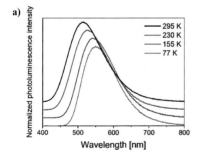
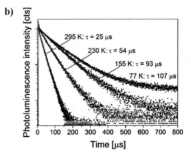

Figure 78. a) Normalized emission spectra of the complex [(PyrTet)Cu(DPEPhos)] (**7-Cu-10**) as powder sample at different temperatures when excited at 350 nm. b) Emission decays at 295 K, 230 K, 155 K, 77 K and monoexponential fit curves (red line). The instrument-response function is not depicted here, but is convolved with the recorded decays for fitting.

When plotting the emission decay times versus the temperature, a steady increase from ambient temperature until a plateau of $\tau \approx 103$ μs below 125 K is observed (Figure 79). Under the assumption that the excited singlet and triplet state are in fast thermal equilibrium (with the energy gap ΔE_{ST}) and that the emission lifetimes of the singlet and the triplet (τ_{S1} and τ_{T1}) are temperature-independent, the observed overall decay time can be expressed by Equation 8, which was derived in the introductory chapter 1.2.3.

$$\tau_{obs} = \frac{3+\exp\left(-\frac{\Delta E}{k_B T}\right)}{\frac{3}{\tau_1}+\frac{1}{\tau_{S1}}\exp\left(-\frac{\Delta E}{k_B T}\right)} \qquad (8)$$

The experimental data is fitted by Equation 8 determining the free parameters to $\Delta E_{ST} = 693$ cm^{-1}, $\tau_{S1} = 395.4$ ns and $\tau_{T1} = 103.6$ μs. However, the fit is not very sensitive to the decay time of the singlet τ_{S1}, i.e. different sets of relatively good fits show a large variety of times from 100–500 ns for τ_{S1}, strongly dependent on minor changes of the triplet lifetime τ_{T1}. Due to the weak slope and only few data points, it is difficult to quantify the quality of the depicted fit. Additionally, the non-radiative transitions of the excited triplet state are not completely temperature-independent, since different non-radiative rates are found for the TADF and phosphorescence emission. This could be assigned to the additional non-radiative channel from the excited singlet state at 293 K, but keeping in mind that about 99% of the excitons sit in the triplet, the main channel of quenching processes is assumed from the triplet state. Nevertheless, the fitting procedure gives an estimation on the transition rates of the spontaneous fluorescence and phosphorescence in the overall TADF picture, although the value of τ_{S1} is probably lower and the energy splitting ΔE_{ST} might be underestimated considering the strong emission red-shift from 293 to 77 K.

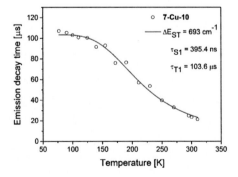

Figure 79. Emission decay times of the complex [(PyrTet)Cu(DPEPhos)] (**7-Cu-10**) as powder sample at different temperatures. Fitting the data by Equation 8 gives the free parameters $\Delta E_{ST} = 693$ cm^{-1}, $\tau_{S1} = 395.4$ ns and $\tau_{T1} = 103.6$ μs.

The radiative rate of the excited singlet state is furthermore approximated from the absorption spectrum according to S. J. STRICKLER and ROBERT A. BERG.[234] Thereby the absorption spectrum is displayed versus energy in eV, and three Gaussians fitted to the absorption spectrum reflecting two ligand-centered bands and a charge-transfer band (Figure 99 in 5.3.1). The integral of the charge-transfer absorption band (in M^{-1} versus energy in cm^{-1}) is a measure for the transition dipole moment μ of the ^{1}MLCT transition. Using the Equation 9, where n_0 is the refractive index of the solvent ($n_0(CH_2Cl_2) = 1.42$) and the integral is the intensity weighted average of the identified absorption area,[235] the transition dipole moment is obtained as $\mu = 2.55$ D.

$$\mu^2 = 9.186 \times 10^{-3} n_0 \int \left[\frac{\varepsilon(\upsilon)}{\upsilon}\right] d\upsilon \qquad (9)$$

The transition dipole moment can further be used as estimation for the radiative decay constant of the same transition (Equation 10), namely $S_1 \rightarrow S_0$, assuming that no geometrical changes occur in the molecule after excitation and thus excitation and emission channels are the same:

$$\mu^2 = \frac{3\pi\varepsilon_0 h^4 c^3 <E^{-3}>}{n_0 \tau_r} \qquad (10)$$

where ε_0 is the vacuum dielectric constant, h is the Planck's constant divided by 2π, c is the speed of light, τ_r is the radiative lifetime and $<E^{-3}>$ is the intensity weighted average of the luminescence spectrum, defined by:

$$< E^{-3} > = \frac{\int E^{-3} I(E) dE}{\int I(E) dE} \qquad (11)$$

with E as the energy of the luminescence of intensity I in joules. Using the emission spectrum at 293 K in powder (where geometrical changes after excitation should be minimized) for the calculation of $<E^{-3}>$, a radiative lifetime of the excited singlet state of $\tau_r = 56$ ns corresponding to a rate constant of $k_r^{S1} = 1.79 \times 10^7 s^{-1}$ is derived. This value is an order of magnitude smaller than the sum of radiative and non-radiative rates of the singlet $k_{S1} = 2.5 \times 10^6 s^{-1}$ from the fitting procedure above. Presuming a significantly lower non-radiative than radiative rate for the excited singlet ($k_r^{S1} \approx k_{S1}$), the obtained values can be understood as upper and lower limit of the radiative singlet rate: $1.79 \times 10^7 s^{-1} > k_r^{S1} > 2.5 \times 10^6 s^{-1}$. Thus, the order of magnitude of k_r^{S1} can be determined as $10^6 s^{-1}$.

With this in hand, the processes in the TADF mechanism can be illustrated in a state scheme with experimentally determined or approximated rate constants (Scheme 9). After photoexcitation to the excited singlet state (vibrational relaxation can be neglected when exciting at the absorption edge of 400 nm) only 0.01% of the excitons relax to the ground state under emission of spontaneous fluorescence, which is observed in the time-resolved emission measurements on a picosecond time-range. All other excitons intersystem cross to the triplet state and either emit phosphorescence or transfer back to the excited singlet state. Taking Boltzmann statistics into account (Equation 6), roughly 98.8% of the excitons reside in the triplet state after establishment of the thermal equilibrium. The dwell time of the excitons in this thermal equilibrium is mainly controlled by the ratio of the intersystem crossing to the lifetime of the singlet state k_{S1}/k_{ISC}, and the rate-limiting step RISC elongates the overall TADF lifetimes into the microsecond regime. The reverse intersystem crossing rate is dependent on the temperature and is close to 0 at 77 K, i.e. no further back transfer of the excitons occurs and deactivation is only possible via phosphorescence. Using the Boltzmann statistics again, k_{RISC} at 293 K can be estimated to $4.1 \times 10^8 \, s^{-1}$ from the ISC rate and ΔE_{ST}, which fits very well to the experimental estimation of $< 5 \times 10^8 \, s^{-1}$. The triplet emission decay time can further be resolved into the radiative rate of $9.0 \times 10^3 \, s^{-1}$ and the non-radiative rate constant of $7.7 \times 10^2 \, s^{-1}$. Since the main part of excitons sit in the excited triplet state, it is assumed that non-radiative processes from the triplet state are accounted for loss channels in the TADF system. For the design of emitting materials with short overall lifetimes, high transition probabilities of the fluorescence $S_1 \rightarrow S_0$ and fast reverse intersystem crossing are required. The latter is controlled by the energy splitting between the excited singlet and triplet states and indirectly by the rate of ISC and thus spin-orbit coupling of the material. On the other hand, small energy gaps ΔE_{ST} are achieved by a distinct charge-transfer character of the emission, i.e. spatially separated HOMO and LUMO orbitals. However, this does not seem to play a large role as compared to organic TADF materials.[32,236-238]

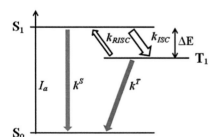

k_{S1} [s^{-1}]	10^6
k_{ISC} [s^{-1}]	3.7×10^{10}
k_{RISC}(293 K) [s^{-1}]	$< 5 \times 10^8$
k_{T1} [s^{-1}]	9.7×10^3
ΔE_{ST} [cm^{-1}]	693

Scheme 9. Schematic illustration of the TADF mechanism in the complex [(PyrTet)Cu(DPEPhos)] (**7-Cu-10**) with attribution of experimental or approximated rate constants to the respective transitions fluorescence, intersystem and reverse intersystem crossing, and phosphorescence.

Additionally, the individual contributions of thermally activated delayed fluorescence and phosphorescence to the overall luminescence can be simulated for different temperatures. The fractional intensities of the phosphorescence or fluorescence, respectively, are given as: [144,149]

$$\frac{I(T_1)}{I_{Total}} = \left(1 + \frac{g_S k_r^S}{g_T k_r^T}\exp\left(-\frac{\Delta E_{ST}}{k_B T}\right)\right)^{-1} \qquad (12)$$

$$\frac{I(S_1)}{I_{Total}} = 1 - \frac{I(T_1)}{I_{Total}} \qquad (13)$$

where g_i denote the degeneracy factors and are given as $g_T = 3$ and $g_S = 1$. If the values obtained from fitting the temperature-dependent lifetime graph are used, and it is assumed that no losses occur via the excited singlet state, i.e. $k_r^{S1} = k_{S1}$, a TADF component of 76% can be estimated for the model complex [(PyrTet)Cu(DPEPhos)] (7-Cu-10) at 293 K (Figure 80). From Equation 13 the influence of the energy gap ΔE_{ST} on the overall TADF emission lifetimes becomes clearer: When keeping the lifetimes of the excited singlet and triplet state constant, a larger energy splitting between the excited states leads to a lower contribution of thermally activated delayed fluorescence to the overall luminescence and thus a longer luminescence decay time is expected.

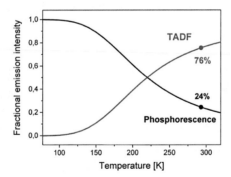

Figure 80. Simulation of the fractional intensities of thermally activated delayed fluorescence and phosphorescence to the overall luminescence by Equations 12 and 13 for the complex [(PyrTet)Cu(DPEPhos)] (7-Cu-10) at different temperatures. The parameters $k_r^S = 2.5 \times 10^6$ s^{-1}, $k_r^T = 3.0 \times 10^3$ s^{-1}, $\Delta E_{ST} = 693$ cm^{-1} are used. The highlighted points on the simulated curve depict the fractional intensities at 293 K.

A further interesting copper(I) complex **46-Cu-10** based on 5-(2-methoxypyridin-6-yl)tetrazolate (**46**) exhibits only a very small red-shift of the emission band and a larger increase of the emission decay time from 293 K to 77 K compared to the model complex **7-Cu-10** (Table 48, see also Figure 70 in 3.2.5.2). The non-radiative rates are almost by two orders of magnitude lower at 77 k, indicating significantly different extents of vibrational or rotational quenching.

Table 48. Photophysical characterization of the complex [(5-(2-methoxypyridin-6-yl)tetrazolate)Cu(DPEPhos)] (**48-Cu-10**) at different temperatures as powder sample.

Temperature	λ_{em} [nm]	ϕ_{em}	τ_{ave} [µs]	k_r [s^{-1}]	k_{nr} [s^{-1}]
293 K	485	0.76	14.3	5.3×10^4	1.7×10^4
77 K	489	0.96[a]	144.9	6.6×10^3	2.8×10^2

[a] Measured in the group of Christian Strassert, University Münster, with an integrating sphere equipped with a Dewar.

Temperature-dependent emission decay times from 2–300 K are shown in Figure 81. The observed increase of decay times with decreasing temperature is fitted by Equation 9 and gives $\Delta E_{ST} = 844$ cm^{-1}, $\tau_{S1} = 106.4$ ns and $\tau_{T1} = 142.9$ µs. The quality of the fit is much more reliable than for the above discussed complex **7-Cu-10** and thus direct comparisons are difficult. However, the values are in a similar range as for complex **7-Cu-10** and confirm the approximated rate constants for the model complex. Since a much smaller shift to longer wavelengths is observed for **46-Cu-10**, the energy splitting ΔE_{ST} of complex **7-Cu-10** might be larger than obtained from the fit (693 cm^{-1}). Interestingly, the triplet lifetime of the sterically stronger hindered complex **46-Cu-10** is longer than of the model complex **7-Cu-10**, which means that the introduction of the methoxy-group in ortho-position on the pyridine moiety results in slightly lower transition probabilities T$_1 \rightarrow$ S$_0$ despite similar coordination geometries. At temperatures below 20 K the triplet substates are not thermally equilibrated anymore and a strong increase up to 324 µs at 2 K is observed, which corresponds to a combined lifetime of the lower triplet substates T$_I$ and T$_{II}$. The emission spectra in the temperature range 5–100 K are consistent due to the small zero-field splitting between the triplet substates, but of slightly lower intensity at 5 K. At temperatures below 20 K, where the excitons reside in the lower substate for a long time, non-radiative processes become more important.[144] Since the lifetimes of the three substates cannot be resolved by a fitting procedure, the energy gap between the substates is estimated to 2–4 cm^{-1}.[108] This characterizes the mononuclear neutral copper(I) complexes **7-Cu-10** and **46-Cu-10** with tetrazolate ligands as materials of moderate spin-orbit coupling when compared to other copper(I) complexes.[144,148]

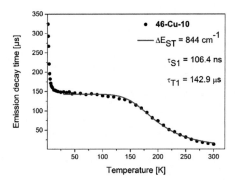

Figure 81. Emission decay times of the complex [(5-(2-methoxypyridin-6-yl)tetrazolate)Cu(DPEPhos)] (**46-Cu-10**) as powder sample at different temperatures. Fitting the data by Equation 9 gives the parameters $\Delta E_{ST} = 844$ cm^{-1}, $\tau_{S1} = 106.4$ ns and $\tau_{T1} = 142.9$ μs.[XIII]

To sum up the investigation of the thermally activated delayed fluorescence mechanism in the model complex [(PyrTet)Cu(DPEPhos)] (**7-Cu-10**) by time-resolved spectroscopy, a method to determine the intersystem crossing rate and to approximate the reverse intersystem crossing rate is developed. Since transient absorption spectroscopy, as typically applied for non-luminescent copper(I) complexes, can lead to ambiguous results when trying to determine intersystem crossing, luminescence decay dynamics on the picosecond time-range are recorded by time-resolved emission spectroscopy and fluorescence upconversion to enable the singlet to triplet crossing to be measured. From temperature-dependent emission decay times in the microsecond regime and a fit to the experimental data, an estimation on the energy gap ΔE_{ST} and the lifetime of the excited singlet state k_{S1} are obtained, while the fitted triplet lifetime is similar to the overall lifetime at 77 K. With these numbers in hand, a general scheme on the mechanism of TADF could be drawn: For highly efficient emitting materials based on copper(I) with short emission decay times, large photoluminescence quantum yields are achieved by not only reducing vibrational or rotational quenching processes but also maintaining a high radiative rate, which is mainly controlled by transition probabilities of the triplet excited state (chapter 3.2.4.3), while small singlet-triplet splitting and strong spin-orbit coupling ensures efficient back-population of the excited singlet state and thus short emission lifetimes.

[XIII] Temperature-dependent emission decay times from 2–300 K are generously measured by Markus Leitl, University of Regensburg.

3.4 Optoelectronic Devices

The herein developed and investigated copper(I) complexes are further applied in organic light-emitting diodes to gain first results on their performance in optoelectronic devices. Several copper(I) complexes have already been used as emitting materials in OLED devices and are usually processed by vapor deposition techniques or from solution. The latter technique being cost- and material-efficient is especially suitable for luminescent copper(I) complexes. In this material class no triplet-triplet annihilation is observed, which ensures high emission efficiency even when aggregation is present in the solution phase. The absence of emission quenching by Dexter processes between individual complex molecules also makes higher concentrations of the dopant in the emitting layer possible. Recently, an OLED with an external quantum efficiency of 23% (maximum current efficiency of 73 cd/A) has been reported for a solution-processed device with a dinuclear copper(I) halide complex of the type [Cu$_2$I$_2$(P^N)(P^P)] at a doping concentration of 30%.[100] This high EQE proves the concept of singlet harvesting, i.e. the use of all singlet and triplet excitons for the generation of light by thermally activated delayed fluorescence. Copper(I) complexes are gaining increasing interest both in academic and industrial research as more abundant alternatives to iridium(III) and platinum(II) emitting materials, while still maintaining high external quantum efficiencies.

In this chapter, quenching processes after photoexcitation are investigated for the tetrazolate copper(I) complex class, and the model complex [(PyrTet)Cu(DPEPhos)] (**7-Cu-10**) will be exemplarily applied in OLED devices.

3.4.1 Quenching Processes in Copper(I) Complexes

Luminescent materials with a participation of the triplet state in the emission mechanism usually suffer from quenching processes via the long-lived triplet state. A major loss channel for phosphorescent materials is triplet-triplet quenching at high concentrations of the dopant in a matrix,[42] and also in organic TADF materials a concentration quenching is found, which reduces the photoluminescence quantum efficiency and results in strong roll-off in organic light-emitting devices.[31,239,240] Furthermore, the emission intensity of organic TADF compounds is strongly decreased in air-equilibrated solutions compared to deaterated solutions due to the non-radiative deactivation channel of the long-lived triplet excitons with triplet oxygen.[32,143,160,241] Although the luminescence of copper(I) complexes in solution seems to be quenched by a third, highly disputed, mechanism,[118,119,242,243] triplet-triplet annihilation or self-quenching play only a minor role since photoluminescence quantum yields near unity are recorded for powder and neat films of copper(I) materials.[136,212,244]

The process of triplet-triplet annihilation can be illustrated as follows: When two triplets encounter and the energy gap between the triplet and the ground state ΔE_{T1-S0} is larger than between the excited singlet and triplet states ΔE_{S1-T1}, one of the excitons is promoted to a higher lying singlet state S_n, while the second one is deactivated to the ground state.[42] One photon is lost for the emission of light and the photoluminescence quantum yield reduced by a factor of 2. This Dexter process occurs via the exchange interaction of overlapping molecule orbitals and is range-limited. Thus, at high dilution of the dopant in a matrix (or solution) strong luminescence is observed, whereas increasing the concentrations leads to enhanced quenching. A typical triplet emitter, for which strong triplet-triplet annihilation is known, is Ir(ppy)$_3$ (4),[245] and its concentration-dependent emission intensity depicted in Figure 82. The emitter is doped into the polymer matrix PMMA at different fractions from 1–50% and a strong loss of photoluminescence quantum yield is found when emitting molecules are in close proximity so that their molecular orbitals can interact. In contrast, for the copper(I) complex 7-Cu-10 a minor decrease by 6% is determined for the emitter concentration range of 5–75% in PMMA (details in Table 49).

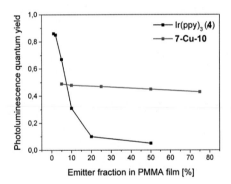

Figure 82. Dependence of the photoluminescence quantum yield on the emitter concentration in a PMMA film for Ir(ppy)$_3$ (4) and the model copper(I) complex 7-Cu-10. The films are fabricated by spin-coating dichloromethane solutions of the host-guest system.

Table 49. Photophysical properties of PMMA films doped with Ir(ppy)$_3$ (4) or the copper(I) complex 7-Cu-10 in different concentrations.

Fraction of Ir(ppy)$_3$	λ_{em} [nm]	ϕ_{em}	Fraction of 7-Cu-10	λ_{em} [nm]	ϕ_{em}
1%	507	0.86	5%	522	0.49
2%	509	0.85	10%	523	0.48
5%	513	0.67	25%	526	0.47
10%	512	0.31	50%	530	0.45
20%	511	0.10	75%	533	0.43
50%	511	0.05			

Obviously, triplet-triplet quenching is not effective for copper(I) complexes after photoexcitation, here exemplarily shown for a mononuclear neutral coordination motif, but also tested for dinuclear copper(I) halide complexes. The slight decrease of the PLQYs for **7-Cu-10** might be attributed to effects of the molecular environment or ground state distortion. Nevertheless, no significant drop is observed, which would be already expected at emitter fractions of 10–20% in PMMA. The insufficient Dexter transfer hints at low orbital overlap of adjacent complex molecules or other processes being faster than quenching. Although the TADF mechanism induces back-transfer of the excitons from the triplet state to the excited singlet state, the major proportion resides in the triplet, and thus reverse intersystem crossing is not a competitive process to triplet quenching. Furthermore, the copper(I) complexes exhibit high emission efficiencies even at low temperatures, when TADF is disabled. Hence, the reason for the absence of triplet quenching is assigned to low orbital overlap of adjacent molecules, and seems to be an intrinsic feature of this material class, as this phenomenon is not observed for organic TADF materials. HARTMUT YERSIN proposed a self-trapping mechanism, where the excited molecule forms a trap and thus does not interact with the surrounding complex molecules.[47,144] Independent of the fundamental reasons of the low orbital overlap with adjacent molecules, the experimental results draw some conclusions for the energy-transfer from an optically active host to the copper(I) complex: If a weak orbital overlap and insufficient Dexter energy transfer is also assumed for host-guest systems, efficient Förster transfer by dipole-dipole interactions has to be ensured. This is usually approximated from the overlap integral of the emission spectrum of the host and absorption bands of the dopant. In the following section, the model complex **7-Cu-10** is tested in organic light-emitting devices without a host, since no triplet annihilation processes are expected from the above results.

3.4.2 Application of the Model Complex [(PyrTet)Cu(DPEPhos)] in an OLED

In the previous chapters it was shown, that the model complex [(PyrTet)Cu(DPEPhos)] (**7-Cu-10**), as example for the mononuclear neutral copper(I) tetrazolate class, exhibits high photoluminescence quantum yields in powder and moderate emission efficiency in film (PLQY powder: 78%, neat film: 35%), which is crucial for the efficiency of an OLED. The thermally activated delayed fluorescence mechanism ensures the potential use of all generated singlet and triplet excitons in the running device. Furthermore, the material has proven to be thermally stable up to 310 °C without decomposition or loss of ligands, and the electrochemical stability could be strongly improved by introduction of the tetrazolate moiety in the mononuclear neutral complexes (chapter 3.2.1). Despite a moderate tendency to crystallize, the material does not show melting points or glass transitions in differential scanning calorimetry measurements either in powder or neat film, which rules out crystallization of the emission layer during annealing procedures as part of the device fabrication (Figures 108 and 109 in 5.3.4). Crystalline areas could cause leakage currents in the running device and thus decrease the performance of the device. However, a drawback of the material might be its long emission decay times and thus accumulation of excitons on the emitter molecules at high current densities, which can lead to significant quenching processes in electrical bias. Triplet-polaron quenching (TPQ) occurs from the interaction of a long-lived triplet exciton with an electrical charge (= polaron) under radiationless deactivation.[52]

Al	55 nm
LiF	1 nm
TPBi	50 nm
7-Cu-10	30 nm
PEDOT:PSS	40 nm
ITO	120 nm
Glass substrate	

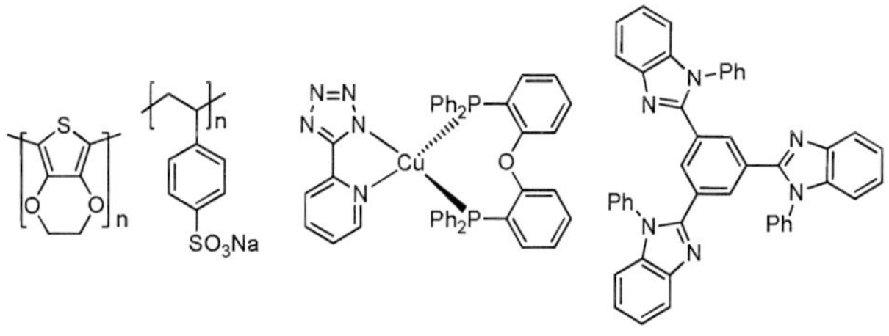

PEDOT:PSS (**53**)   **7-Cu-10**   TPBi (**54**)

Figure 83. Stack design of organic light-emitting diodes with the model complex [(PyrTet)Cu(DPEPhos)] (**7-Cu-10**), and the materials PEDOT:PSS (**53**) and 1,3,5-Tris(1-phenyl-1H-benzimidazol-2-yl)benzene (TPBi, **54**) used as supporting layers.[XIV]

For the OLED devices a simple stack design is chosen to focus on the emitter's properties in the running device. However, despite the development and improvement of the emitting material, also the design and optimization of the device and a good adjustment of transport layers and host is important to fully exploit the potential of the emitting material, which is not focus of this preliminary study. Since triplet-triplet annihilation is insufficient for the copper(I) complex [(PyrTet)Cu(DPEPhos)] (**7-Cu-10**), as was shown in the previous chapter 3.4.1, an emission layer without host is used. The stack design is depicted in Figure 83. PEDOT:PSS (**53**) is used as hole-injection layer adjacent to the anode (ITO), while LiF is applied as electron injection layer for the aluminum cathode. The electron transport layer TPBi further acts as blocking layer to minimize leakage currents towards the cathode. Two OLED devices with the copper(I) complex **7-Cu-10** as emitting material are fabricated by deposition of the individual layers on

---

[XIV] OLED devices were fabricated and characterized by Dr. Harald Flügge, CYNORA GmbH (see also 5.1.6).

a glass substrate with ITO structures from solution up to the emission layer, and subsequently TPBi, LiF and the cathode are processed by thermal evaporation.

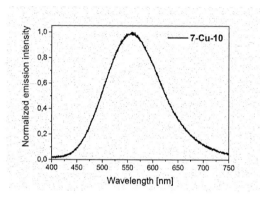

Figure 84. Electroluminescence spectrum of device B with the mononuclear copper(I) complex **7-Cu-10** as emitting material.

The characterization of the two OLED devices, herein denoted as device A and B, are shown in Table 50. Three of the four pixels of the devices are used for general IVL (I = current, V = voltage, L = luminance) characterization, while the electroluminescence spectrum of the device is recorded from the remaining pixel (Figure 84). Compared to the green photoluminescence of the neat film at 535 nm, the electroluminescence is red-shifted to 556 nm, but exciplex quenching can be ruled out as the spectra do not change with increasing voltage.

Despite similar overall characteristics as device B, the three pixels of device A show differing luminance and current (and power) efficiencies due to an inhomogeneous emission layer, and thus the device B with more reliable data will be discussed in the following. In Figure 85, the current density and luminance of the device B versus applied voltage is depicted. The relatively high turn-on voltage of 4.3 V and the shape of the current density graph indicate a strong ohmic resistance and poor charge transport in the device. Furthermore the luminance of the device amounts only to 269 Cd/m at 10 V and low external quantum efficiencies of around 1% are recorded. The emitting material in the running OLED device seems to have a high electrical resistance and thus reduces current and power efficiencies, while on the same time the electroluminescence intensity of the complex is much lower than expected from the photoluminescence results. The relatively long lifetime of the copper(I) complex **7-Cu-10** of 11.5 µs leads to an accumulation of triplet excitons in the emission layer and induces strong quenching processes such as triplet-polaron quenching in the running device. This might be even more enhanced by the low conducting properties of the emitting material, which results in the formation of excitons solely on the small band of interface to adjacent layers.

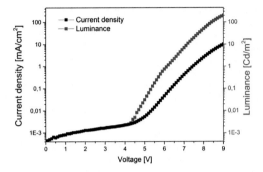

Figure 85. Current density and luminance of device B at different voltages. A turn-on voltage of 4.3 V is observed.

Table 50. OLED characteristics for the device A and B based on the copper(I) complex **7-Cu-10** for the voltage range 0–9 V. The averaged values are obtained from 3 pixels per device.

	U_{turnon} [V]	$L_{max}{}^a$ [Cd/m^2]	P_{eff} [lm/W]	C_{eff} [Cd/A]	current density[b] [Cd/cm^2]	EQE [%]	CIE$_x$, CIE$_y$
device A	4.8	273	3.7	3.8	5.4	1.2	0.402, 0.501
device B	4.3	269	1.0	2.5	5.0	0.8	0.398, 0.501

[a] At 10 V; [b] at 100 cd/m^2.

To conclude, the copper(I) complex [(PyrTet)Cu(DPEPhos)] (**7-Cu-10**) shows only moderate performance data in the tested OLED devices due to its relatively long TADF lifetime and poor conducting properties. Despite the absence of self-quenching mechanisms between adjacent copper(I) complexes based on TTA, other quenching processes like triplet-polaron quenching become important under electrical bias, when the excited molecule deactivates with a long decay time. However, these preliminary studies mainly concentrated on the emitting material and further investigations with different hosts and properly adjusted transport layers are to be performed. First OLED devices with mononuclear neutral copper(I) complexes have been reported in literature, which show similar performance data for the non-optimized devices as discussed above, but could be improved significantly by doping the copper(I) complexes in a host and adjusting the charge transport in the device to avoid the accumulation of charge carriers at the EML/ETL interface.[101,126,129] This was shown to be the main loss channel of excitons and results in poor current efficiencies, thus confirming the above made conclusions. The luminance could be increased from 415 cd/m^2 for a device with neat emission layer to almost 5000 cd/m^2 for a host-guest system with an optimized ETL,[101] while the external quantum efficiency of a device based on a different copper(I) complex was improved from 3.6% to 7.8% at a maximum current efficiency of 9.7 cd/A or 21.3 cd/A, respectively, by simply using a different host.[129] These results confirm the necessity of a well adjusted host-guest system as well as optimization of the charge transport in the devices to avoid exciton quenching processes.

4. Conclusion

Luminescent copper(I) complexes present a promising class of emitting materials for organic light-emitting diodes and have attracted enormous academic and industrial interest due to their high emission efficiencies, color tunability by ligand variation, and their abundancy compared to iridium(III) or platinum(II) compounds. Especially the assessment of thermally activated delayed fluorescence to copper(I) compounds has accelerated the development and investigation of several complex classes, and ligand (and halide) variation helped to understand the fundamental photophysical properties of the different coordination motifs. Despite thorough investigations on the mononuclear cationic copper(I) complexes by emission and absorption spectroscopy as well as time-resolved methods, only few examples of their neutral counterparts have been reported up to the beginning of this research work.

In this thesis, a synthetic route to luminescent, mononuclear neutral copper(I) complexes of the type [(N^N)Cu(P^P)] has been developed, and the excited states scheme of the most promising complex [(PyrTet)Cu(DPEPhos)] – based on 5-(pyridin-2-yl)tetrazolate and a chelating bis(phosphine) ligand – and modifications thereof investigated by spectroscopical methods in combination with DFT calculations. Furthermore the mechanism of thermally activated delayed fluorescence could be illustrated by means of time-resolved emission spectroscopy in the pico-, nano-, and microsecond regime.

In detail, a systematic study was performed on the coordination and photochemistry of copper(I) with bidentate pyridine-amines or their deprotonated amide analogues, respectively, and various phosphine ligands of different denticity and bite angle to form cationic or neutral complexes (3.1). A set of general rules to predict the complex's coordination motif (mononuclear or cluster) with the given ligands could be derived and a synthetic route to mononuclear neutral copper(I) complexes was developed (Scheme 10). While mononuclear cationic complexes based on the neutral pyridine-amine ligands are formed both with DPEPhos (**10**) and triphenylphosphine (**8**) (3.1.2.1), the right combination of pyridine-amide and phosphine ligands has to be chosen to obtain mononuclear neutral complexes. The complexation reactions of copper(I) with pyridine-amide ligands and monodentate triphenylphosphine (**8**) or the narrow bite angle dppm (**9**) afforded cluster-type structures, but the bis(phosphine) DPEPhos (**10**) has proven as a good candidate to yield the desired mononuclear neutral complex motif (3.1.2.2).

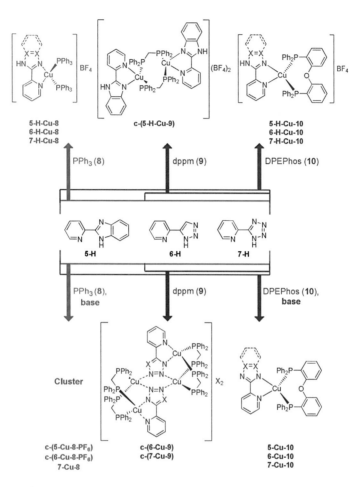

Scheme 10. The coordination chemistry of copper(I) with the pyridine-amine ligands PyrBimH (**5-H**), PyrTriH (**6-H**), PyrTetH (**7-H**) or their anionic forms **5**, **6**, **7**, respectively, and monodentate or bidentate phosphines **8–10** (x means C or N, while X denotes the counterions BF_4^- or PF_6^-).

The photoluminescence of the mononuclear neutral complexes of the type [(N^N)Cu(DPEPhos)] is characterized by a (ML+IL)CT transition and strongly depends on the electronic properties of the applied N^N ligands. For the highly luminescent complex [(PyrTet)Cu(DPEPhos)] (**7-Cu-10**) the HOMO is localized on the copper(I) center and the two phosphorous atoms of the bis(phosphine), and to a lesser extent on the amide moiety of the N^N ligand, while the LUMO lies on the whole pyridine-amide ligand (3.1.3.2). By comparison with their cationic counterparts [(N^N)Cu(DPEPhos)]$^+$ better understanding of the emission properties of the neutral mononuclear complexes [(N^N)Cu(DPEPhos)] was achieved: Deprotonation of the pyridine-amine ligand to the pyridine-amide usually blue-shifts the emission of the corresponding complexes due to an additional intraligand charge transfer character, and a stronger coordination of the copper(I) atom in the neutral complexes and less vibrational quenching after loss of the counterion enhance the photoluminescence quantum yields, thus proving that mononuclear neutral copper complexes are superior emitting materials compared to their cationic counterparts.

The mononuclear neutral complex [(PyrTet)Cu(DPEPhos)] (**7-Cu-10**) based on pyridine-tetrazolate has been shown to be advantageous over the developed pyridine-benzimidazolate or -triazolate complexes in terms of emission efficiency, low sensitivity against oxygen, and thermal stability. It was then varied by different modifications of the bis(phosphine) and the 5-(pyridin-2-yl)tetrazolate ligands (Figure 86), and the electronic effects on the complexes were systematically studied by absorption and emission spectroscopy and further related to DFT calculations (3.2).

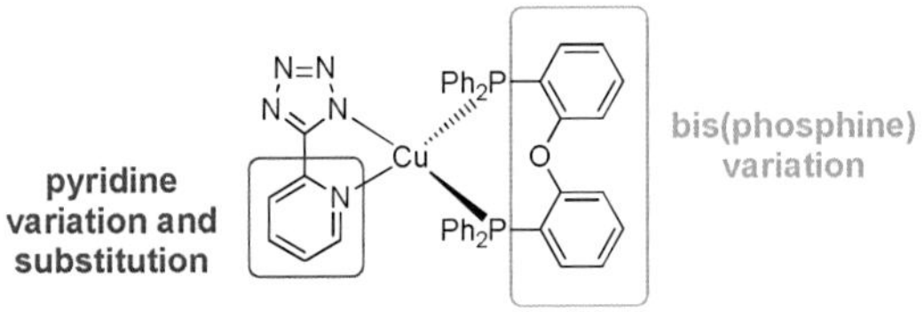

Figure 86. Variation possibilities of the model compound [(PyrTet)Cu(DPEPhos)] (**7-Cu-10**) to investigate the ligands' steric and electronic influence on the photophysical properties.

Bis(phosphines) of different P-Cu-P bite angles mainly control the formation of the complex (3.2.3 and 3.2.5.1), while different electronic effects of the P^P ligands alter the energy levels of the complexes' excited states. This is most relevant for the bis(phosphine) R-(+)-BINAP (**23**) of lower electron donating ability than DPEPhos (**10**) and its derivatives PTEPhos (**18**) and XantPhos (**19**), and leads to a shift of the complex's LUMO orbital to the P^P instead of the N^N ligand (3.2.3). Furthermore, a radiationless deactivation could be observed for the complexes based on the thioether analogue DPTPhos (**21**), which form a pentacoordinated copper(I) center in the excited state (3.2.2). The variation of the pyridine moiety of the N^N ligand in the complexes by enlargement of the aromatic system (3.2.4.1), substitution against five-membered ring heterocycles (3.2.4.3) and introduction of electron donating groups (3.2.5) led to several conclusions: Since both the LUMO and to a smaller extent the HOMO energy levels of the respective complexes are affected by ligand modulation, the color tunabiliy is limited when altering the electronic effects of the ligands, e.g. only a moderate red-shift is observed for the enlargement of the aromatic system in quinoline-tetrazolate ligands (3.2.4.1).

Additionally, the LUMO levels of the bis(phosphine) and the pyridine-tetrazolate ligand are energetically close to each other in the complex, so that the complex's LUMO orbital can easily shift to the bis(phosphine) even for slight electronic modifications of the N^N ligand, thus reducing the tunability of the emission color to shorter wavelengths (3.2.4.3 and 3.2.5.2). This also explains the differing absorption and emission properties for complexes with various electron donating substituents in ortho- or para-position on the pyridine moiety, and is further supported by DFT calculations (3.2.5.2). In addition, the emission efficiency of the complexes is the higher, the lower the molecular movements and vibrational quenching (low k_{nr}) but also the higher the radiative rate (k_r), which was found to be controlled by transition probabilities of the relaxed excited state to the ground state. This was especially observed for complexes with imidazole- or benzothiazole-tetrazolate ligands (3.2.4.3), but also for isoquinoline structures (3.2.4.1). To conclude, by a systematic variation of the tetrazole and also the bis(phosphine) ligand in herein developed mononuclear neutral complexes, and by their spectroscopical and theoretical examination, a thorough understanding of the structure-property relationships was achieved.

The mechanism of thermally activated delayed fluorescence and its interplay of radiative and non-radiative transitions was further investigated for the model complex [(PyrTet)Cu(DPEPhos)] (**7-Cu-10**) by time-resolved spectroscopy (3.3). Since the commonly applied transient absorption technique is not sensitive to intersystem crossing in highly luminescent copper(I) complexes in the solid state, a method to determine the non-radiative ISC (and reverse intersystem crossing) in these systems had to be established (3.3.1). By time-resolved emission spectroscopy and fluorescence upconversion in the picosecond time-range, a time constant of 27 ps was obtained for the intersystem crossing from the excited singlet to triplet states (Scheme 11).

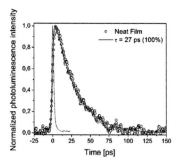
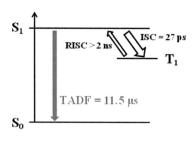

Scheme 11. Schematic illustration of the TADF mechanism with experimental time constants for intersystem and reverse intersystem crossing (right) from picosecond luminescence decay dynamics of [(PyrTet)Cu(DPEPhos)] (**7-Cu-10**) in neat film (left).

Further parameters of the TADF mechanism were determined from a fit to temperature-dependent overall emission decay times (3.3.2), giving rise to the following picture: Luminescent copper(I) complexes with short emission lifetimes, which are crucial for low quenching in the OLED device, are achieved by high transition probabilities of the fluorescence $S_1 \rightarrow S_0$ and fast reverse intersystem crossing, which is controlled by the energy splitting ΔE_{ST} between the excited singlet and triplet states and indirectly by the rate of ISC and thus spin-orbit coupling of the material.

Finally, the exemplary complex [(PyrTet)Cu(DPEPhos)] (**7-Cu-10**) was applied in an organic light-emitting diode from solution processing to gain first results on its performance in optoelectronic devices (3.4). Despite the absence of self-quenching mechanisms by triplet-triplet quenching for copper(I) complexes (3.4.1), other quenching processes like triplet-polaron quenching become important under electrical bias, when the excited molecule deactivates with a long decay time (3.4.2). The optimization of the devices and a good adjustment of transport layers and different hosts to fully exploit the potential of the emitting material will be subject to further investigations.

Consequently, the promising class of mononuclear neutral copper(I) complexes, in particular based on pyridine-tetrazolate ligands and bis(phosphines), shows outstanding luminescence properties and exhibits great potential as emitting materials for optoelectronic applications.

5. Experimental Part

5.1 General

5.1.1 Analytical Methods and Instrumentation

Nuclear Magnetic Resonance Spectroscopy (NMR)

^{1}H NMR spectra were recorded at room temperature on *Bruker* AC 250 (250 MHz), *Bruker* Avance 300 (300 MHz), *Bruker* Avance 400 (400 MHz), *Bruker* Avance III 500 (500 MHz) and *Bruker* Avance 600 (600 MHz), ^{13}C NMR spectra were recorded on *Bruker* AC 250 (63 MHz), *Bruker* Avance 300 (75 MHz), *Bruker* AM 400 (100 MHz), *Bruker* Avance III 500 (125 MHz), and ^{31}P NMR spectra were recorded on a *Bruker* AC 250 (250 MHz), *Bruker* Avance III 500 and *Bruker* Avance 600 (600 MHz). Deuterated solvents (chloroform-d1, dimethylsulfoxide-d6, methanol-d4) were purchased from *Sigma-Aldrich* and *Deutero*. Chemical shifts of all compounds are expressed in parts per million (ppm) downfield from tetramethylsilane (TMS) and referenced to the residual protons or carbons of the solvents: 7.26 ppm (^{1}H) or 77.16 ppm (^{13}C) in chloroform-d1, 2.50 ppm (^{1}H) or 39.43 ppm (^{13}C) in dimethylsulfoxide-d6, 3.31 ppm (^{1}H) or 49.1 ppm (^{13}C) in methanol-d4, and 4.790 ppm (^{1}H) in deuterium oxide-d2. Chemical shifts in ^{31}P spectra were calculated without reference. Multiplicities of signals in ^{1}H NMR spectra are denoted as follows: s = singlet, bs = broad singlet, d = doublet, t = triplet, q = quartet, quin = quintet, m = multiplet, dd = doublet of doublets, td = doublet of triplet, quind = doublet of quintet, ddd = doublet of doublet of doublet, tt = triplet of triplets. All spectra were analyzed according to first order and coupling constants (J) are given in Hertz (Hz). Abbreviations for signal assignment include: H_{Ar} = aromatic protons, X-H = protons at position X, C_{Ar} = aromatic carbons, X-C: C-atom at position X. Multiplicities of signals in ^{13}C NMR spectra are assigned by the DEPT method (Distortionless Enhancement by Polarization Transfer): Primary or tertiary C-atoms give positive signals (+), secondary C-atoms give negative signals (−), and quaternary C-atoms don't show signals (C_q = quaternary carbon).

Mass Spectrometry (EI-MS, FAB-MS)

Electron ionization (EI, 70 eV) and fast atom bombardment (FAB) mass spectra were measured on a *MasCom* MAT 95 spectrometer. Molecule fragments and clusters are expressed in mass to charge ratio (m/z) and the intensities are given as a percentage relative to the base signal (100%). The molecular ion (EI) and protonated molecule ion (FAB) are referred to as $[M^+]$ and $[M^+ + H]$, respectively. For FAB-MS the matrix 3-Nitrobenzylalcohol (3-NBA) or glycerol was used. For copper(I) complexes the abbreviations N^N refers to the pyridine-amine/tetrazole ligand or its anionic form, and P^P or P refer to the bis(phosphine) or phosphine ligand, respectively.

Infrared Spectroscopy (IR)

IR spectra were recorded on a *Bruker* IFS 88 and *Bruker* Alpha. Samples were measured as film between KBr plates (oils) or with DRIFT (Diffused Reflectance Infrared Fourier Transform Spectroscopy) or ATR (Attenuated Total Reflection) technique (solid samples). Wavenumbers ($\tilde{v}$) of the absorption bands are given in cm^{-1} and intensities are described as follows: vs = very strong (0–10% T), s = strong (11–40% T), m = middle (41–70% T), w = weak (71–90% T) and vw = very weak (91–100% T).

Elemental Analysis (EA)

Elemental analysis was measured on *Elementar* model vario MICRO CUBE with *Sartorius* analytical scale model M2P, and *HEKAtech* Euro EA-CHNS. Values for carbon (C), hydrogen (H), nitrogen (N) and sulfur (S) are given in mass percent. Used abbreviations include: calc.: calculated theoretical value, found: experimental value.

X-ray Diffraction Analysis

Single crystal X-ray diffraction was measured at 123 K using MoKα radiation ($\lambda = 0.71073$ Å) on a *Bruker-Nonius* APEXII and *Bruker-Nonius* KappaCCD, or using CuKα radiation ($\lambda = 1.54178$ Å) on a Bruker D8 Venture Diffractometer with Photon100 detector at 123(2) K by Dr. Martin Nieger from the University of Helsinki. Direct (SHELXS-97[246]) or Patterson methods were used to interpret the data, and refinement was carried out using SHELXL-97/2013/2014[247] (full-matrix least-squares on F_2). Non-hydrogen atoms were refined anisotropically, and hydrogen atoms were localized by difference electron density determination. Semi-absorption corrections were applied for all structures.

Crystallographic data of some compounds has already been deposited with the Cambridge Crystallographic Data Centre (see 5.4 for details).

Thin-layer Chromatography (TLC)

Ready-to-use TLC plates with fluorescence indicator from *Merck* (Silica gel 60, F_{254}, thickness 0.2 mm) were used for thin layer chromatography (TLC). Substances were detected by fluorescence quenching in UV light of $\lambda = 254$ nm or by color reagents (potassium permanganate solution, [XV] vanillin solution[XVI]) and subsequent heating in hot airflow. Retardation factors R_f are given as ratios of the migration distances of the substance to the solvent front.

Gas Chromatography – Mass spectrometry (GC-MS)

For reaction control and substance identification a combined GC-MS system of *Perkin Elmer* with a Clarus 680 gas chromatograph equipped with an autosampler and a Elite-5ms column (lengths: 30 m, internal diameter: 0.25 mm, film thickness: 0.25 μm, temperature limit: –60–325/350 °C) and a Clarus SQ8C mass spectrometer was used. Samples were applied in low concentrations of 1 mg/ml in dichloromethane or ethyl acetate.

5.1.2 Photophysical Methods and Instrumentation

Absorption Spectroscopy (UV-Vis)

UV-Vis spectra were recorded on a *Thermo Scientific* Evolution 201. Samples were measured as dichloromethane solutions (spectroscopic grade) at concentrations of 10^{-5} M, or as films on quartz substrates (spincoated from 20 mg/ml dichloromethane solutions). The spectra were recalculated by

$$A = \varepsilon \cdot c \cdot l$$

with A: absorbance, ε: molar extinction coefficient, c: concentration, l: path length, and are displayed in extinction coefficients ε vs. wavelengths if not otherwise noted.

[XV] 1.00 g potassium premanganate, 2.00 g acetic acid, 5.00 g sodium hydrogen carbonate in 100 ml water
[XVI] 10.0 g vanillin in 90 ml ethanol and 10 ml concentrated sulfuric acid

Steady-state Photoluminescence Spectroscopy (PL)

Steady-state photoluminescence spectra were recorded on a *Horiba Scientific* FluoroMax-4 with a 150 W xenon-arc lamp, excitation and emission monochromators and a *Hamamatsu* R928 photomultiplier tube. Emission and excitation spectra were corrected by standard calibration curves.

Steady-state emission spectra of the complex **7-Cu-10** in the temperature range 77–310 K (chapter 3.3.2) were recorded on an *Edinburgh Instruments FLS980* equipped with an xenon arc lamp as excitation source and a NIR photomultiplier tube as detector, in the group of Ifor Samuel, University of St. Andrews. Emission and excitation spectra are corrected by standard calibration curves. The sample was placed in an *Oxford Instruments* Optistat DN-V cryostat in vacuum (10^{-4} mbar), equipped with a liquid nitrogen reservoir and an external temperature controller.

Time-resolved Emission Spectroscopy

Emission decay times on microsecond timescales were measured on the same system using the TCSPC (time-correlated single photon counting) method with a flash light (λ = 250–2000 nm, 3 µs pulse) or SpectraLED (λ = 314/355 nm, pulse FWHM = 493 ns) as excitation sources. Decay curves were analyzed with the software DAS-6 and DataStation provided by Horiba Yvon Jobin.

Emission decay times of the complex **7-Cu-10** in the temperature range 77–310 K (chapter 3.3.2) were recorded on an *Edinburgh Instruments FLS980* using the technique of time-resolved single photon counting – multi-channel scaling (MCS), with a MCP photomultiplier tube as detector. The sample is placed in the cryostat, as specified above, and excited by an external SpectraLED (375 nm) in manual control by a delay generator.

Emission decay times and time-resolved PL spectra on a ps time-scale were recorded on a *Hamamatsu* synchroscan universal streak camera with an instrument response of 2.5 ps (FWHM) in the group of Ifor Samuel, University of St. Andrews. Measurements were completed in vacuum at 10^{-5} mbar. The excitation source for streak camera measurements was at 375 nm, which was derived from the frequency doubled output of the signal of a parametric amplifier with pulses at 20 kHz and 300 fs (full-width half-maximum), pumped by a Pharos regenerative amplifier from Light Conversion. Care was taken to keep peak excitation powers low to avoid exciton annihilation or degradation effects. As prompt fluorescence in copper(I) complexes is very weak, the signals had to be averaged from several scans (no change was observed during integrations).

Streak camera (*Hamamatsu* universal streak camera C6860) measurements of THF solution samples (in a rotating cell) were taken at a range of wavelengths from 480 to 600 nm. The excitation source was the frequency doubled output from a Ti:Sapphire oscillator with pulses of 100 fs (full-width half-maximum) at 80 MHz. The instrument response function (IRF) of the setup was recorded by measuring excitation scatter on a blank fused silica disk, and was found to be ~ 4 ps (full-width half-maximum). As prompt fluorescence in copper(I) complexes is very weak, the signals had to be averaged from several scans (no change was observed during integrations).

Emission decay times on an fs time-scale were recorded with the fluorescence optical gating technique (FOG100 by *CDP Systems*). The excitation source was the frequency doubled output from a Ti:Sapphire oscillator with pulses of 100 fs (FWHM) at 80 MHz. The 400 nm excitation was focused on the rotating cuvette containing the solution and the residual 800 nm was sent onto a delay-line mounted reteroreflector to enable it to act as the gating pulse. Optical gating that enabled femtosecond temporal resolution was achieved by sum-frequency mixing of the luminescence photons from the sample and the gating pulse in a BBO nonlinear crystal. The upconverted luminescence was filtered before being detected with a photomultiplier tube in a photon counting regime. Sum-frequency generation between the luminescence and the 800 nm gate in the crystal was a weak process, therefore many scans were averaged, with no change in kinetics observed during integration. Dynamics were measured at a range of wavelengths from 550 to 650 nm in THF. The accumulated background of the long-lived phosphorescence was subtracted from the kinetics before analysis. The instrument response function of the setup was recorded by observing Raman scattering of a sample cell filled with water and an IRF of 360 fs (FWHM) was found.

Time-resolved Absorption Spectroscopy (TA)

Transient absorption studies were undertaken in the group of Ifor Samuel, University of St. Andrews, by the pump/probe technique with a Spectra Physics Hurricane regenerative amplifier, producing pulses of 100 fs at 5 kHz and 800 nm. The pump (400 nm) was obtained by frequency doubling the fundamental, while the probe was a white light continuum (420 to 760 nm) generated with 800 nm amplified light focused onto calcium fluoride. As dynamics >> 2 ps were of interest, the setup was not corrected for spectral chirp. Detection of the continuum probe was made with a silicon photodiode array from Stressing, with transient absorption measured by chopping the pump at 2.5 kHz. The measurements were carried out under active vacuum of 10^{-5} mbar to reduce photodegradation with the amplified laser pump.

Photoluminescence Quantum Yields (PLQY)

Absolute photoluminescence quantum yields were measured using the integrating sphere method[248] on a *Hamamatsu Photonics* C9920-02G system. PLQYs are given as ratio of emitted photons to absorbed photons. Compounds were measured as powder or film samples with excitation at 350 nm. For film measurements the sphere was purged with nitrogen to prevent oxygen quenching. Calculation of the PLQYs and CIE coordinates was carried out with the software C9 920-02 V3.4.1. PLQY values within an accuracy of ±0.02.

5.1.3 Cyclovoltammetry (CV)

Cyclovoltammetric measurements in dichloromethane were recorded at room temperature under nitrogen flow with a *CH Instruments* Model 600D with Workstation by Dr. Georgios Liaptsis and Dr. Sebastian Dück, CYNORA GmbH. The measurements were performed with a glassy carbon working electrode, a platinum wire counter electrode and a platinum wire reference electrode at a scan rate of 100 mV/s. The supporting electrolyte was 0.1 M NBu_4PF_6. The measurement was calibrated against $FeCp_2/FeCp_2^+$, and corrected versus SCE.[249]

5.1.4 Thermogravimetric Analysis (TGA)

Thermogravimetric analyses were measured on a *Mettler Toledo* TGA/SDTA 851 by Dr. Mathias Schwotzer, Karlsruhe Institute of Technology (KIT) in the temperature range 30–1000 °C. The powder samples in an Alox crucible were heated up under nitrogen atmosphere with a rate of 10 °C/min, and the weight loss recorded against temperature.

5.1.5 Differential Scanning Calorimetry (DSC)

Thermal analysis was performed with differential scanning calorimetry (DSC) by Felix Limberg, Fraunhofer Institute for Applied Polymer Research (IAP), using a *Netzsch* DSC 204 Phoenix with a scanning rate of 10 K/min. For this purpose 3–5 mg of the sample were weighed in an aluminum pan, closed with a lid and the closed crucible perforated with a needle. Glass transitions were obtained from the second heating cycle.

5.1.6 Fabrication and IVL-Characterization of OLEDs

OLED devices have been prepared by Dr. Harald Flügge, CYNORA GmbH, starting from indium-tin-oxide (ITO) covered glass substrates with a sheet resistance of 16 Ω/m^2. The substrates were cleaned in a 2% aqueous solution of Helmanex, washed with deionized water, and treated with ozone for 10 min prior to spin-coating with PEDOT:PSS (Heraeus AL4083) filtered through a 0.45 μm syringe filter. This layer was dried at 200 °C for 5 min in air. The following emitting layer, consisting of compound **7-Cu-10**, was deposited from a chlorobenzene solution passed through 0.45 μm syringe filters under an inert atmosphere, resulting in a 30 nm thick layer. After annealing at 80 °C for 10 min under vacuum, the following layers were deposited by evaporation under vacuum: TPBi (50 nm), LiF (1 nm), and Al (55 nm). The active area of the resulting device was 24 mm^2. Finally, the device was encapsulated using 3M barrier foil. Current (I), voltage (V), and luminance (L) characteristics (IVL) were measured with a BoTest System.

5.1.7 Density-functional theory calculations (DFT)

Selected complexes were studied using density functional theory (DFT) by Dr. Jana Friedrichs, CYNORA GmbH. Initial geometries were obtained from single-crystal X-ray diffraction data, or from crystal structures of similar complexes and adding/exchanging respective atoms. These structures were optimized in the ground state and in the lowest triplet state using the BP86 functional [250,251] with the resolution-of-identity (RI) approximation. [252, 253, 254] Harmonic vibrational frequency analysis was employed to ensure that the found stationary points are true minima on the potential energy hypersurface. Excitation and phosphorescence energies were calculated as energy differences between the closed-shell singlet ground-state and the lowest (unrestricted) triplet state. For particular complexes (denoted in the main text), frontier orbitals and TD-DFT excitation energies were computed at the BP86 ground state geometry using the B3LYP functional [255, 256, 257, 258] without the RI-approximation. For all calculations, the def2-SV(P) basis set [259,260] and the m4 grid for numerical integration was used. All calculations were performed with the *Turbomole* program package (version 6.4). [261] Orbital plots were produced by Jmol.

5.1.8 Synthetic Methods and Materials

General

Starting materials and solvents were purchased from commercial suppliers (*Sigma-Aldrich, VWR, TCI Europe, ABCR, Chempur, Roth, Acros*) and used without further purification. Analytical grade solvents (ethyl acetate, cyclohexane, diethyl ether, dichloromethane) for extraction and chromatography, and dry solvents (dichloromethane, ethanol, tetrahydrofurane, toluene, diethyl ether, *n*-hexane; < 50 mm H_2O) for reactions were also used as purchased without further purification.

Solid reagents were used as powders and reactions carried out at room temperature if not otherwise noted. For low temperature reactions the following freezing mixtures in Dewar vessels were used:

–0 °C	ice/water
–10 °C	ice/water/sodium chloride
–78 °C	dry ice/*iso*-propanol

Conversion of the reactions was controlled by TLC or GC-MS (see 5.1.1). Solvents were removed from the reaction mixture under reduced pressure by a rotary evaporator at lowest possible temperature (40 °C). Solutions of inorganic salts were always used as saturated salt-water solutions. Crude products were purified by flash column chromatography as described by Still[262] with silica 60 from *Merck* (0.040 – 0.063 mm) as stationary phase, or on a *Büchi* Sepacore flash chromatography equipment if not otherwise noted.

Reactions with oxygen and/or water sensitive reagents were carried out under nitrogen atmosphere according the Schlenk technique.[263] Thus, glassware was sealed with a septum and evacuated under heating and subsequently purged with nitrogen several times before usage. Solvents or liquids were transferred via syringes and V2A-steel needles. Degassed solvents were obtained by purging the respective solvent with nitrogen (or by the pump-freeze-thaw technique).

5.2 Synthetic Procedures and Analytical Characterization

5.2.1 General Procedures

GP1: Tetrazole synthesis

Warning! HN_3 gas evolves during tetrazole synthesis, thus only closed reaction setups were used and released gases were transferred to washing bottles with potassium hydroxide solution.

Nitrogen-rich compounds such as tetrazole derivatives are used as components for explosive mixtures.[264] The reactions described here were run on only a few grams scale, and no problems were encountered. However, great caution should be exercised when handling or heating compounds of this type.

Method A:[174,175] In a flask with reflux condenser and subsequent washing bottle (filled with saturated potassium hydroxide solution) the respective carbonitrile (1.00 equiv.), sodium azide (1.30 equiv.) and ammonium chloride (1.30 equiv.) were dissolved in DMF (3–5 ml/mmol) and refluxed. After cooling down the reaction, water was added (15 ml/mmol). The aqueous phase was then acidified by adding hydrochloric acid (5 M) and the precipitating solid filtered off, washed with water (pH lower than needed for precipitation of the tetrazole) and dried in vacuum.

Method B:[201] Same equipment as in Method A. The respective carbonitrile (1.00 equiv.), sodium azide (1.40 equiv.), and acetic acid (4.00 equiv.) were dissolved in *tert*-butanol (4 ml/mmol). After stirring at 80 °C for 2 h, the reaction mixture was cooled down, the same amount of acetic acid (4.00 equiv.) added and stirred overnight at 80 °C. Water (10 ml/mmol) was added to the reaction mixture and stirred for 1 h. Acidification of the aqueous phase with 5 M hydrochloric acid led to the precipitation of a solid, which is filtered off, washed with water (pH lower than needed for precipitation of the tetrazole) and dried in vacuum.

Method C:[202] Same equipment as in Method A. The respective carbonitrile (1.00 equiv.), sodium azide (1.20 equiv.), and pyridine hydrochloride[XVII] (1.00 equiv.) were suspended in DMF (2 ml/mmol) and refluxed. After cooling down, potassium hydroxide solution (5 M, 0.4 ml/mmol) was added, stirred for 30 min, and solvents evaporated. The solid residue was dissolved in few water (1–2 ml/mmol),

---

[XVII] Pyridine hydrochloride was obtained from pyridine (5.00 ml, 61.9 mmol, 1.00 equiv.) and hydrochloric acid (1 M, 62.5 ml, 62.5 mmol, 1.01 equiv.) in equivalent parts and removal of the solvent.

extracted with ethyl acetate twice (15 ml/mmol), and the aqueous phase separated from the organic layers. Hydrochloric acid (5 M) was added to the aqueous phase until a precipitate formed. This solid was filtered off, washed with water (pH lower than needed for precipitation of the tetrazole) and dried in vacuum.

GP2: Synthesis of cationic copper(I) complexes from tetrazole ligands

The copper(I) salt [Cu(CH$_3$CN)$_4$]BF$_4$ (1.00 equiv.), the tetrazole ligand (1.00 equiv.) and phosphine PPh$_3$ (2.00 equiv.) or bis(phosphine) (1.00 equiv.), respectively, were dissolved in a dry dichloromethane/ethanol mixture (3:1, 10 ml/mmol) under nitrogen atmosphere and stirred for 12 h at room temperature. The solvents were evaporated and the residue extracted with few dichloromethane (6 ml/mmol). After filtering off possible residuals, the filtrate was precipitated in diethyl ether (100 ml/mmol), the solid filtered off, washed with diethyl ether and dried in vacuum.

GP3: Synthesis of neutral copper(I) complexes from their cationic analogues

The analogous cationic complex (1.00 equiv.) was suspended in dry methanol (6 ml/mmol) with potassium hydroxide (2.50 equiv.) under nitrogen atmosphere and stirred for 12 h at room temperature. The solvent was evaporated and the residue dissolved in dichloromethane (6 ml/mmol), possible residuals were filtered off. The filtrate was then precipitated in diethyl ether (100 ml/mmol), the solid filtered off, washed with diethyl ether several times and dried in vacuum.

GP4: Synthesis of neutral copper(I) complexes from tetrazole ligands

[Cu(CH$_3$CN)$_4$]BF$_4$ salt (1.00 equiv.), the tetrazole ligand (1.00 equiv.), potassium hydroxide (2.00 equiv.) and phosphine PPh$_3$ (2.00 equiv.) or bis(phosphine) (1.00 equiv.), respectively, were suspended in a dry dichloromethane/ethanol mixture (3:1, 10 ml/mmol) under nitrogen atmosphere. After stirring for 12 h at room temperature, the solvents were evaporated and the residue extracted with few dichloromethane (6 ml/mmol). Precipitation of the filtrate in diethyl ether gave a powder, which was collected by filtration, washed with diethyl ether and dried in vacuum.

5.2.2 Synthesis of Ligands

Pyridine-imines

2-(1,2,3-Triazol-5-yl)pyridine (**6-H**).[173] After dissolving 2-ethynylpyridine (0.93 ml, 0.947 g,

9.00 mmol, 1.00 equiv.) in 10 ml dry toluene under nitrogen atmosphere, azido(trimethyl)silane (2.25 ml, 1.97 g, 17.1 mmol, 1.90 equiv.) was added and the reaction mixture refluxed for 2 d. Subsequently the reaction was hydrolyzed with 2 ml water, and all solvents evaporated. The title compound was obtained as yellow solid (431 mg, 2.95 mmol, 37% yield) from purification by column chromatography (cH/EE 1:1). – R_f = 0.29 (cH/EE 1:1). – ^{1}H NMR (300 MHz, DMSO): δ = 7.43 (ddd, J = 7.9, 4.3, 1.1 Hz, 1 H, H_{Ar}), 7.96 (td, J = 7.7, 1.7 Hz, 1 H, H_{Ar}), 8.04 (d, J = 7.8 Hz, 1 H, H_{Ar}), 8.43 (bs, 1 H, $H_{Triazole}$), 8.68 (d, J = 4.7 Hz, 1 H, H_{Ar}), 15.35 (bs, 1 H, NH) ppm. – MS (70 eV, EI, 40 °C), m/z (%): 146 (100) [M$^+$], 118 (27) [M$^+$– N$_2$], 105 (9) [M$^+$– N$_3$], 97 (11), 91 (21) [M$^+$– CHN$_3$], 78 (43) [M$^+$– CCHN$_3$].

2-(Tetrazol-5-yl)pyridine (**7-H**).[202] Compound **7-H** was synthesized according to **GP1** method **A** from 2-pyridinecarbonitrile (2.08 g, 20.0 mmol, 1.00 equiv.). The reaction mixture was stirred at 110 °C for 48 h. Acidification of the water phase to pH = 5 afforded the product as white precipitate. Yield: 1.87 g, 12.7 mmol, 64%; white solid. – ^{1}H NMR (300 MHz, MeOD): δ = 7.55–7.59 (m, 1 H, H_{Ar}), 8.03 (td, J = 7.8, 1.7 Hz, 1 H, H_{Ar}), 8.23–8.26 (m, 1 H, H_{Ar}), 8.75–8.77 (m, 1 H, H_{Ar}) ppm. – ^{13}C NMR (63 MHz, MeOD): δ = 123.6 (+, CH), 127.9 (+, CH), 139.0 (+, CH), 145.0 (C$_q$, 2-C$_{Ar}$), 151.2 (+, CH), 156.7 (C$_q$, C$_{TetH}$) ppm. – MS (FAB), m/z (%): 148 (86) [M$^+$+ H], 120 (23) [M$^+$– N$_2$]. – C$_6$H$_5$N$_5$ (147.14): calc. C 48.98, H 3.43, N 47.60; found C 48.94, H 3.43, N 47.28.

Pyridine-2-carboxamidrazone (**15**).[189] To a closed flask filled with 2-pyridinecarbonitrile (0.833 g, 8.00 mmol, 1.00 equiv.) in 10 ml ethanol was added hydrazine monohydrate (0.792 ml, 16.0 mmol, 2.00 equiv.) and stirred at room temperature for 72 h. The solvent of the clear, yellow solution was evaporated and the yellow residue suspended in pentane at 0 °C for 30 min. The insoluble solid was filtered off, washed with cold pentane and dried in vacuum giving 1.02 g product (7.50 mmol, 94%), which could be directly used for synthesis of the 1,2,4-triazole **16-H** and **17-H**. – ^{1}H NMR (500 MHz, CDCl$_3$): δ = 4.40 (bs, 2 H, NH), 5.28 (bs, 2 H, NH), 7.25–7.27 (m, 1 H, H_{Ar}), 7.69 (td, J = 7.9, 1.7 Hz, 1 H, H_{Ar}), 8.02 (d, J = 8.1 Hz, 1 H, H_{Ar}), 8.52 (d, J = 4.8 Hz, 1 H, H_{Ar}) ppm.

2-(3-Methyl-1,2,4-triazol-5-yl)pyridine (**16-H**).[190] Starting material **15** (1.54 g, 11.3 mmol, 1.00 equiv.)

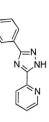 was added to a mixture of 3.40 ml (59.5 mmol, 5.27 mmol) acetic acid and 3.40 ml (36.3 mmol, 3.21 equiv.) acetic anhydride at 0 °C and subsequently stirred at room temperature for 4 h. The solvents of the clear, orange solution were evaporated resulting in the crystallization of a yellow solid. After addition of 10 ml water, which destroyed the acetic anhydride, the solvent was evaporated again. The yellow intermediate product was then heated to 130 °C for 1.5 h in an open flask giving red oil. This was recrystallized twice from MTBE affording 554 mg (3.46 mmol, 31%) of a white solid. – ^{1}H NMR (500 MHz, CDCl$_3$): δ = 2.56 (s, 3 H, CH$_3$), 7.39–7.41 (m, 1 H, H$_{Ar}$), 7.87 (td, J = 7.7, 1.4 Hz, 1 H, H$_{Ar}$), 8.20 (d, J = 7.9 Hz, 1 H, H$_{Ar}$), 8.75 (d, J = 4.6 Hz, 1 H, H$_{Ar}$), 12.19 (bs, 1 H, NH) ppm. – ^{13}C NMR (125 MHz, CDCl$_3$): δ = 13.5 (+, CH$_3$), 121.8 (+, 3-C$_{Ar}$), 124.7 (+, 5-C$_{Ar}$), 137.8 (+, 4-C$_{Ar}$), 146.9 (C$_q$, 2-C$_{Ar}$), 149.2 (+, 6-C$_{Ar}$), 155.8 (C$_q$, 3-C$_{TetH}$), 159.6 (C$_q$, 5-C$_{TetH}$) ppm. – MS (70 eV, EI, 100 °C), m/z (%): 160 (94) [M$^+$], 131 (100) [M$^+$ – N$_2$], 116 (2) [M$^+$ – N$_2$ – Me], 105 (17) [M$^+$ – N$_2$ – CMe], 92 (19) [M$^+$ – N$_2$ – NHCMe], 79 (16) [M$^+$ – C$_2$N$_3$H – Me], 74 (6), 64 (13), 43 (14).

2-(3-(*p*-Tolyl)-1,2,4-triazol-5-yl)pyridine (**17-H**).[189] Starting material **15** (1.02 g, 7.50 mmol, 1.00 equiv.) and sodium carbonate (0.795 g, 7.50 mmol, 1.00 equiv.) were dissolved in dry DMF (15 ml) under nitrogen atmosphere. At 0 °C *p*-toluoyl chloride (1.18 g, 7.50 mmol, 1.00 equiv.) was added dropwise and the reaction mixture was stirred at room temperature. Over night a yellowish precipitate was formed, which was washed with water and ethanol, dried under vacuum (1.08 g), and used without further purification. The crude product was suspended in 10 ml ethylene glycol, heated to 180 °C in an open flask until the suspension cleared and refluxed for another 30 min at 180 °C. While cooling down a white solid crystallized from the solution, which was filtered off, washed with water and dried in vacuum. Yield: 855 mg, 3.62 mmol, 48%. – ^{1}H NMR (500 MHz, CDCl$_3$): δ = 2.41 (s, 3 H, CH$_3$), 7.28 (d, J = 8.0 Hz, 2 H, H$_{Tol}$), 7.45 (ddd, J = 7.8, 5.0, 1.2 Hz, 1 H, H$_{Pyr}$), 7.93 (td, J = 7.8, 1.5 Hz, 1 H, H$_{Pyr}$), 8.11 (d, J = 8.1 Hz, 2 H, H$_{Tol}$), 8.35 (d, J = 7.9 Hz, 1 H, H$_{Pyr}$), 8.80 (d, J = 4.1 Hz, 1 H, H$_{Pyr}$) ppm. – MS (70 eV, EI, 100 °C), m/z (%): 236 (94) [M$^+$], 207 (100) [M$^+$ – N$_2$], 192 (11) [M$^+$ – N$_2$ – Me], 180 (13) [M$^+$ – N$_3$H – Me], 166 (4), 131 (2) [M$^+$ – N$_2$ – Me – C$_5$H$_4$], 119 (2) [M$^+$ – N$_2$ – Tolyl], 103 (7) [M$^+$ – CN$_2$H – Tolyl], 91 (8) [M$^+$ – CN$_3$H], 78 (3) [M$^+$ – C$_2$N$_3$H – Tolyl], 64 (5).

Bis(phosphines)

Bis(2-diphenylphosphino-p-tolyl)ether (**18**).[265] To a flask with n-butyllithium solution (2.5 M in

n-hexanes, 8.00 ml, 20.0 mmol, 2.00 equiv.) under nitrogen atmosphere was added TMEDA (3.02 ml, 20.0 mmol, 2.00 equiv.) dropwise and the reaction mixture warmed up to 50–60 °C. After cooling to room temperature, a solution of di-p-tolyl ether (2.00 g, 10.0 mmol, 1.00 equiv.) in 3 ml dry diethyl ether was added to the reaction mixture and stirred at room temperature for 16 h. Adding a solution of diphenylphosphinous chloride (3.59 ml, 20.0 mmol, 2.00 equiv.) in 3.5 ml dry pentane and stirring another 24 h at room temperature gave a yellow suspension, which cleared up when degassed water (10 ml) and degassed dichloromethane (10 ml) were added. The organic phases were separated, dried over magnesium sulfate and solvents evaporated. The solid residue was purified by dissolving in dry acetone (10 mL) at 40–45 °C and filtering off the precipitating white solid after cooling down. The solid was dried in vacuum affording 1.62 g (2.86 mmol, 29% yield) pure product. – ^{1}H NMR (300 MHz, CDCl$_3$): δ = 2.15 (s, 6 H, CH$_3$), 6.55–6.60 (m, 4 H, H$_{Ar}$), 6.97–6.99 (m, 2 H, H$_{Ar}$), 7.19–7.28 (m, 20 H, H$_{Ar}$) ppm. – ^{31}P NMR (202 MHz, CDCl$_3$): δ = –16.4 ppm. – MS (70 eV, EI, 170 °C), m/z (%): 565 (1) [M$^+$], 536 (<1) [M$^+$ – 2Me], 489 (3) [M$^+$ – Ph], 476 (<1) [M$^+$ – Ph – Me], 457 (<1) [M$^+$ – Ph – 2Me], 445 (1), 411 (1) [M$^+$ – 2Ph], 399 (2) [M$^+$ – 2Ph – Me], 381 (100) [M$^+$ – PPh$_2$], 365 (3) [M$^+$ – PPh$_2$ – Me], 351 () [M$^+$ – PPh$_2$ – 2Me], 304 (10) [M$^+$ – PPh$_2$ – Ph], 292 (6) [M$^+$ – PPh$_2$ – Ph – Me], 262 (4) [M$^+$ – PPh$_2$ – Ph – 2Me], 227 (21) [M$^+$ – PPh$_2$ – 2Ph], 214 (6) [M$^+$ – PPh$_2$ – 2Ph – Me], 197 (4) [M$^+$ – 2PPh$_2$], 183 (28) [PPh$_2$], 165 (5) [M$^+$ – 2PPh$_2$ – 2Me], 152 (8), 108 (7) [PPh], 77 (7) [Ph], 51 (3).

4,6-Bis(diphenylphosphino)dibenzofuran (**20**).[172] Synthesis of compound **20** was performed as described for **18** from dibenzofuran. The residue was recrystallized in dry n-hexanes twice and dry acetone affording a white solid. Yield: 661 mg, 1.23 mmol, 12%. – ^{1}H NMR (500 MHz, CDCl$_3$): δ = 7.00–7.02 (m, 2 H, H$_{Ar}$), 7.24–7.27 (m, 4 H, H$_{Ar}$), 7.31–7.34 (m, 2 H, H$_{Ar}$), 7.38–7.41 (m, 8 H, H$_{Ar}$), 7.45–7.49 (m, 12 H, H$_{Ar}$) ppm. – ^{1}H NMR (125 MHz, CDCl$_3$): δ = 127.2, 128.5 (d, ^{C-P}J = 7.3 Hz, C$_{Ar}$), 128.7, 129.5, 132.2 (d, ^{C-P}J = 3.8 Hz, C$_{Ar}$), 133.5, 134.0, 134.2, 136.6 (d, ^{C-P}J = 11.5 Hz, C$_{Ar}$), 139.9 (d, ^{C-P}J = 10.2 Hz, C$_{Ar}$), 141.7 (d, ^{C-P}J = 5.6 Hz, C$_{Ar}$), 141.9 (d, ^{C-P}J = 5.6 Hz, C$_{Ar}$). – MS (70 eV, EI, 180 °C), m/z (%): 536 (15) [M$^+$], 460 (<1) [M$^+$ – Ph], 367 (6) [M$^+$ – Ph + O], 428 (<1), 382 (<1) [M$^+$ – 2Ph], 352 (11) [M$^+$ – PPh$_2$], 289 (2) [M$^+$ – PPh$_2$ – Ph + O], 273 (10) [M$^+$ – PPh$_2$ – Ph],

244 (7), 215 (3) [M⁺– PPh₂– 2 Ph + O], 201 (6) [M⁺– PPh₂– 2 Ph], 183 (22) [PPh₂], 154 (5), 139 (6), 115 (16), 108 (8) [PPh], 77 (12) [Ph], 72 (16), 58 (100), 42 (49).

Bis(2-diphenylphosphinophenyl)thioether (21).²⁶⁶ Synthesis of compound 21 was performed as

described for 18 from diphenyl sulfide. The crude product was purified by repeated recrystallization in propanol affording a white solid. Yield: 2.21 g, 3.98 mmol, 27%. – ¹H NMR (300 MHz, CDCl₃): δ = 6.74–6.78 (m, 2 H, H_{Ar}), 7.00–7.31 (m, 26 H, H_{Ar}) ppm. – ³¹P NMR (101 MHz, CDCl₃): δ = –11.6 ppm. – IR (KBr): ṽ = 3046 (vw), 1568 (vw), 1475 (w), 1430 (w), 1301 (vw), 1180 (vw), 1155 (vw), 1090 (w), 1067 (vw), 1026 (w), 997 (w), 913 (vw), 767 (w), 744 (m), 694 (m), 540 (w), 504 (m), 451 (w), 429 (w), 396 (vw) cm⁻¹. – MS (70 eV, EI, 200 °C), m/z (%): 554 (<1) [M⁺], 521 (<1), 477 (3) [M⁺ – Ph], 445 (12), 401 (<1) [M⁺ – 2Ph], 369 (100) [M⁺ – PPh₂].

Tetrazoles

2-(Tetrazol-5-yl)pyrazine (28-H).²⁰⁰ Pyrazine-2-carbonitrile (1.58 g, 15.0 mmol, 1.00 equiv.), sodium azide (1.07 g, 16.5 mmol, 1.10 equiv.) and zinc bromide (3.38 g, 15.0 mmol, 1.00 equiv.) were dissolved in 25 ml water and refluxed at 100 °C for 12 h, resulting in a white precipitate. After cooling down, 22.5 ml of 3 M hydrochloric acid (pH = 1) and ethyl acetate (80 ml) were added to dissolve the formed precipitate, and the organic layer was separated from the aqueous phase and the solvent evaporated. Dissolving the residue in 150 ml of 0.25 M potassium hydroxide solution and stirring for 30 min. afforded a white Zn(OH)₂ precipitate, which was filtered off. Acidification of the filtrate with 30 ml 3 M hydrochloric acid gave a white solid. This was filtered off, washed with acidic water (pH = 4–5) and dried in vacuum. Yield: 221 mg, 1.49 mmol, 10%. – MS (FAB), m/z (%): 150 (36) [M⁺ + H], 121 (10) [M⁺ – N₂], 106 (6) [M⁺ – N₃], 92 (9) [M⁺ – N₄], 82 (3) [M⁺ – CN₄]. – C₅H₄N₆ (148.13) · 0.2 H₂O: calc. C 39.58, H 2.92, N 55.39; found C 39.49, H 2.83, N 55.78.

2-(Tetrazol-5-yl)quinoline (29-H).²⁶⁷ The tetrazole 29-H was synthesized according to GP1 method A from quinoline-2-carbonitrile at 130°C for 48 h. A yellow solid was obtained in good yields: 641 mg, 3.25 mmol, 84%. – ¹H NMR (500 MHz, MeOD): δ = 7.66 (t, J = 7.1 Hz, 1 H, H_{Ar}), 7.82 (t, J = 6.9 Hz, 1 H, H_{Ar}), 7.99 (bs, 1 H, H_{Ar}), 8.16 (d, J = 7.2 Hz, 1 H, H_{Ar}), 8.28 (bs, 1 H, H_{Ar}), 8.47 (d, J = 8.3 Hz, 1 H, H_{Ar}) ppm. – ¹³C NMR (125 MHz, MeOD): δ = 119.1, 127.9, 128.0, 129.0, 130.4, 138.01, 141.35, 143.52, 147.6 ppm. – MS (FAB), m/z (%): 198 (100) [M⁺ + H], 170 (23) [M⁺ – N₂], 141 (3) [M⁺ – N₄],

128 (9) [M$^+$ – CN$_4$], 120 (2), 115 (2) [M$^+$ – CN$_4$ – N]. – C$_{10}$H$_7$N$_5$ (197.20) · 0.1 H$_2$O: calc. C 60.36, H 3.65, N 35.19; found C 60.09, H 3.66, N 35.54.

1-(Tetrazol-5-yl)isoquinoline (30-H).[268] The tetrazole 30-H was obtained from reaction of isoquinoline-

1-carbonitrile with sodium azide and ammonium chloride at 130 °C for 48 h following GP1 method A. Yield: 514 mg, 2.61 mmol, 65%; yellow solid. – ^{1}H NMR (500 MHz, CDCl$_3$): δ = 7.84–7.90 (m, 2 H, H$_{Ar}$), 8.01 (d, J = 5.6 Hz, 1 H, H$_{Ar}$), 8.07 (d, J = 8.1 Hz, 1 H, H$_{Ar}$), 8.67 (d, J = 5.6 Hz, 1 H, H$_{Ar}$), 9.41 (d, J = 8.2 Hz, 1 H, H$_{Ar}$) ppm. – MS (FAB), m/z (%): 198 (80) [M$^+$ + H], 170 (15) [M$^+$ – N$_2$]. – C$_{10}$H$_7$N$_5$ (197.20) · 0.1 H$_2$O: calc. C 60.36, H 3.65, N 35.19; found C 60.79, H 3.62, N 33.98.

5-(Furan-2-yl)tetrazole (32-H).[202] Tetrazole 32-H was synthesized according to GP1 method A from

furan-2-carbonitrile. The reaction was refluxed at 100 °C for 24 h. Since no precipitate was formed after acidification of the aqueous phase to pH = 3, it was extracted with ethyl acetate (3 × 50 ml), the combined organic layers dried over magnesium sulfate and the solvent evaporated. The orange oil was again dissolved in water and a few drops of 5 M sodium hydroxide solution (pH > 9) and extracted with ethyl acetate to remove residual reagent. Acidification of the aqueous phase to pH = 3, extraction with ethyl acetate (3 × 30 ml) and evaporation of the organic solvent afforded a yellow solid. Yield: 485 mg, 3.57 mmol, 71%. – ^{1}H NMR (500 MHz, MeOD): δ = 6.71 (q, J = 3.5 Hz, 1 H, 4-H$_{Ar}$), 7.24 (d, J = 3.4 Hz, 1 H, 3-H$_{Ar}$), 7.83 (s, 1 H, 5-H$_{Ar}$) ppm. – ^{13}C NMR (125 MHz, MeOD): δ = 112.0 (+, 4-C$_{Fu}$), 112.8 (+, 3-C$_{Fu}$), 139.9 (C$_q$, 2-C$_{Fu}$), 145.7 (+, 5-C$_{Fu}$), 148.4 (C$_q$, C$_{TetH}$) ppm. – MS (FAB), m/z (%): 137 (56) [M$^+$ + H]. – C$_5$H$_4$N$_4$O (136.11) · 0.5 H$_2$O: calc. C 43.83, H 3.02, N 40.89; found C 43.64, H 2.91, N 41.17.

5-(Thiophen-2-yl)tetrazole (33-H).[269] Compound 33-H was synthesized following GP1 method A from

thiophen-2-carbonitrile at 120 °C for 48 h. Sodium hydroxide solution (5 M, 4 ml) was added to the reaction mixture and the solvents were evaporated. After dissolving the residue in 10 ml water and a few drops of 5 M sodium hydroxide solution (pH > 9) and extracting it with ethyl acetate to remove residual starting material, the aqueous phase was acidified by 5 M hydrochloric acid to reach pH = 4 and a precipitate formed. This was filtered off, washed with acidic water (pH = 3–4) and dried in vacuum. Yield: 954 mg, 6.27 mmol, 78%; yellowish solid. – ^{1}H NMR (500 MHz, MeOD): δ = 7.24 (q, J = 5.1 Hz, 1 H, 4-H$_{Ar}$), 7.73 (dd, J = 5.1, 1.1 Hz, 1 H, 3-H$_{Ar}$), 7.76 (dd, J = 3.7, 1.1 Hz, 1 H, 5-H$_{Ar}$) ppm. – ^{13}C NMR (125 MHz, MeOD): δ = 125.1 (C$_q$, 2-C$_{Ar}$), 128.0 (+, 5-C$_{Ar}$), 128.9 (+, 4-C$_{Ar}$), 129.7 (+, 3-C$_{Ar}$), 151.9 (C$_q$, C$_{TetH}$) ppm.

– MS (FAB), m/z (%): 153 (36) [M$^+$ + H], 110 (3) [M$^+$ – N$_3$]. – C$_5$H$_4$N$_4$S (152.18): calc. C 39.46, H 2.65, N 36.82, S 21.07; found C 39.58, H 2.61, N 36.73, S 21.05.

2-(Tetrazol-5-yl)-*N*-methylimidazole (**34-H**).[270] The tetrazole was synthesized according to **GP1** method **A** from starting material **34-CN** at 120 °C by Dr. Daniel Zink, CYNORA GmbH. Due to its polarity the product precipitated only slowly from the acidic aqueous phase at 5 °C. It was further purified by dissolving in 10 ml water and a few drops of 5 M sodium hydroxide solution (pH > 9), filtration, and acidification of the filtrate to pH = 3. Yield: 735 mg, 4.90 mmol, 44%; yellow solid. – ^{1}H NMR (400 MHz, DMSO/D$_2$O): δ = 4.00 (s, 1 H, CH$_3$), 7.44 (dd, J = 13.5, 1.8 Hz, 2 H, H$_{Ar}$) ppm. – ^{13}C NMR (100 MHz, DMSO): δ = 33.9 (+, CH$_3$), 116.9 (+, C$_{Ar}$), 122.5 (+, C$_{Ar}$), 133.5 (C$_q$, 2-C$_{Ar}$), 147.0 (C$_q$, C$_{TetH}$) ppm. – IR (KBr): ṽ = 3175 (vw), 3137 (vw), 2929 (vw), 2696 (vw), 1856 (vw), 1630 (vw), 1536 (vw), 1520 (vw), 1453 (vw), 1406 (vw), 1365 (w), 1301 (vw), 1273 (vw), 1128 (vw), 1077 (vw), 1041 (vw), 1024 (vw), 1012 (vw), 914 (w), 860 (vw), 771 (w), 749 (vw), 691 (w), 505 (w), 412 (w) cm^{-1}. – MS (FAB), m/z (%): 151 (100) [M$^+$ + H], 137 (51) [M$^+$ – N], 123 (21) [M$^+$ – N$_2$], 120 (9) [M$^+$ – Me – N], 108 (11) [M$^+$ – N$_3$], 95 (3) [M$^+$ – N$_4$]. – C$_5$H$_6$N$_6$ (150.14) · H$_2$O: calc. C: 39.29, H 4.15, N 54.98; found C 39.42, H 4.21, N 54.89.

2-(Tetrazol-5-yl)-*N*-hexylimidazole (**35-H**). The tetrazole **DZ1583** was synthesized according to **GP1** method **A** from starting material **35-CN** at 120 °C by Dr. Daniel Zink, CYNORA GmbH. The compound was further purified by dissolving in 5 ml water and a few drops of 5 M sodium hydroxide solution (pH > 9), filtration, and acidification of the filtrate to pH = 2. Yield: 446 mg, 2.02 mg, 51%; white solid. – ^{1}H NMR (400 MHz, MeOD): δ = 0.88 (t, J = 6.8 Hz, 3 H, CH$_3$), 1.29–1.35 (m, 6 H, CH$_2$), 1.88–1.91 (m, 2 H, CH$_2$), 4.72 (t, J = 7.4 Hz, 2 H, N-CH$_2$), 7.61 (bs, 1 H, H$_{Ar}$), 7.76 (bs, 1 H, H$_{Ar}$) ppm. – ^{13}C NMR (100 MHz, MeOD): δ = 14.2 (+, CH$_3$), 23.5 (–, CH$_2$), 27.0 (–, CH$_2$), 31.2 (–, CH$_2$), 32.3 (–, CH$_2$), 124.8 (+, C$_{Ar}$), 132.7 (+, C$_{Ar}$), 138.2 (C$_q$, 2-C$_{Ar}$), 150.8 (C$_q$, C$_{TetH}$) ppm. – IR (KBr): ṽ = 3128 (w), 2951 (w), 2922 (m), 2855 (w), 1849 (vw), 1627 (w), 1537 (w), 1518 (w), 1456 (m), 1367 (m), 1297 (w), 1275 (w), 1134 (w), 1106 (w), 1031 (w), 1012 (w), 944 (w), 912 (m), 862 (w), 771 (m), 751 (w), 693 (m), 663 (vw), 548 (vw), 497 (w), 409 (m) cm^{-1}. – MS (FAB), m/z (%): 221 (100) [M$^+$ + H], 193 (10) [M$^+$ – N$_2$], 178 (4) [M$^+$ – N$_3$], 165 (2) [M$^+$ – N$_4$], 121 (6) [M$^+$ – C$_3$H$_7$ – N$_4$]. – C$_{10}$H$_{16}$N$_6$ (220.28) · 0.85 H$_2$O: calc. C: 50.98, H 7.57, N 35.67; found C 50.83, H 6.98, N 35.23.

2-(Tetrazol-5-yl)-N-methyl-benzimidazole (**36-H**). The tetrazole **36-H** was synthesized following **GP1** method **A** at 130 °C for 85 h from starting material **36-CN** by Dr. Daniel Zink, CYNORA GmbH. The product was further purified by dissolving in 15 ml alkaline water (pH > 9), filtration, and acidification of the filtrate to pH = 5. Yield: 908 mg, 4.53 mmol, 79%, white solid. – ^{1}H NMR (400 MHz, DMSO/D$_2$O): δ = 4.26 (s, 3 H, CH$_3$), 7.56–7.59 (m, 2 H, H$_{Ar}$), 7.74–7.77 (m, 1 H, H$_{Ar}$), 7.81–7.84 (m, 1 H, H$_{Ar}$) ppm. – ^{13}C NMR (100 MHz, DMSO): δ = 32.7 (+, CH$_3$), 112. 5 (+, C$_{Ar}$), 114.1 (+, C$_{Ar}$), 126.2 (+, C$_{Ar}$), 126.7 (+, C$_{Ar}$), 130.6 (C$_q$, C$_{Ar}$), 132.9 (C$_q$, C$_{Ar}$), 140.7 (C$_q$, 2-C$_{Ar}$), 149.7 (C$_q$, C$_{TetH}$) ppm. – IR (KBr): ṽ = 2878 (w), 2782 (w), 2662 (w), 1717 (vw), 1628 (w), 1598 (w), 1528 (w), 1495 (w), 1468 (w), 1431 (w), 1362 (m), 1348 (m), 1265 (w), 1160 (w), 1131 (w), 1105 (w), 1037 (m), 1012 (w), 962 (vw), 903 (w), 873 (m), 810 (w), 759 (m), 747 (s), 683 (w), 600 (m), 542 (m), 453 (w) cm^{-1}. – MS (FAB), m/z (%): 202 (12) [M$^+$ + H], 189 (14) [M$^+$ – N], 178 (12) [M$^+$ – N$_2$], 165 (19) [M$^+$ – N$_3$], 133 (100) [M$^+$ – CN$_4$], 115 (21) [M$^+$ – Me – CN$_4$]. – C$_9$H$_8$N$_6$ (200.20) · 0.25 H$_2$O: calc. C: 52.81, H 4.19, N 41.05; found C 52.82, H 4.03, N 40.71.

2-Methyl-4-(tetrazol-5-yl)oxazole (**37-H**). Tetrazole **37-H** was synthesized according to **GP1** method **C** from starting material **37-CN** with 4.00 equiv. of sodium azide and 4.00 equiv. of pyridine hydrochloride. The reaction mixture was refluxed at 100 °C for 40 h. Acidification of the aqueous phase to pH = 2 afforded a yellowish precipitate, which was filtered off, washed with few acidic water (pH = 1–2) and dried in vacuum. Yield: 217 mg, 1.44 mmol, 63%. – ^{1}H NMR (300 MHz, MeOD): δ = 2.55 (s, 3 H, CH$_3$), 8.50 (s, 1 H, H$_{Ar}$) ppm. – ^{13}C NMR (100 MHz, MeOD): δ = 13.5 (+, CH$_3$), 128.4 (C$_q$, 4-C$_{Ar}$), 140.6 (+, 5-C$_{Ar}$), 150.9 (C$_q$, C$_{TetH}$), 165.0 (C$_q$, 2-C$_{Ar}$) ppm. – IR (KBr): ṽ = 3149 (w), 3050 (vw), 2959 (w), 2851 (w), 2720 (w), 1650 (w), 1579 (w), 1516 (w), 1442 (vw), 1410 (w), 1381 (w), 1309 (w), 1240 (w), 1228 (w), 1178 (vw), 1111 (m), 1082 (w), 1039 (m), 998 (w), 952 (w), 938 (w), 924 (m), 859 (w), 798 (m), 743 (m), 710 (w), 700 (w), 679 (w), 666 (vw), 626 (w), 490 (w) cm^{-1}. – MS (FAB), m/z (%): 153 (100) [M$^+$ + H], 138 (44) [M$^+$ – Me], 125 (13) [M$^+$ – N$_2$], 110 (9) [M$^+$ – Me – N$_2$], 96 (11) [M$^+$ – N$_4$], 91 (8), 85 (5) [M$^+$ – CN$_4$]. – C$_5$H$_5$N$_5$O (151.13) · 0.1 H$_2$O: calc. C 39.27, H 3.43, N 45.80; found C 39.17, H 3.32, N 45.97.

5-Methyl-3-(tetrazol-5-yl)isoxazole (**38-H**). Tetrazole **38-H** was obtained from reaction of the carbonitrile **38-CN** with 4.00 equiv. of sodium azide and 4.00 equiv. of pyridine hydrochloride following **GP1** method **C**. The reaction temperature was kept at 100 °C for 48 h. Acidification of the aqueous phase yielded no precipitate, thus it was stored at –20 °C overnight. After warming up to room temperature, a yellowish precipitate was formed, which was collected by filtration, washed with few acidic water (pH = 1–2) and dried in vacuum. Yield: 181 mg, 1.20 mmol, 44%. – ^{1}H NMR (400 MHz, MeOD): δ = 2.53 (s, 3 H, CH$_3$), 6.67 (s, 1 H, H$_{Ar}$) ppm. – ^{13}C NMR (100 MHz, MeOD): δ = 12.0 (+, CH$_3$), 102.0 (+, 4-C$_{Ar}$), 152.5 (C$_q$, 3-C$_{Ar}$), 154.3 (C$_q$, C$_{TetH}$), 172.7 (C$_q$, 5-C$_{Ar}$) ppm. – IR (KBr): ṽ = 3459 (w), 3129 (vw), 2349 (vw), 1932 (vw), 1663 (vw), 1636 (vw), 1607 (w), 1464 (w), 1366 (w), 1339 (vw), 1238 (w), 1202 (vw), 1102 (vw), 1028 (vw), 1011 (w), 976 (vw), 959 (w), 924 (w), 812 (w), 801 (w), 760 (vw), 670 (vw), 536 (w), 480 (w) cm^{-1}. – MS (70 eV, EI, 70 °C), m/z (%): 151 (47) [M$^+$], 123 (34) [M$^+$– N$_2$], 109 (20) [M$^+$– N$_2$], 95 (6) [M$^+$– Me – N$_2$], 88 (17), 80 (8) [M$^+$– Me – N$_4$], 52 (16) [CCCN], 43 (100) [C(Me)O].

2-(Tetrazol-5-yl)benzoxazole (**39-H**).[267] Compound **39-H** was obtained from reagent **39-CN** by

reflux with 4.25 equiv. of sodium azide and 4.25 equiv. of pyridine hydrochloride at 100 °C for 48 h following **GP1** method **C**. From acidification of the water phase to pH = 4 a reddish solid was obtained. Yield: 574 mg, 3.07 mmol, 77%. – ^{1}H NMR (300 MHz, MeOD): δ = 7.45 (quind, J = 7.5 , 1.7 Hz, 2 H, H$_{Ar}$), 7.67 (d, J = 7.5 Hz, 1 H, H$_{Ar}$), 7.81 (d, J = 7.1 Hz, 1 H, H$_{Ar}$) ppm. – ^{13}C NMR (75 MHz, MeOD): δ = 112.8 (+, C$_{Ar}$), 122.3 (+, C$_{Ar}$), 127.0 (+, C$_{Ar}$), 128.6 (+, C$_{Ar}$), 142.1 (C$_q$, C$_{Ar}$), 152.0 (C$_q$, C$_{TetH}$), 153.3 (C$_q$, C$_{Ar}$), 154.7 (C$_q$, C$_{Ar}$) ppm. – IR (KBr): ṽ = 3074 (vw), 2865 (vw), 2532 (w), 1879 (w), 1630 (w), 1615 (w), 1527 (w), 1466 (w), 1447 (m), 1341 (w), 1323 (m), 1247 (m), 1196 (w), 1180 (w), 1164 (w), 1108 (w), 1035 (m), 1020 (m), 948 (w), 891 (w), 801 (w), 746 (s), 687 (w), 629 (w), 437 (w) cm^{-1}. – MS (FAB), m/z (%): 188 (100) [M$^+$+ H], 176 (9) [M$^+$– N], 160 (19) [M$^+$– N$_2$], 146 (2) [M$^+$– N$_3$], 121 (10) [M$^+$– CN$_4$H]. – C$_8$H$_5$N$_5$O (187.16) · 0.1 H$_2$O: calc. C 50.85, H 2.77, N 37.06; found C 50.85, H 2.54, N 36.97.

2-(Tetrazol-5-yl)benzothiazole (**40-H**). Tetrazole **40-H** was synthesized according to **GP1** method **C**

with starting material **40-CN** and 4.25 equiv. of sodium azide and 4.25 equiv. of pyridine hydrochloride. The reaction was refluxed at 100 °C for 48 h. Acidification of the aqueous phase to pH = 4 and storage at 5 °C overnight afforded 500 mg white solid (2.46 mmol, 82%). – ^{1}H NMR (400 MHz, MeOD): δ = 7.54 (td, J = 7.7, 1.2 Hz, 1 H, H$_{Ar}$), 7.60 (td, J = 7.7, 1.1 Hz, 1 H, H$_{Ar}$), 8.06 (d, J = 7.2 Hz, 1 H, H$_{Ar}$), 8.14 (d, J = 7.7 Hz, 1 H, H$_{Ar}$) ppm. – ^{13}C NMR (100 MHz, MeOD): δ = 123.5 (+, C$_{Ar}$), 125.3 (+, C$_{Ar}$), 128.3 (+, C$_{Ar}$), 128.4 (+, C$_{Ar}$), 136.7 (C$_q$, 7a-C$_{Ar}$), 150.5 (C$_q$, C$_{TetH}$), 153.9 (C$_q$, 3a-C$_{Ar}$), 154.7 (C$_q$, 2-C$_{Ar}$) ppm. – IR (KBr): ṽ = 3087 (vw), 2976 (w), 2853 (w), 2755 (w), 2692 (w), 1703 (vw), 1560 (w), 1573 (m), 1557 (w), 1492 (vw), 1454 (w), 1436 (w), 1402 (w), 1362 (m), 1320 (m), 1283 (w), 1225 (vw), 1124 (w), 1080 (m), 1050 (w), 1025 (w), 1002 (w), 977 (s), 890 (w), 855 (w), 767 (s), 742 (m), 726 (m),

708 (w) cm^{-1}. – MS (FAB), m/z (%): 204 (30) [M$^+$+ H], 184 (2), 176 (6) [M$^+$ – N$_2$], 167 (6), 138 (65) [M$^+$ – CN$_4$], 125 (10) [M$^+$ – CN$_4$ – C], 121 (9) [M$^+$ – CN$_4$ – N], 108 (13) [M$^+$ – CN$_4$ – C=N], 90 (12) [M$^+$ – CN$_4$ – CS]. – C$_8$H$_5$N$_5$S (203.23): calc. C: 47.28, H 2.48, N 34.46, S 15.78; found C 47.22, H 2.47, N 34.11, S 15.81.

4-Phenyl-2-(tetrazol-5-yl)pyridine (41-H). a) Tetrazole **41-H** was synthesized from 4-phenylpyridine-2-carbonitrile following **GP1** method **A** and refluxed at 120 °C for 48 h. Sodium hydroxide solution (5 M, 4 ml) was added to the reaction mixture and stirred for 30 min, and subsequently all solvents evaporated. The residue was dissolved in 10 ml water (pH > 9) and extracted with ethyl acetate (2 × 50 ml). When acidifying the aqueous phase to pH = 4, a precipitate formed, which was filtered off, washed with acidic water (pH = 3–4) and dried in vacuum. Yield: 2.05 g, 9.17 mmol, 92%. – ^{1}H NMR (400 MHz, MeOD): δ = 7.48–7.56 (m, 3 H, H$_{Ar}$), 7.80–7.83 (m, 3 H, H$_{Ar}$), 8.48 (s, 1 H, H$_{Ar}$), 8.76 (d, J = 5.2 Hz, 1 H, H$_{Ar}$) ppm. – ^{13}C NMR (100 MHz, MeOD): δ = 121.2 (+, C$_{Pyr}$), 124.6 (+, C$_{Pyr}$), 128.1 (+, C$_{Ar}$), 130.5 (+, C$_{Ar}$), 131.0 (+, C$_{Ar}$), 138.1(C$_q$, C$_{Ar}$), 145.7 (C$_q$, 4-C$_{Ar}$), 151.6 (+, 6-C$_{Ar}$), 151.8 (C$_q$, 2-C$_{Ar}$), 156.9 (C$_q$, C$_{TetH}$) ppm. – IR (KBr): $\tilde{v}$ = 3098 (vw), 2570 (w), 1984 (w), 1623 (m), 1465 (m), 1440 (w), 1376 (w), 1347 (w), 1240 (w), 1180 (w), 1130 (w), 1080 (w), 1037 (w), 1025 (w), 990 (w), 966 (w), 906 (w), 845 (w), 802 (w), 773 (m), 755 (m), 726 (w), 696 (m), 616 (w), 517 (w), 490 (w), 445 (m) cm^{-1}. – MS (FAB), m/z (%): 224 (98) [M$^+$+ H], 196 (20) [M$^+$ – N$_2$], 181 (7) [M$^+$ – N$_3$], 165 (6) [M$^+$ – N$_4$], 120 (9) [M$^+$ – Ph – N$_2$]. – C$_{12}$H$_9$N$_5$ (223.24) · 0.25 H$_2$O: calc. C 63.29, H 4.20, N 30.75; found C 63.01, H 4.26, N 30.78. b) Compound **41-H** was synthesized following **GP1** method **B** from 4-phenylpyridine-2-carbonitrile. The product was obtained by acidification of the water phase to pH = 1. The precipitate was washed with acidic water (pH = 1) and dried in vacuum giving 2.14 g (9.60 mmol, 64% yield) product. – ^{1}H NMR (300 MHz, DMSO): δ = 7.54–7.56 (m, 3 H, H$_{Ar}$), 7.89–7.92 (m, 2 H, H$_{Ar}$), 8.07 (dd, J = 5.2 Hz, 1 H, H$_{Ar}$), 8.43 (s, 1 H, H$_{Ar}$), 8.78 (d, J = 5.2 Hz, 1 H, H$_{Ar}$) ppm. – ^{13}C NMR (62.5 MHz, MeOD): δ = 124.1 (+, 3-C$_{Ar}$), 127.4 (+, 5-C$_{Ar}$), 130.9 (+, Ph-C$_{Ar}$), 133.2 (+, Ph-C$_{Ar}$), 133.9 (+, Ph-C$_{Ar}$), 140.5 (C$_q$, Ph-C$_{Ar}$), 147.9 (C$_q$, 4-C$_{Ar}$), 153.6 (+, 6-C$_{Ar}$), 155.2 (C$_q$, 2-C$_{Ar}$), 159.7 (C$_q$, C$_{TetH}$) ppm. – IR (KBr): $\tilde{v}$ = 3364 (vw), 2487 (w), 1963 (vw), 1612 (w), 1555 (w), 1468 (w), 1439 (w), 1379 (w), 1283 (vw), 1266 (vw), 1247 (vw), 1205 (vw), 1097 (vw), 1055 (vw), 1033 (w), 1007 (w), 895 (w), 848 (w), 802 (vw), 773 (w), 754 (m), 733 (w), 711 (vw), 688 (w), 620 (w), 513 (vw), 487 (vw), 445 (w) cm^{-1}. – MS (FAB), m/z (%): 224 (29) [M$^+$+ H], 207 (6) [M$^+$ – N], 196 (10) [M$^+$ – N$_2$], 181 (7) [M$^+$ – N$_3$], 165 (6) [M$^+$ – N$_4$], 121(14) [M$^+$ – N$_2$ – Ph].

4-Methyl-2-(tetrazol-5-yl)pyridine (**42-H**).[271] Ligand **42-H** was synthesized from 4- methylpyridine-2-carbonitrile according to **GP1** method **A** at 130 °C for 48 h by Dr. Daniel Zink, CYNORA GmbH. From acidification of the aqueous phase to pH = 4 with hydrochloric acid, a white solid precipitated after 30 min and after storage at 5 °C overnight, the solid was filtered off, washed with acidic water (pH = 3–4) and dried in vacuum. Yield: 4.83 g, 30.0 mmol, 71%; off-white solid. – ^{1}H NMR (500 MHz, MeOD): δ = 2.51 (s, 3 H, CH$_3$), 7.56 (bs, 1 H, H$_{Ar}$), 8.19 (bs, 1 H, H$_{Ar}$), 8.64 (bs, 1 H, H$_{Ar}$) ppm. – MS (FAB), m/z (%): 178 (15) [M$^+$ + H$_2$O], 162 (83) [M$^+$ + H], 134 (15) [M$^+$ – N$_2$], 120 (12) [M$^+$ – Me – N$_2$], 107 (19) [M$^+$ – N$_4$], 89 (15) [M$^+$ – CN$_4$H]. – C$_7$H$_7$N$_5$ · 0,05 H$_2$O (161.17): calc. C 51.88, H 4.42, N 43.21; found C 51.55, H 4.56, N 43.25.

4-Methoxy-2-(tetrazol-5-yl)pyridine (**43-H**). The tetrazole **43-H** was synthesized from **43-CN** following **GP1** method **B** by refluxing at 80 °C for 120 h. Due to the very polar character of the tetrazole the workup and purification had to be adjusted: The solvents from the reaction mixture were evaporated, the residue dissolved in 5–6 ml water with a few drops of 5 M sodium hydroxide solution (pH > 9), extracted with ethyl acetate (2 × 50 ml) and the aqueous phase acidified to pH = 4 and after storage at 5 °C overnight, a precipitate was formed. This was filtered off, washed with acidic water (pH = 3–4) and dried in vacuum. Yield: 746 mg, 4.21 mmol, 54%; white solid. – IR (KBr): ṽ = 3042 (w), 2713 (w), 1643 (m), 1621 (m), 1537 (w), 1495 (m), 1428 (w), 1388 (w), 1373 (w), 1311 (m), 1239 (m), 1186 (m), 1145 (w), 1116 (w), 1044 (w), 1029 (w), 1014 (m), 987 (m), 959 (w), 897 (m), 852 (m), 762 (w), 627 (w), 581 (m), 504 (m), 465 (w), 419 (w) cm^{-1}. – MS (FAB), m/z (%): 176 (18) [M$^+$ + H], 152 (8) [M$^+$ – N$_2$], 120 (11) [M$^+$ – N$_4$], 95 (3) [M$^+$ – CN$_4$ – Me], 89 (12) [M$^+$ – Me – N$_4$]. – C$_7$H$_7$N$_5$O (177.17) · 0.2 H$_2$O: calc. C 46.51, H 4.13, N 38.74; found C 46.42, H 3.92, N 39.05. NMR data cannot be recorded here due to the insolubility of **553** in deuterated solvents.

4-Piperidyl-2-(tetrazol-5-yl)pyridine (**44-H**). The tetrazole **44-H** was synthesized from **44-CN** according to **GP1** method **A** at 130 °C for 48 h. After cooling down the reaction, sodium hydroxide solution (5 M, 4 ml) was added, the solvents evaporated and the residue dissolved in 5 ml water (pH > 9). The aqueous phase was extracted with ethyl acetate (2 × 50 ml) and then acidified to reach pH = 4 by addition of 5 M hydrochloric acid. A white solid precipitated, which was filtered off, washed with few acidic water (pH = 3–4) and dried in vacuum. Yield: 713 mg, 3.10 mmol, 77%. – ^{1}H NMR (500 MHz, DMSO/D$_2$O): δ = 1.52–1.57 (m, 6 H, CH$_2$), 3.50 (bs, 4 H, CH$_2$), 6.82 (dd, J = 7.6, 2.8 Hz, 1 H, H$_{Ar}$),

7.37 (d, J = 2.8 Hz, 1 H, H$_{Ar}$), 7.82 (d, J = 7.5 Hz, 1 H, H$_{Ar}$) ppm. – ^{13}C NMR (125 MHz, DMSO): δ = 23.0 (–, CH$_2$), 24.8 (–, CH$_2$), 47.5 (–, CH$_2$), 103.2 (+, C$_{Ar}$), 106.2 (+, C$_{Ar}$), 138.7 (+, C$_{Ar}$), 139.3 (C$_q$, 2-C$_{Ar}$), 155.6 (C$_q$, C$_{TetH}$), 155.8 (C$_q$, 4-C$_{Ar}$) ppm. – IR (KBr): ṽ = 3636 (vw), 3082 (vw), 2851 (w), 1623 (m), 1538 (m), 1456 (w), 1407 (w), 1367 (w), 1264 (w), 1245 (m), 1106 (vw), 1043(vw), 1023 (w), 984 (w), 941 (w), 915 (w), 864 (w), 849 (w), 820 (m), 759 (w), 680 (w), 625 (vw), 582 (vw), 527 (w), 435 (w) cm^{-1}. – MS (FAB), m/z (%): 231 (100) [M$^+$ + H], 203 (31) [M$^+$ – N$_2$], 188 (9) [M$^+$ – N$_3$]. – C$_{11}$H$_{14}$N$_6$ (230.27) · 0.45 H$_2$O: calc. C 55.42, H 6.30, N 35.26; found C 55.16, H 6.30, N 35.66.

2-Methyl-6-(tetrazol-5-yl)pyridine (45-H).[271] Ligand **45-H** was synthesized from 6-methylpyridine-2-carbonitrile following **GP1** method **A** at 120 °C for 48 h. After the reaction was

cooled down, 5 ml sodium hydroxide solution (5 M) was added and the solvents evaporated. After dissolving the residue in few ml water (pH > 9), the aqueous phase was extracted with ethyl acetate (2 × 50 ml) to remove unreacted starting material, and then acidified with 5 M hydrochloric acid to reach pH = 5. A white solid precipitated, which was filtered off and washed with acidic water (pH = 4–5). Yield: 1.08 g, 6.71 mmol, 79%. – ^{1}H NMR (500 MHz, MeOD): δ = 2.62 (s, 3 H, CH$_3$), 7.40 (d, J = 7.8 Hz, 1 H, H$_{Ar}$), 7.87 (t, J = 7.8 Hz, 1 H, H$_{Ar}$), 8.01 (d, J = 7.7 Hz, 1 H, H$_{Ar}$) ppm. – ^{13}C NMR (125 MHz, MeOD): δ = 22.7 (+, CH$_3$), 119.4 (+, 3-C$_{Ar}$), 125.4 (+, 5-C$_{Ar}$), 137.9 (+, 4-C$_{Ar}$), 142.9 (C$_q$, 2-C$_{Ar}$), 155.4 (C$_q$, C$_{TetH}$), 159.4 (C$_q$, 6-C$_{Ar}$) ppm. – MS (FAB), m/z (%): 162 (100) [M$^+$ + H], 134 (17) [M$^+$ – N$_2$], 119 (5) [M$^+$ – Me – N$_2$], 92 (6) [M$^+$ – CN$_4$]. – C$_7$H$_7$N$_5$ (161.17)·0.4 H$_2$O: calc. C: 49.93, H 4.67, N 41.59; found C 49.60, H 4.55, N 41.57.

2-Methoxy-6-(tetrazol-5-yl)pyridine (46-H). The tetrazole **46-H** was synthesized from 2-methoxypyridine-6-carbonitrile following **GP1** method **B**. Due to the very polar character of the tetrazole, it could not be obtained by precipitation of the acidified aqueous phase. Thus, the solvents were evaporated from the reaction mixture, the residue dissolved in 6 ml water with a few drops of 5 M sodium hydroxide solution (pH > 9), extracted with ethyl acetate (2 × 50 ml) and the aqueous phase acidified to pH = 3 and after storage at 5 °C overnight a precipitate was formed. Yield: 194 mg, 1.41 mmol, 43%; white solid. – ^{1}H NMR (500 MHz, DMSO): δ = 4.03 (s, 3 H, OCH$_3$), 6.94 (dd, J = 8.2 Hz, 1.0 1 H, H$_{Ar}$), 7.79 (dd, J = 7.3, 1.0 Hz, 1 H, H$_{Ar}$), 7.83–7.86 (m, 1 H, H$_{Ar}$) ppm. – ^{13}C NMR (125 MHz, DMSO): δ = 53.0 (+, CH$_3$), 113.6 (+, 3-C$_{Ar}$), 115.3 (+, 5-C$_{Ar}$), 139.9 (+, 4-C$_{Ar}$), 140.7 (C$_q$, 6-C$_{Ar}$), 155.0 (C$_q$, C$_{TetH}$), 164.4 (C$_q$, 2-C$_{Ar}$) ppm. – IR (KBr): ṽ = 2783 (w), 1607 (w), 1577 (w), 1556 (w),

1474 (w), 1424 (w), 1382 (w), 1370 (w), 1303 (w), 1274 (w), 1243 (w), 1165 (w), 1095 (w), 1079 (w), 1049 (w), 1010 (w), 985 (w), 896 (w), 870 (w), 806 (m), 749 (m), 726 (w), 669 (w), 612 (vw), 468 (vw), 446 (w) cm^{-1}. – MS (FAB), m/z (%): 178 (100) [M$^+$ + H], 165 (2) [M$^+$ – N], 150 (35) [M$^+$ – N$_2$], 136 (34) [M$^+$ – N$_3$], 120 (9) [M$^+$ – N$_4$], 108 (30) [M$^+$ – CN$_4$], 93 (7) [M$^+$ – CN$_4$ – Me], 89 (22) [M$^+$ – Me – N$_4$]. – C$_7$H$_7$N$_5$O (177.17): calc. C 47.46, H 3.98, N 39.53; found C 47.39, H 3.81, N 39.67.

Carbonitriles and Precursors

N-Methylimidazole-2-carbonitrile (**34-CN**).[272] The synthesis of compound **34-CN** was performed by Dr. Daniel Zink, CYNORA GmbH. Copper(I) cyanide (436 mg, 4.87 mmol, 0.100 equiv.), 1,10-phenanthroline (1.76 g, 9.74 mmol, 0.200 equiv.), sodium cyanide (3.10 g, 63.3 mmol, 1.30 equiv.) and potassium *tert*-butoxide (10.9 g, 97.4 mmol, 2.00 equiv.) were slowly added to a flask filled with 20 ml dry 1,4-dioxane/m-xylene (2:1) under nitrogen atmosphere. Then first N-methylimidazole (4.00 g, 48.7 mmol, 1.00 equiv.) and subsequently iodine (18.5 g, 73.1 mmol,1.50 equiv.) were added to the reaction mixture, which resulted in an exothermic reaction. The reaction mixture was stirred at 110 °C for 24 h. After filtration of the reaction mixture over silica gel and washing with ethyl acetate, the solvents were evaporated and the raw product purified by column chromatography (cH/EE 1:1). Yield: 1.63 g, 15.2 mmol, 31%; yellow oil. – R_f = 0.19 (cH/EE 1:1). – ^{1}H NMR (500 MHz, CDCl$_3$): δ = 3.87 (s, 3 H, CH$_3$), 7.09 (s, 1 H, H$_{Ar}$), 7.19 (s, 1 H, H$_{Ar}$) ppm. – MS (70 eV, EI, 70 °C), m/z (%): 132 (80) [M$^+$ + CN], 107 (100) [M$^+$], 87 (26), 67 (8) [M$^+$ – Me – CN], 57 (13) [M$^+$ – Me – CN – N].

N-hexylimidazole (**35**). The compound **35** was synthesized in a slightly modified route of Wang et al by Dr. Daniel Zink, CYNORA GmbH.[273] 1H-Imidazole (5.00 g, 73.4 mmol, 1.00 equiv.) and sodium hydroxide (3.09 g, 77.1 mmol, 1.05 equiv.) were dissolved in 40 ml DMSO and stirred at room temperature for 1 h. Then 1-bromohexane (12.4 ml, 88.1 mmol, 1.20 equiv.) was added and after stirring for 30 min at room temperature, another 20 ml of DMSO were added and heated to 50 °C overnight. After dissolving the reaction mixture in 1 l water, it was extracted with ethyl acetate (3 × 200 ml), the organic phase washed with water (400 ml) and brine (300 ml), dried over magnesium sulfate and the solvents evaporated. The yellow oil was directly used for synthesis of compound **DZ 1557** without further purification. Yield: 8.55 g, 56.19 mmol, 77%, yellow oil. – 1H NMR (500 MHz, CDCl$_3$): δ = 0.86–0.88 (m, 3 H, CH$_3$), 1.29 (bs, 6 H, CH$_2$), 1.76–1.78 (m, 2 H, CH$_2$), 3.94 (t, J = 7.2 Hz, 2 H, CH$_2$), 6.91 (s, 1 H, H$_{Ar}$), 7.07 (s, 1 H, H$_{Ar}$), 7.57 (s, 1 H, H$_{Ar}$) ppm. – MS (70 eV, EI, 50 °C), m/z (%): 152 (51) [M$^+$], 137 (4) [M$^+$ – CH$_3$], 125 (73)

$[M^+ - C_2H_5]$, 109 (9) $[M^+ - C_3H_7]$, 96 (39) $[M^+ - C_4H_9]$, 82 (100) $[M^+ - C_5H_{11}]$, 69 (19) $[M^+ - C_6H_{13}]$, 55 (29) $[M^+ - C_6H_{11} - N]$.

N-Hexylimidazole-2-carbonitrile (**35-CN**). Compound **35-CN** was synthesized as described for **34-CN** from *N*-hexylimidazole (**35**) by Dr. Daniel Zink, CYNORA GmbH. The reaction mixture was stirred at 120 °C for 36 h. After filtration of the reaction mixture over silica gel and washing with ethyl acetate, the solvents were evaporated and the residue purified by column chromatography (cH/EE 10:1). Yield: 747 mg, 4.21 mmol, 16%. – $R_f = 0.49$ (cH/EE 1:1). – ^{1}H NMR (500 MHz, CDCl₃): δ = 0.89 (t, J = 7.0 Hz, 3 H, CH₃), 1.29–1.33 (m, 6 H, CH₂), 1.84 (quin, J = 7.2 Hz, 2 H, CH₂), 4.12 (t, J = 7.2 Hz, 2 H, NCH₂), 7.10 (s, 1 H, H$_{Ar}$), 7.20 (s, 1 H, H$_{Ar}$) ppm.

N-Methylbenzimidazole-2-carbonitrile (**36-CN**).[204] The carbonitrile **36-CN** was synthesized as described for **34-CN** from *N*-methylbenzimidazole by Dr. Daniel Zink, CYNORA GmbH. The reaction mixture was stirred at 120 °C for 12 h. After filtration of the reaction mixture over silica gel and washing with ethyl acetate, the solvents were evaporated and the residue purified by column chromatography (cH/EE 5:1). Yield: 1.21 g, 7.71 mmol, 25%; yellow solid. – ^{1}H NMR (500 MHz, CDCl₃): δ = 4.02 (s, 3 H, CH₃), 7.40–7.52 (m, 2 H, H$_{Ar}$), 7.85 (d, J = 8.2 Hz, 2 H, H$_{Ar}$) ppm. – MS (70 eV, EI, 80 °C), *m/z* (%): 157 (100) [M⁺], 148 (8), 129 (7) $[M^+ - CN]$, 119 (4) $[M^+ - Me - CN]$, 105 (4) $[M^+ - Me - CN - N]$, 90 (4) $[M^+ - Me - CN - C≡N]$, 77 (5) [Ph].

2-Methyloxazole-4-carboxamide (**37-CONH₂**). From reaction of 2-methyloxazole-4-(methyl)carboxylate (582 mg, 4.00 mmol, 1.00 equiv.) in an excess of ammonium hydroxide solution (28–30% in water, 12.0 ml, 10.8 mg) for 12 h at room temperature a white solid precipitated, which was filtered off, washed with water and dried in vacuum giving 409 mg (3.24 mmol, 81%) pure product **37-CONH₂**. – ^{1}H NMR (400 MHz, CDCl₃): δ = 2.48 (s, 3 H, CH₃), 5.53 (bs, 1 H, NH), 6.75 (bs, 1 H, NH), 8.11 (s, 1 H, H$_{Ar}$) ppm. – ^{13}C NMR (100 MHz, CDCl₃): δ = 13.8 (+, CH₃), 135.6 (C$_q$, 4-C$_{Ar}$), 141.5 (+, 5-C$_{Ar}$), 153.5 (C$_q$, CONH₂), 161.5 (C$_q$, 2-C$_{Ar}$) ppm. – IR (KBr): $\tilde{v}$ = 3422 (vw), 3109 (vw), 1707 (vw), 1651 (w), 1404 (w), 1379 (vw), 1324 (vw), 1238 (vw), 1192 (vw), 1110 (w), 1083 (vw), 988 (vw), 921 (vw), 807 (w), 727 (vw), 678 (vw), 618 (w), 533 (w), 442 (vw) cm⁻¹. – MS (70 eV, EI, 30°C), *m/z* (%): 126 (100) [M⁺], 110 (12) $[M^+ - Me]$, 98 (6) $[M^+ - C(Me)]$, 85 (37) $[M^+ - NC(Me)]$, 82 (8) $[M^+ - CONH_2]$, 69 (11) $[M^+ - NC(Me)O]$, 54 (12) [CCONH₂], 43 (20) [CONH₂].

2-Methyloxazole-4-carbonitrile (**37-CN**). To a solution of starting material **37-CONH₂** (409 mg, 3.24 mmol, 1.00 equiv.) in 5 ml pyridine was added 0.42 ml phosphoryl trichloride (4.54 mmol, 1.40 equiv.) and the reaction mixture stirred for 3 h at room temperature. After adding ice water, hydrochloric acid (5 M) is added to reach pH = 3. Extraction with diethyl ether (3 × 50 ml), washing with brine (1 × 80 ml), drying over magnesium sulfate and evaporation of the solvent gave an orange oil. Yield: 269 mg, 2.49 mmol, 77%. – ^{1}H NMR (400 MHz, CDCl₃): δ = 2.52 (s, 3 H, CH₃), 8.07 (s, 1 H, H$_{Ar}$) ppm. – ^{13}C NMR (100 MHz, CDCl₃): δ = 13.8 (+, CH₃), 111.8 (C$_q$, CN), 114.9 (C$_q$, 4-C$_{Ar}$), 146.1 (+, 5-C$_{Ar}$), 163.2 (C$_q$, 2-C$_{Ar}$) ppm.

5-Methylisoxazole-3-carboxamide (**38-CONH₂**).[205, 274] Compound **38-CONH₂** was synthesized as described for **37-CONH₂** from 5-methylisoxazole-3-(ethyl)carboxylate. Yield: 457 mg, 3.62 mg, 63%; white solid. – ^{1}H NMR (400 MHz, CDCl₃): δ = 2.49 (s, 3 H, CH₃), 5.72 (bs, 1 H, NH), 6.45 (s, 1 H, H$_{Ar}$), 6.71 (bs, 1 H, NH) ppm. – ^{13}C NMR (100 MHz, CDCl₃): δ = 12.4 (+, CH₃), 101.4 (+, 4-C$_{Ar}$), 158.2 (C$_q$, 3-C$_{Ar}$), 160.9 (C$_q$, CONH₂), 171.5 (C$_q$, 5-C$_{Ar}$) ppm. – IR (KBr): ṽ = 3358 (w), 3178 (w), 2932 (vw), 2781 (vw), 1665 (m), 1597 (m), 1490 (w), 1458 (m), 1437 (w), 1333 (m), 1234 (w), 1110 (w), 1038 (vw), 1009 (w), 995 (w), 980 (vw), 915 (w), 828 (vw), 787 (w), 727 (w), 637 (m), 606 (m), 523 (w) cm^{-1}. – MS (70 eV, EI, 25 °C), m/z (%): 126 (86) [M⁺], 110 (2) [M⁺ – Me], 83 (54) [M⁺ – CONH₂], 68 (68) [M⁺ – Me – CONH₂], 51 (3), 44 (100) [CONH₂].

5-Methylisoxazole-3-carbonitrile (**38-CN**).[205,274] Compound **38-CN** was obtained by reaction of starting material **38-CONH₂** as described for **37-CN** and stirring overnight at room temperature. Yield: 257 mg, 2.38 mmol, 66%; yellow oil. The product could be directly used for synthesis of compound **38-H** without further purification. – ^{1}H NMR (400 MHz, CDCl₃): δ = 2.53 (s, 3 H, CH₃), 6.36 (s, 1 H, H$_{Ar}$) ppm. – ^{13}C NMR (100 MHz, CDCl₃): δ = 12.3 (+, CH₃), 104.1 (+, 4-C$_{Ar}$), 110.2 (C$_q$, CN), 139.6 (C$_q$, 3-C$_{Ar}$), 172.5 (C$_q$, 5-C$_{Ar}$) ppm. – IR (KBr): ṽ = 3442 (vw), 3143 (vw), 2933 (vw), 2258 (vw) [CN], 1767 (vw), 1713 (vw), 1593 (w), 1436 (w), 1409 (w), 1246 (vw), 1129 (vw), 1041 (vw), 1006 (w), 932 (vw), 805 (vw), 700 (vw), 550 (vw), 485 (w) cm^{-1}. – MS (70 eV, EI, 30 °C), m/z (%): 188 (35), 108 (71) [M⁺], 97 (6), 95 (4) [M⁺ – N], 91 (5), 83 (9) [M⁺ – CN], 78 (50) [M⁺ – NO], 68 (9) [M⁺ – Me – CN], 57 (9) [C(Me)ON], 51 (33) [CC(CN)], 43 (100) [C(Me)O].

Benzoxazole-2-carbaldehyde oxime (**39-NOH**).[206] *o*-Aminophenol (1.64 g, 15.0 mmol, 1.00 equiv.) was
treated with 10 M hydrochloric acid to reach pH = 3. At 60 °C hydroxylamine
hydrochloride (1.56 g, 22.5 mmol, 1.50 equiv.) in 4 ml water and chloral hydrate
(3.72, 22.5 mmol, 1.50 equiv.) in 4 ml water were added simultaneously, and
with portions of sodium acetate (7.69 g, 93.8 mmol, 6.25) the pH was kept at 4. While stirring at 60 °C
for 2 h the product precipitated as orange solid. The product was obtained by filtering off from the
reaction mixture, washing with water and drying under vacuum at 50 °C. Yield: 1.68 g, 10.4 mmol,
69%; orange solid. – ^{1}H NMR (300 MHz, CDCl$_3$): δ = 7.37–7.46 (m, 2 H, H$_{Ar}$), 7.58–7.63 (m, 1 H, H$_{Ar}$),
7.79 (d, J = 7.6 Hz, 1 H, H$_{Ar}$), 8.27 (s, 1H, C(NOH)*H*) ppm.

Benzoxazole-2-carbonitrile (**39-CN**). [204,206] The dehydration was performed by adding thionyl chloride
(945 μl, 13.0 mmol, 1.25 equiv.) over 30 min to a solution of oxime **39-NOH**
(1.68 g, 10.4 mmol, 1.00 equiv.) in toluene (60 ml). The reaction mixture was refluxed
at 110 °C until the gas evolution deceased (3 h). After cooling down, the reaction
mixture was stirred with charcoal for 10 min, filtrated and solvents evaporated giving a yellow solid.
This was further purified by sublimation (10^{-2} mbar, 80 °C): 1.22 g, 8.45 mmol, 81%; yellow-white
solid. – ^{1}H NMR (400 MHz, CDCl$_3$): δ = 7.24–7.26 (m, 1 H, H$_{Ar}$), 7.32–7.41 (m, 2 H, H$_{Ar}$),
7.62–7.46 (m, 1 H, H$_{Ar}$) ppm. – ^{13}C NMR (100 MHz, CDCl$_3$): δ = 109.1 (C$_q$, CN), 111.5 (+, 7-C$_{Ar}$),
122.0 (+, 4-C$_{Ar}$), 126.6 (+, 5-C$_{Ar}$), 129.1 (+, 6-C$_{Ar}$), 137.3 (C$_q$, 2-C$_{Ar}$), 139.5 (C$_q$, 3a-C$_{Ar}$), 150.4 (C$_q$,
7a-C$_{Ar}$) ppm. – IR (KBr): ṽ = 3194 (vw), 2249 (vw) [CN], 1723 (vw), 1673 (w), 1604 (vw), 1531 (w),
1475 (w), 1443 (w), 1370 (vw), 1338 (w), 1274 (vw), 1256 (w), 1219 (vw), 1167 (w), 1135 (w),
1102 (w), 992 (w), 950 (w), 893 (w), 816 (w), 760 (m), 747 (m), 685 (w), 630 (vw), 619 (w), 575 (w),
450 (m), 417 (m) cm^{-1}. – MS (70 eV, EI, 30 °C), *m/z* (%): 144 (100) [M$^+$], 116 (6) [M$^+$– CN], 92 (8)
[PhO], 77 (1) [Ph], 64 (26) [M$^+$– Ph], 38 (5) [C-CN].

Benzothiazole-2-carboxamide (**40-CONH₂**).[275,276] The amide **40-CONH₂** was synthesized according to
the method described for compound **37-CONH₂** from benzothiazole-
2-(methyl)carboxylate and stirred at room temperature for 24 h. The formed
precipitate was collected by filtration, washed with water and dried in vacuum.
Purity and substance identity was confirmed by GCMS and ^{1}H NMR, and the product used for synthesis
of compound **40-CN** without further purification. Yield: 751 mg, 4.21 mmol, 84%, off-white powder.
– ^{1}H NMR (250 MHz, DMSO): δ = 7.48 (quind, J = 7.3 Hz, 2 H, H$_{Ar}$), 7.90 (bs, NH$_2$),
8.00–8.05 (m, 2 H, H$_{Ar}$) ppm. – MS (GCMS), *m/z* (%): 178 (62) [M$^+$], 135 (100) [M$^+$– CONH$_2$],
108 (41) [M$^+$– CONH$_2$– CN], 91 (9) [M$^+$– CONH$_2$– CS], 82 (11).

Benzothiazole-2-carbonitrile (**40-CN**).[204,277] The amide reduction to carbonitrile **40-CN** was performed as described for **37-CN** from the starting material **40-CONH₂**. Yield: 534 mg, 3.33 mmol, 95%; orange crystals. – ^{1}H NMR (250 MHz, CDCl₃): δ = 7.60–7.70 (m, 2 H, H$_{Ar}$), 7.96–8.02 (m, 1 H, H$_{Ar}$), 8.20–8.27 (m, 1 H, H$_{Ar}$) ppm. – MS (70 eV, EI, 20 °C), m/z (%): 160 (100) [M⁺], 133 (1) [M⁺ – CN], 116 (<1), 108 (15) [M⁺ – CN – N≡C], 93 (<1) [M⁺ – CN – SC], 82 (2), 69 (3) [SC≡NC], 58 (1) [SC≡N].

4-Methoxypyridine-2-carbonitrile (**43-CN**).[207,278,279] The synthesis of compound **43-CN** was performed by Dr. Daniel Zink, CYNORA GmbH. 4-Methoxypyridine (2.60 g, 23.3 mmol, 1.00 equiv.) was dissolved in 50 ml chloroform, cooled to 0 °C and 3-chloroperoxybenzoic acid (11.5 g, 46.6 mmol, 2.00 equiv.) added portionwise. The reaction was stirred at room temperature for 24 h, diluted with 100 ml chloroform and then potassium carbonate (9.93 g, 93.2 mmol, 4.00 equiv.) was added. After stirring for another 15 min, the reaction mixture was filtrated and the solvent of the filtrate evaporated. The obtained 4-methoxypyridine-*N*-oxide (**43-O**) (2.059 g, 16.5 mmol, 71%) was directly used for synthesis of the carbonitrile **43-CN** without further purification. The intermediate product (2.20 g, 17.6 mmol, 1.00 equiv.) was dissolved in 40 ml dry dichloromethane under nitrogen atmosphere. Trimethylsilyl cyanide (2.49 ml, 18.7 mmol, 1.06 equiv.) and dimethylcarbamoyl chloride (1.73 ml, 18.5 mmol, 1.05 equiv.) were added in small portions and the reaction temperature kept at 25 °C with a water bath. After stirring the reaction for 4 d at room temperature, the solvents were evaporated and the raw product purified by column chromatography (cH/EE 1:1). Yield: 1.83 g, 13.6 mmol, 77%; violet solid. – R_f = 0.49 (cH/EE 1:1). – ^{1}H NMR (500 MHz, CDCl₃): δ = 3.91 (s, 3 H, CH₃), 7.00 (dd, J = 5.8 Hz, 1 H, H$_{Ar}$), 7.21 (d, J = 2.5 Hz, 1 H, H$_{Ar}$), 8.50 (d, J = 5.8 Hz, 1 H, H$_{Ar}$) ppm. – MS (70 eV, EI, 80 °C), m/z (%): 134 (100) [M⁺], 104 (30) [M⁺ – OMe], 91 (7) [M⁺ – OMe – N], 77 (11) [M⁺ – OMe – CN].

4-Piperidiylpyridine-2-carbonitrile (**44-CN**). The synthesis of compound **44-CN** was performed by Dr. Michael Danz, CYNORA GmbH. The starting material 4-chloropyridine-2-carbonitrile (**52-CN**) (1.11 g, 8.00 mmol, 1.00 equiv.) was dissolved in piperidine (10.0 ml, 101 mmol, 12.7 equiv.) and refluxed at 100 °C for 1 h. Sodium carbonate solution (30 ml) was added to the reaction mixture and then extracted with dichloromethane (2 × 40 ml). The combined organic layers were dried over magnesium sulfate and the solvent evaporated. A yellowish-white solid was obtained from purification by column chromatography (cH/EE 1:1). Yield: 894 mg, 4.77 mmol, 60%; yellowish solid. – ^{1}H NMR (500 MHz, CDCl₃): δ = 1.65–1.67 (m, 6 H, CH₂), 3.36–3.38 (m, 4 H, CH₂), 6.72 (dd, J = 6.1, 2.7 Hz, 1 H, H$_{Ar}$),

6.99 (d, J = 2.5 Hz, 1 H, H_{Ar}), 8.22 (d, J = 6.1 Hz, 1 H, H_{Ar}) ppm. – ^{13}C NMR (125 MHz, CDCl$_3$): δ = 24.1 (–, CH$_2$), 25.0 (–, CH$_2$), 47.1 (–, CH$_2$), 110.0 (+, 3-C_{Ar}), 113.0 (+, 5-C_{Ar}), 118.2 (C_q, CN), 134.5 (C_q, 2-C_{Ar}), 151.2 (+, 6-C_{Ar}), 154.4 (C_q, 7-C_{Ar}) ppm. – MS (70 eV, EI, 90 °C), m/z (%): 186 (100) [M$^+$], 172 (3) [M$^+$ – CH$_2$], 158 (7) [M$^+$ – C$_2$H$_4$], 146 (24) [M$^+$ – C$_3$H$_6$], 131 (26) [M$^+$ – C$_4$H$_8$], 117 (2) [M$^+$ – C$_5$H$_{10}$], 103 (11) [M$^+$ – NC$_5$H$_{10}$], 76 (2) [M$^+$ – NC$_5$H$_{10}$ – CN]. – C$_{11}$H$_{13}$N$_3$ (187.11): calc. C 69.89, H 7.04, N 22.23; found C 70.01, H 7.04, N 22.53.

4-Dimethylaminopyridine-2-carbonitrile (47-CN).[210] The starting material 49-CN (498 mg, 3.59 mmol, 1.00 equiv.) and an excess of dimethylamine (40% in water, 2 ml, 15.8 mmol, 4.40 equiv.) were heated to 100 °C for 2 h in a closed vial. After cooling down, the reaction mixture was hydrolyzed with 5 ml water and extracted with dichloromethane (3 × 15 ml). The combined organic phase was dried over magnesium sulfate and solvents evaporated giving the pure product as brownish solid. Yield: 414 mg, 2.81 mmol, 78%. – R_f = 0.07 (cH/EE 1:1). – ^{1}H NMR (500 MHz, CDCl$_3$): δ = 3.06 (s, 6 H, CH$_3$), 6.56 (dd, J = 5.9, 2.8 Hz, 1 H, 5-H_{Ar}), 7.46 (d, J = 2.8 Hz, 1 H, 3-H_{Ar}), 8.15 (d, J = 5.9 Hz, 1 H, 6-H_{Ar}) ppm. MS (GCMS), m/z (%): 10.39: 165 (64) [M$^+$ + H$_2$O], 146 (10) [M$^+$], 122 (100) [M$^+$ – CN], 105 (13) [M$^+$ – NMe$_2$], 94 (14) [M$^+$ – CN – 2 Me], 80 (13) [M$^+$ – CN – NMe$_2$].

6-Dimethylaminopyridine-2-carbonitrile (48-CN). The carbonitrile 48-CN was synthesized from starting material 6-chloropyridine-2-carbonitrile (50-CN) (1.0 g, 7.22 mmol, 1.00 equiv.) and an excess of dimethylamine solution (40% in water, 4.00 ml, 31.6 mmol, 4.38 equiv.) as described for compound 44-CN. After stirring at 100 °C for 4 h, the reaction was cooled down yielding yellow crystals, which were filtered off, washed with diethyl ether and dried in vacuum. Yield: 687 mg, 4.67 mmol, 65%. – R_f = 0.30 (cH/EE 1:1). – ^{1}H NMR (500 MHz, CDCl$_3$): δ = 3.10 (s, 6 H, CH$_3$), 6.66 (d, J = 8.5 Hz, 1 H, 3-H_{Ar}), 7.44 (d, J = 7.2 Hz, 1 H, 5-H_{Ar}), 7.57 (dd, J = 8.5, 7.3 Hz, 1 H, 4-H_{Ar}) ppm. – ^{13}C NMR (125 MHz, CDCl$_3$): δ = 38.1 (+, CH$_3$), 109.2 (+, C_{Ar}), 110.2 (+, C_{Ar}), 138.2 (+, C_{Ar}), 147.4 (C_q, C_{Ar}), 157.9 (C_q, C_{Ar}), 167.9 (C_q, 2-C_{Ar}) ppm. – MS (70 eV, EI), m/z (%): 165 (100) [M$^+$ + H$_2$O], 150 (12) [M$^+$ – Me + H$_2$O], 136 (41) [M$^+$ – 2 Me + H$_2$O], 133 (22) [M$^+$ – Me], 120 (40) [M$^+$ – CN], 105 (48) [M$^+$ – NMe$_2$], 93 (11) [M$^+$ – CN – 2 Me], 78 (14) [M$^+$ – CN – NMe$_2$]. – MS (FAB), m/z (%): 166 (100) [M$^+$ + H + H$_2$O], 149 (6) [M$^+$ – Me + H$_2$O], 133 (4) [M$^+$ – Me], 121 (23) [M$^+$ – CN], 105 (4) [M$^+$ – NMe$_2$], 91 (6) [M$^+$ – CN – 2 Me].

4-Chloropyridine-*N*-oxide (**49-O**).[285, 280] The N-oxide **49-O** was obtained from reaction of 4-chloropyridine hydrochloride (2.25 g, 15.0 mmol, 1.00 equiv.) with 3-chloroperoxybenzoic acid (5.18 g, 30.0 mmol, 2.00 equiv.) as described for **43-O**. The crude product was obtained as yellow solid, which was directly used for synthesis of compound **49-CN** without further purification. Yield: 937 mg, 7.23 mmol, 48%. – MS (GCMS), *m/z* (%): 129 (46) [M$^+$], 113 (100) [M$^+$– O], 78 (86) [M$^+$– O –Cl].

4-Chloropyridine-2-carbonitrile (**49-CN**).[278,210] The N-oxide **49-O** (937 mg, 7.23 mmol, 1.00 equiv.) was converted to the carbonitrile **49-CN** as described for **43-CN**. The reaction was hydrolyzed and the reaction mixture extracted with ethyl acetate (3 × 40 ml). The combined organic phase was dried over magnesium sulfate and solvents evaporated. Purification by column chromatography (cH/EE 5:1) yielded the product as orange solid. Yield: 498 mg, 3.59 mmol, 50%. – R_f = 0.41 (cH/EE 5:1). – ^{1}H NMR (500 MHz, CDCl$_3$): δ = 7.54 (dd, *J* = 5.3, 2.0 Hz, 1 H, 5-H$_{Ar}$), 7.71 (d, *J* = 1.9 Hz, 1 H, 3-H$_{Ar}$), 8.63 (d, *J* = 5.3 Hz, 1 H, 6-H$_{Ar}$).

2-Chloropyridine-*N*-oxide (**50-O**).[281] The N-oxide **50-O** was obtained by reaction of 2-chloropyridine hydrochloride (2.37 ml, 25.0 mmol, 1.00 equiv.) with 3-chloroperoxybenzoic acid (8.63 g, 50.0 mmol, 2.00 equiv.) as described for **43-O**. The crude product was obtained as yellow, oily solid, which was directly used for synthesis of compound **50-CN** without further purification. Yield: 2.05g, 15.8 mmol, 63%. – MS (GCMS), *m/z* (%): 129 (100) [M$^+$], 113 (70) [M$^+$– O], 78 (96) [M$^+$– O –Cl].

6-Chloropyridine-2-carbonitrile (**50-CN**).[278,282] Compound **50-CN** was synthesized as described above for **43-CN** from 2-chloropyridine-*N*-oxide (**50-O**). After reacting for 7 d, the solvent was evaporated and the crude product purified by filtration over Alox (cH/EE 5:1) affording a yellow solid, which still contained dimethylcarbamoyl chloride. Yield: 1.0 g brsm, 7.22 mmol, 46%. – R_f = 0.93 (cH/EE 5:1). – ^{1}H NMR (500 MHz, CDCl$_3$): δ = 7.57 (d, *J* = 8.1 Hz, 1 H, 3-H$_{Ar}$), 7.65 (d, *J* = 7.5 Hz, 1 H, 5-H$_{Ar}$), 7.82 (t, *J* = 7.9 Hz, 1 H, 4-H$_{Ar}$) ppm. – MS (GCMS), *m/z* (%): 138 (87) [M$^+$], 103 (100) [M$^+$– Cl], 76 (32) [M$^+$– CN –Cl].

4-Trifluoromethylpyridine-*N*-oxide (**51-O**).[283] The compound **51-O** was synthesized as described for **43-O** from 4-trifluoromethylpyridine. It could be directly used for the reaction to **51-CN** without further purification. Yield: 811 mg, 4.97 mmol, 73%; white solid. – MS (GCMS), *m/z* (%): 163 (100) [M$^+$], 147 (51) [M$^+$– O], 144 (19) [M$^+$– F], 127 (23) [M$^+$– O – F], 107 (10) [M$^+$– 3 F], 88 (11) [M$^+$– O – 3 F], 78 (17) [M$^+$– O – CF$_3$].

2-Trifluoromethylpyridine-*N*-oxide (**52-O**).[284,285] The N-oxide **52-O** was synthesized as described above

for compound **43-O** from 2-trifluoromethylpyridine. The raw product **52-O** could be directly used for the reaction to **52-CN** without further purification. Yield: 560 mg, 3.43 mmol, 51%; yellow, oily solid. – MS (GCMS), *m/z* (%): 163 (100) [M$^+$], 147 (58) [M$^+$– O], 128 (11) [M$^+$– O – F], 84 (54) [M$^+$– O – 3 F], 78 (99) [M$^+$– O – CF$_3$].

5.2.3 Synthesis of Complexes

The synthesized complexes **7-H-Cu-8**, **7-H-Cu-10**, **7-Cu-8**, **7-Cu-10**, **7-Cu-18**, **7-H-Cu-18**, **7-Cu-19**, **7-H-Cu-19** were already published,[167] and the complexes **5-H-Cu-8**, **6-H-Cu-8**, **6-H-Cu-8-PF$_6$**, **c-(5-H-Cu-9-PF$_6$)**, **5-H-Cu-10**, **6-H-Cu-10**, **c-(5-Cu-8-PF$_6$)**, **c-(6-Cu-8-PF$_6$)**, **c-(6-Cu-9)**, **c-(6-Cu-9-PF$_6$)**, **c-(7-Cu-9)**, **c-(7-Cu-9-PF$_6$)**, **5-Cu-10**, **6-Cu-10** are prepared for publication.[102]

[(2-(Pyridin-2-yl)benzimidazole)Cu(PPh$_3$)$_2$]BF$_4$ (**5-H-Cu-8**). The complex **5-H-Cu-8** was synthesized following **GP2** from ligands 2-(pyridin-2-yl)benzimidazole (**5-H**) and triphenylphosphine (**8**). Yield: 366 mg, 421 μmol, 84%; yellow powder. – MS (FAB), m/z (%): 1128 (<1) [N^N$_2$Cu$_2$P$_2$BF$_4$], 844 (1) [N^NCu$_2$P$_2$], 782 (6) [N^NCuP$_2$], 587 (7) [CuP$_2$], 520 (100) [N^NCuP], 325 (8) [CuP]. – C$_{48}$H$_{39}$BCuF$_4$N$_3$P$_2$ · 0.1 CH$_2$Cl$_2$ (870.16) (precipitated): calc. C 65.75, H 4.50, N 4.78; found C 65.77, H 4.74, N 4.89. Yellow crystals were obtained by layering diethyl ether upon a solution of **5-H-Cu-8** in dichloromethane under nitrogen atmosphere. Found (crystal) C 65.83, H 4.48, N, 4.76. – X-ray diffraction data of single crystals can be found in section 5.4.

[(2-(1,2,3-Triazol-5-yl)pyridine)Cu(PPh$_3$)$_2$]BF$_4$ (**6-H-Cu-8**). The complex **6-H-Cu-8** was synthesized following **GP2** from ligand 2-(1,2,3-triazol-5-yl)pyridine (**6-H**) and triphenylphosphine (**8**). Yield: 299 mg, 364 μmol, 61%, brownish powder. – MS (FAB), m/z (%): 1320 (1) [N^NCu$_2$P$_4$], 1290 (<1) [N^N$_2$Cu$_2$P$_3$BF$_4$], 1264 (<1) [N^N$_2$Cu$_3$P$_3$], 1057 (1) [N^NCu$_2$P$_3$], 1031 (<1) [N^N$_2$Cu$_2$P$_2$BF$_4$], 1006 (<1) [N^N$_2$Cu$_3$P$_2$], 886 (<1) [N^NCu$_2$P$_2$BF$_4$], 796 (9) [N^NCu$_2$P$_2$], 733 (11) [N^NCuP$_2$], 587 (32) [CuP$_2$], 533 (6) [N^NCu$_2$P], 471 (100) [N^NCuP], 325 (23) [CuP]. – C$_{43}$H$_{36}$BCuF$_4$N$_4$P$_2$· 0.1 CH$_2$Cl$_2$ (821.08) (precipitated): calc. C 62.40, H 4.40, N 6.75; found C 62.51, H 4.42, N 6.72. Brown crystals were obtained by layering diethyl ether upon a solution of **6-H-Cu-8** in dichloromethane under nitrogen atmosphere. Found (crystal) C 62.38, H 4.44, N 6.80%.

[(2-(1,2,3-Triazol-5-yl)pyridine)Cu(PPh₃)₂]PF₆ **(6-H-Cu-8-PF₆)**. The synthesis was performed as recorded for **6-H-Cu-8** with the copper salt [Cu(CH₃CN)₄]PF₆. Yield: 360 mg, 409 µmol, 68%; brownish powder. – MS (FAB), m/z (%): 1087 (<1) [N^N₃Cu₂P₂], 1057 (<1) [N^NCu₂P₃], 1004 (<1) [N^N₂Cu₃P₂], 972 (<1) [N^N₃Cu₂P(PF₆)], 849 (1) [Cu₃P], 797 (6) [N^NCu₂P₂], 733 (6) [N^NCuP₂], 587 (38) [CuP₂], 533 (7) [N^NCu₂P], 471 (100) [N^NCuP], 325 (45) [CuP]. – C₄₃H₃₆CuF₆N₄P₃· 0.6 CH₂Cl₂ (879.23): calc. C 56.30, H 4.03, N 6.02; Found C 55.97, H 4.11, N 6.04. – X-ray diffraction data of single crystals can be found in section 5.4.

[(2-(Tetrazol-5-yl)pyridine)Cu(PPh₃)₂]BF₄ **(7-H-Cu-8)**. The complex **7-H-Cu-8** was synthesized following **GP2** from ligand 2-(tetrazol-5-yl)pyridine **(7-H)** and triphenylphosphine **(8)**. Yield: 575 mg, 699 µmol, 70%, white powder. – MS (FAB), *m/z* (%): 1792 (<1) [N^N₂Cu₃P₅], 1628 (<1) [N^N₄Cu₄P₃], 1577 (<1), 1532 (<1) [N^N₂Cu₃P₄], 1420 (<1) [N^N₃Cu₃P₃], 1322 (<1) [N^N₄Cu₂P₂BF₄], 1272 (<1) [N^N₂Cu₃P₃], 1216 (<1) [N^N₃Cu₄P₂], 1159 (<1) [N^N₂Cu₄P₂BF₄], 1060 (1) [N^NCu₂P₃], 1007 (2) [N^N₂Cu₃P₂], 896 (1) [N^N₂Cu₃P₂− 2N₄], 799 (43) [N^NCu₂P₂], 741 (2) [N^NCu₂P₂ – N₄], 735 (2) [N^NCuP₂], 588 (82) [CuP₂], 473 (18) [N^NCuP], 327 (50) [CuP], 264 (15) [P^P]. – C₄₂H₃₅BCuF₄N₅P₂ (822.07): calc. C 61.36, H 4.29, N 8.52; found C 61.53, H 4.31, N 8.45. – X-ray diffraction data of single crystals can be found in section 5.4.

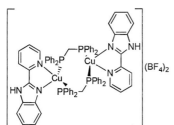

[(2-(Pyridin-2-yl)benzimidazole)₂Cu₂(dppm)₂](BF₄)₂ **(c-(5-H-Cu-9))**. Complex **c-(5-H-Cu-9)** was synthesized according to the general method **GP2** from 2-(pyridin-2-yl)benzimidazole **(5-H)** and dppm **(9)** affording a white suspension when stirring overnight. When suspending in dichloromethane after evaporation of the solvents, only a small fraction was dissolved. Instead the white solid was filtered off, washed with diethyl ether and dichloromethane and dried in vacuum. Yield: 353 mg, 242 µmol, 81%; white solid. – MS (FAB), m/z (%): 1780 (1), 1596 (1) [N^NCu₄P^P₃], 1546 (1) [N^N₃Cu₃P^P₂], 1498 (<1) [N^N₂Cu₄P^P₂BF₄], 1413 (2) [N^N₂Cu₄P^P₂], 1371 (1) [N^N₂Cu₂P^P₂BF₄], 1179 (4) [N^N₃Cu₂P^PBF₄], 1155 (4) [N^N₃Cu₃P^P], 1123 (13) [N^N₃CuP^PBF₄], 1090 (6) [N^N₃Cu₂P^P], 1062 (8), 1048 (4) [N^N₂Cu₃P^PBF₄], 1029 (3) [N^N₂Cu₄P^P], 989 (11) [N^N₂Cu₂P^PBF₄], 931 (3), 915 (11), 901 (12) [N^N₂Cu₂P^P], 831 (8) [CuP^P₂], 795 (7) [N^NCu₂P^PBF₄], 768 (7) [N^NCu₃P^P], 737 (24) [N^NCuP^PBF₄], 704 (41) [N^NCu₂P^P], 678 (13),

642 (34) [N^NCuP^P], 447 (100) [CuP^P]. – $C_{74}H_{62}B_2Cu_2F_8N_6P_4 \cdot CH_2Cl_2$ (1459.94): calc. C 60.61, H 4.27, N 5.72; found C 60.42, H 4.24, N 5.71. – X-ray diffraction data of single crystals can be found in section 5.4.

[(2-(Pyridin-2-yl)benzimidazole)Cu(DPEPhos)]BF$_4$ (**5-H-Cu-10**). Compound **5-H-Cu-10** was synthesized following **GP2** from 2-(pyridin-2-yl)-benzimidazole (**5-H**) and DPEPhos (**10**). Yield: 400 mg, 452 μmol, 90%; yellow powder. – MS (FAB), m/z (%): 2190 (<1) [N^N$_2$Cu$_3$P^P$_3$], 1982 (<1) [N^N$_3$Cu$_5$P^P$_2$], 1679 (<1) [N^N$_2$Cu$_2$P^P$_2$BF$_4$], 1591 (<1) [N^N$_2$Cu$_2$P^P$_2$], 1397 (1) [N^NCu$_2$P^P$_2$], 1154 (<1), 796 (15) [N^NCuP^P], 601 (100) [CuP^P]. – $C_{48}H_{37}BCuF_4N_3OP_2 \cdot 0.1\ CH_2Cl_2$ (884.14): calc. C 64.72, H 4.20, N 4.71; found: C 64.67, H 4.45, N 4.55. – X-ray diffraction data of single crystals can be found in section 5.4.

[(2-(1,2,3-Triazol-5-yl)pyridine)Cu(DPEPhos)]BF$_4$ (**6-H-Cu-10**). The complex **6-H-Cu-10** was synthesized following **GP2** from ligand 2-(1,2,3-triazol-5-yl)-pyridine (**6-H**) and DPEPhos (**10**). The compound was obtained by recrystallization from a dichloromethane solution layered with diethyl ether under nitrogen atmosphere affording brownish crystals. Yield: 349 mg, 418 μmol, 84%; brownish powder. – MS (FAB), m/z (%): 1582 (1) [N^N$_2$Cu$_3$P^P$_2$BF$_4$], 1556 (<1) [N^N$_2$Cu$_3$P^P$_2$], 1349 (5) [N^NCu$_2$P^P$_2$], 1290 (<1) [N^NCuP^P$_2$], 1273 (<1), 1223 (<1) [N^N$_3$Cu$_4$P^P], 955 (<1) [N^N$_2$Cu$_2$P^P], 811 (5) [N^NCu$_2$P^P], 747 (18) [N^NCuP^P], 601 (100) [CuP^P]. – $C_{43}H_{34}BCuF_4N_4OP_2 \cdot 0.2\ CH_2Cl_2$ (835.05): calc. C 60.90, H 4.07, N 6.58; found C 61.05, H 4.42, N 6.37. – X-ray diffraction data of single crystals can be found in section 5.4.

[(2-(Tetrazol-5-yl)pyridine)Cu(DPEPhos)][(5-(pyridin-2-yl)tetrazolate)Cu(DPEPhos)]BF$_4$ (**7-H-Cu-10**). The complex **7-H-Cu-10** was synthesized following **GP2** from ligand 2-(tetrazol-5-yl)pyridine (**7-H**) and DPEPhos (**10**). Yield: 300 mg, 189 μmol, 57%, light yellow powder. – MS (FAB), m/z (%): 2097 (<1) [N^N$_2$Cu$_3$P^P$_3$], 1559 (1) [N^N$_2$Cu$_3$P^P$_2$], 1350 (18) [N^NCu$_2$P^P$_2$], 1291 (<1) [N^NCuP^P$_2$], 1139 (<1) [CuP^P$_2$], 997 (<1), 812 (4) [N^NCu$_2$P^P], 749 (1) [N^NCuP^P], 691 (1) [N^NCuP^P – N$_4$], 602 (100) [CuP^P]. – $C_{84}H_{65}BCu_2F_4N_{10}O_2P_4 \cdot 2\ H_2O$ (1584.27): calc. C 62.27,

H 4.29, N 8.64, found C 62.27, H 4.31, N 8.11. – X-ray diffraction data of single crystals can be found in section 5.4.

[(2-(Pyridin-2-yl)benzimidazolate)$_3$Cu$_4$(PPh$_3$)$_6$]PF$_6$ (c-(5-Cu-8-PF$_6$)). The pyridine-amine ligand 2-(2-pyridyl)benzimidazole (**5-H**) (195 mg, 1.00 mmol, 1.00 equiv.) was dissolved in 5 ml dry THF under nitrogen atmosphere and 0.4 ml *n*-butyllithium solution (2.5 M in *n*-hexanes, 1.00 mmol, 1.00 equiv.) was added dropwise at 0 °C. After stirring for 30 min, a second solution made from [Cu(CH$_3$CN)$_4$]PF$_6$ (373 mg, 1.00 mmol, 1.00 equiv.) and PPh$_3$ (**8**) (525 mg, 2.00 mmol, 2.00 equiv.) in 10 ml dry THF under nitrogen was added. The reaction mixture was stirred overnight at room temperature, and afterwards precipitated in 100 ml diethyl ether, washed and filtrated. The solid was suspended in dichloromethane and filtrated. The solvents of the filtrate were evaporated under reduced pressure giving a yellow solid. Yield: 525 mg, 205 µmol, 82%, yellow powder. The compound was recrystallized from a dichloromethane solution by layering with diethyl ether under nitrogen atmosphere affording yellow crystals. – MS (FAB), m/z (%): 1620 (1) [N$^{\wedge}$N$_3$Cu$_4$P$_3$], 1361 (5) [N$^{\wedge}$N$_3$Cu$_4$P$_2$], 1104 (70) [N$^{\wedge}$N$_2$Cu$_3$P$_2$], 845 (53) [N$^{\wedge}$NCu$_2$P$_2$], 780 (5) [N$^{\wedge}$NCuP$_2$], 579 (91) [N$^{\wedge}$NCu$_2$P], 520 (100) [N$^{\wedge}$NCuP], 325 (70) [CuP], 258 (98) [N$^{\wedge}$NCu], 183 (51). – C$_{144}$H$_{114}$Cu$_4$F$_6$N$_9$P$_7$ · 0.3 CH$_2$Cl$_2$ (2555.50): calc. C 67.15, H 4.48, N 4.88; found C 67.06, H 4.56, N 4.83. – X-ray diffraction data of single crystals can be found in section 5.4.[XVIII]

[(5-(Pyridin-2-yl)1,2,3-triazolate)$_2$Cu$_3$(PPh$_3$)$_6$]PF$_6$ (c-(6-Cu-8-PF$_6$)). The complex **c-(6-Cu-8-PF$_6$)** was synthesized following **GP3** from complex **6-H-Cu-8-PF$_6$**. Yield: 152 mg, 69.1 µmol, 62%; white powder. The compound was further purified by recrystallization of a dichloromethane solution layered with diethyl ether under nitrogen atmosphere.

– MS (FAB), m/z (%): 1946 (<1) [N$^{\wedge}$N$_4$Cu$_5$P$_4$], 1694 (<1), 1685 (<1) [N$^{\wedge}$N$_4$Cu$_5$P$_3$], 1635 (1), 1561 (<1) [N$^{\wedge}$N$_4$Cu$_3$P$_3$], 1476 (<1) [N$^{\wedge}$N$_3$Cu$_4$P$_3$], 1268 (2) [N$^{\wedge}$N$_3$Cu$_3$P$_3$], 1215 (1) [N$^{\wedge}$N$_3$Cu$_4$P$_2$], 1160 (<1) [N$^{\wedge}$N$_3$Cu$_3$P$_2$], 1060 (2) [N$^{\wedge}$NCu$_2$P$_3$], 1006 (16) [N$^{\wedge}$N$_2$Cu$_3$P$_2$], 950 (2) [N$^{\wedge}$N$_2$Cu$_2$P$_2$], 898 (3), 796 (52) [N$^{\wedge}$NCu$_2$P$_2$], 744 (56) [N$^{\wedge}$N$_2$Cu$_3$P], 588 (36) [CuP$_2$], 533 (100) [N$^{\wedge}$NCu$_2$P], 471 (18) [N$^{\wedge}$NCuP], 325 (67) [CuP].

[XVIII] The pyridine-amide ligand acts as tridentate ligand here, but is denoted as N$^{\wedge}$N in the MS-FAB analysis due to consistency.

– $C_{122}H_{102}Cu_3F_6N_8P_7 \cdot 0.5$ CH_2Cl_2 (2199.60): calc. C 65.56, H 4.63, N 4.99; found C 65.48, H 4.72, N 5.08. – X-ray diffraction data of single crystals can be found in section 5.4.[XVIII]

[(5-(Pyridin-2-yl)tetrazolate)Cu(PPh$_3$)$_2$]BF$_4$ (**7-Cu-8**). Synthesis of the complex **7-Cu-8** was accomplished with the ligands 2-(tetrazol-5-yl)pyridine (**7-H**) and triphenylphosphine (**8**). Purification of **7-Cu-8** was difficult due to ligand exchange reactions during precipitation and stirring in diethyl ether, which resulted in the reproducible formation of the cluster [Cu$_2$(PPh$_3$)$_2$(PyrTet)$_2$], proven by elemental analysis. Therefore, complex **7-Cu-8** had to be obtained by reaction of the cationic species **7-H-Cu-8** with potassium hydroxide following **GP3**, evaporation of the solvent, extraction with dichloromethane and evaporation of dichloromethane. However, small impurities of **7-Cu-8** with PPh$_3$ were not preventable with this method. Yield: 113 mg, 138 µmol, 70%, white powder. – MS (FAB), *m/z* (%): 1742 (<1) [N^N$_3$Cu$_4$P^P$_4$], 1530 (<1) [N^N$_2$CuP^P$_{43}$], 1271 (<1) [N^N$_2$Cu$_3$P^P$_3$], 1218 (<1), 1098 (<1), 1061 (1) [N^NCu$_2$P^P$_3$], 1008 (3) [N^N$_2$Cu$_3$P^P$_2$], 956 (<1) [N^N$_2$Cu$_4$P^P$_2$ – 2 N$_4$], 891 (<1) [N^N$_2$Cu$_3$P^P$_2$ – 2 N$_4$], 799 (14) [N^NCu$_2$P^P$_2$], 746 (1) [N^NCu$_2$P^P$_2$ – N$_4$], 706 (2) [N^NCuP^P$_2$ – N$_2$], 691 (2) [N^NCuP^P$_2$ – N$_3$], 681 (2) [N^NCu$_3$P^P$_2$ – N$_4$], 588 (58) [CuP^P$_2$], 473 (18) [N^NCuP^P], 327 (80) [CuP^P], 264 (100) [P^P]. – C$_{42}$H$_{34}$CuN$_5$P$_2$ · 0.33 PPh$_3$ (734.26): calc. C 70.16, H 4.78, N 8.52; found C 70.22, H 4.75, N 8.38. – X-ray diffraction data of single crystals can be found in section 5.4.

[(5-(Pyridin-2-yl)1,2,3-triazolate)$_2$Cu$_4$(dppm)$_4$](BF$_4$)$_2$ (**c-(6-Cu-9)**). The complex **c-(6-Cu-9)** was synthesized following **GP2** from ligand 2-(1,2,3-triazol-5-yl)pyridine (**6-H**) and dppm (**9**). Yield: 177 mg, 78.5 µmol, 78%; white powder. White crystals were obtained from layering diethyl ether upon a solution of **c-(6-Cu-9)** in dichloromethane under nitrogen atmosphere. – MS (FAB), m/z (%): 2167 (1) [N^N$_3$Cu$_3$P^P$_4$], 1850 (2) [N^N$_3$Cu$_4$P^P$_3$], 1784 (18) [N^N$_3$Cu$_3$P^P$_3$], 1465 (3) [N^N$_3$Cu$_4$P^P$_2$], 1399 (30) [N^N$_3$Cu$_3$P^P$_2$], 1248 (3) [N^N$_2$Cu$_3$P^P$_2$], 1041 (8) [N^NCu$_2$P^P$_2$], 948 (7) [N^N$_3$Cu$_2$P^P], 927 (9) [N^N$_2$Cu$_4$P^P], 865 (70) [N^N$_2$Cu$_3$P^P], 657 (100) [N^NCu$_2$P^P], 447 (34) [CuP^P].

– $C_{114}H_{98}B_2Cu_4F_8N_8P_8 \cdot 0.5$ CH_2Cl_2 (2255.64): calc. C 59.84, H 4.34, N 4.88, found C 60.06, H 4.40, N 4.98. – X-ray diffraction data of single crystals can be found in section 5.4.[XIX]

[(5-(Pyridin-2-yl)1,2,3-triazolate)₂Cu₄(dppm)₄](PF₆)₂ (c-(6-Cu-9-PF₆)). The synthesis was performed

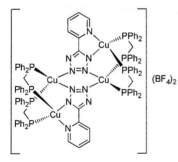

as recorded for **c-(6-Cu-9)** with the copper salt [Cu(CH₃CN)₄]PF₆. Yield: 180 mg, 75.9 μmol, 76%; white powder. Crystallization of its dichloromethane solution by slow diffusion of diethyl ether gave colorless crystals. – MS (FAB), m/z (%): 2234 (<1) [N^N₃Cu₄P^P₄], 2166 (<1) [N^N₃Cu₃P^P₄], 1850 (4) [N^N₃Cu₄P^P₃], 1819 (1) [N^N₃Cu₄P^P₃ – N₂], 1784 (2) [N^N₃Cu₃P^P₃], 1743 (<1), 1697 (<1) [N^N₂Cu₄P^P₃], 1465 (4) [N^N₃Cu₄P^P₂], 1400 (3) [N^N₃Cu₃P^P₂], 1331 (3) [N^N₃Cu₂P^P₂], 1312 (6) [N^N₂Cu₄P^P₂], 1249 (2) [N^N₂Cu₃P^P₂], 1041 (15) [N^NCu₂P^P₂], 929 (6) [N^N₂Cu₄P^P], 865 (27) [N^N₂Cu₃P^P], 849 (6) [N^N₂Cu₃P^P – N], 831 (3) [N^N₂Cu₃P^P – N₂], 655 (54) [N^NCu₂P^P], 447 (16) [CuP^P]. – $C_{114}H_{98}Cu_4F_{12}N_8P_{10} \cdot 0.5$ CH_2Cl_2 (2371.96): calc. C 56.96, H 4.13, N 4.64, found C 56.83, H 4.44, N 4.74. – X-ray diffraction data of single crystals can be found in section 5.4.[XIX]

[(5-(Pyridin-2-yl)tetrazolate)₂Cu₄(dppm)₄](BF₄)₂ (c-(7-Cu-9)). The complex **c-(7-Cu-9)** was

synthesized following **GP2** from ligand 2-(tetrazol-5-yl)-pyridine (**7-H**) and dppm (**9**). Yield: 176 mg, 78.0 μmol, 78%; white powder. Crystallization by layering diethyl ether upon a solution of **x** in dichloromethane under nitrogen atmosphere gave greenish crystals. – MS (FAB), m/z (%): 2890 (<1) [N^N₄Cu₆P^P₅], 2356 (<1) [N^N₃Cu₆P^P₄], 2172 (<1) [N^N₃Cu₃P^P₄], 1852 (1) [N^N₃Cu₄P^P₃], 1786 (18) [N^N₃Cu₃P^P₃], 1545 (1) [N^N₄Cu₃P^P₂], 1467 (3) [N^N₃Cu₄P^P₂], 1402 (9) [N^N₃Cu₃P^P₂], 1333 (23) [N^N₃Cu₂P^P₂], 1250 (6) [N^N₂Cu₃P^P₂], 1043 (13) [N^NCu₂P^P₂], 867 (52) [N^N₃Cu₃P^P], 658 (63) [N^NCu₂P^P], 552 (72) [N^NCuP^P – N₃], 447 (100) [CuP^P].

[XIX] The pyridine-amide ligand acts as tetradentate ligand here, but is denoted as N^N in the MS-FAB analysis due to consistency.

– $C_{112}H_{96}B_2Cu_4F_8N_{10}P_8 \cdot 2\ CH_2Cl_2$ (2257.61): calc. C 56.41, H 4.15, N 5.77; found C 56.45, H 4.21, N 5.78. – X-ray diffraction data of single crystals can be found in section 5.4.[XIX]

[(5-(Pyridin-2-yl)tetrazolate)$_2$Cu$_4$(dppm)$_4$](PF$_6$)$_2$ (c-(7-Cu-9-PF$_6$)). The synthesis was performed as

recorded for c-(7-Cu-9) with the copper salt [Cu(CH$_3$CN)$_4$]PF$_6$. Yield: 363 mg, 153 μmol, 87%; greenish powder. Colorless crystals were obtained by layering diethyl ether upon a solution of c-(7-Cu-9-PF$_6$) in dichloromethane under nitrogen atmosphere. – MS (FAB), m/z (%): 2228 (2) [N^N$_3$Cu$_4$P^P$_4$], 1844 (12) [N^N$_3$Cu$_4$P^P$_3$], 1459 (13) [N^N$_3$Cu$_4$P^P$_2$], 1333 (13) [N^N$_3$Cu$_2$P^P$_2$], 1313 (4) [N^N$_2$Cu$_4$P^P$_2$], 1249 (4) [N^N$_2$Cu$_3$P^P$_2$], 1148 (1), 1128 (2) [N^N$_2$CuP^P$_2$], 1040 (20) [N^NCu$_2$P^P$_2$], 867 (39) [N^N$_2$Cu$_3$P^P], 656 (72) [N^NCu$_2$P^P], 552 (73) [N^NCuP^P – N$_3$], 447 (100) [CuP^P]. – $C_{112}H_{96}Cu_4F_{12}N_{10}P_{10} \cdot CH_2Cl_2$ (2373.93): calc. C 55.20, H 4.02, N 5.70; found C 54.98, H 4.29, N, 5.71. – X-ray diffraction data of single crystals can be found in section 5.4.[XIX]

[(2-(Pyridin-2-yl)benzimidazolate)Cu(DPEPhos)] (5-Cu-10). The complex 5-Cu-10 was synthesized

following GP4 from 2-(2-pyridyl)benzimidazole (5-H) and DPEPhos (10). Yield: 136 mg, 171 μmol, 34%; yellow powder. – MS (FAB), m/z (%): 1398 (6) [N^NCu$_2$P^P$_2$], 1054 (1) [N^N$_2$Cu$_2$P^P], 861 (1) [N^NCuP^P], 795 (15) [N^NCuP^P], 718 (1) [(N^N – Pyr)CuP^P], 601 (100) [CuP^P]. – $C_{48}H_{36}CuN_3OP_2 \cdot 0.2\ CH_2Cl_2$ (796.31): calc. C 71.17, H 4.51, N 5.17; found C 71.12, H 4.66, N 5.01. – X-ray diffraction data of single crystals can be found in section 5.4.

[(5-(Pyridin-2-yl)1,2,3-triazolate)Cu(DPEPhos)] (6-Cu-10). The complex 6-Cu-10 was synthesized

following GP4 from ligand 2-(1,2,3-triazol-5-yl)pyridine (6-H) and DPEPhos (10). Yield: 218 mg, 292 μmol, 73%; off-white powder. – MS (FAB), m/z (%): 1349 (3) [N^NCu$_2$P^P$_2$], 811 (4) [N^NCuP^P], 747 (9) [N^NCuP^P], 601 (65) [CuP^P]. – $C_{43}H_{33}CuN_4OP_2 \cdot 0.4\ CH_2Cl_2$ (747.24): calc. C 66.72, H 4.36, N 7.17; found C 66.60, H 4.70, N 6.95.

[(5-(Pyridin-2-yl)tetrazolate)Cu(DPEPhos)] (**7-Cu-10**). a) The complex **7-Cu-10** was synthesized

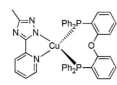

following **GP3** from **7-H-Cu-10**. Yield: 203 mg, 271 μmol, 68%, white powder. – MS (FAB), m/z (%): 2096 (<1) [N^N$_2$Cu$_3$P^P$_3$], 1559 (1) [N^N $_2$Cu$_3$P^P$_2$], 1350 (6) [N^NCu$_2$P^P$_2$], 1139 (2) [CuP^P$_2$], 812 (1) [N^NCu$_2$P^P], 749 (4) [N^NCuP^P], 691 (3) [N^NCuP^P – N$_4$], 602 (100) [CuP^P]. – C$_{42}$H$_{32}$CuN$_5$OP$_2$ (748.28): calc. C 67.42, H 4.31, N 9.36; found C 67.26, H 4.36, N 9.28. b) The complex **7-Cu-10** was synthesized following **GP4** from ligand 2-(tetrazol-5-yl)pyridine (**7-H**) and DPEPhos (**10**). Yield: 1.82 g, 2.43 mmol, 61%; white powder. – C$_{42}$H$_{32}$CuN$_5$OP$_2$ (748.28): calc. C 67.42, H 4.31, N 9.36; found C 67.25, H 4.40, N 9.06. – X-ray diffraction data of single crystals can be found in section 5.4.

[(5-(Pyridin-2-yl)-3-methyl-1,2,4-triazolate)Cu(DPEPhos)] (**16-Cu-10**). The complex **16-Cu-10** was

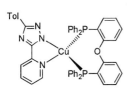

synthesized from ligand **16** and DPEPhos (**10**) following **GP4**, for further purification its dichloromethane solution was precipitated again in diethyl ether/n-hexane (70 ml, 4:3) under nitrogen atmosphere giving a yellow solid. Yield: 704 mg, 924 μmol, 37%. – MS (FAB), m/z (%): 1362 (5) [N^NCu$_2$P^P$_2$], 1046 (<1) [N^N$_2$Cu$_3$P^P], 982 (1) [N^N$_2$Cu$_2$P^P], 886 (<1) [N^NCu$_3$P^P], 825 (10) [N^NCu$_2$P^P], 761 (21) [N^NCuP^P], 682 (1) [N^NCuP^P – C$_2$N$_3$CH$_3$], 601 (100) [CuP^P]. – C$_{44}$H$_{35}$CuN$_4$OP$_2$ · 0.2 CH$_2$Cl$_2$ (761.27): calc. C 68.21, H 4.58, N 7.20; found C 68.20, H 4.64, N 6.98.

[(3-(p-Tolyl)-5-(pyridin-2-yl)triazolate)Cu(DPEPhos)] (**17-Cu-10**). The complex was synthesized following **GP4** from ligand **17** and DPEPhos (**10**). The pure complex was obtained by precipitation of its dichloromethane solution in diethyl ether/n-hexane (120 ml, 5:1). Yield: 535 mg, 639 μmol, 64%; yellow powder. – MS (FAB), m/z (%): 1439 (3) [N^NCu$_2$P^P$_2$], 1199 (<1) [N^N$_2$Cu$_3$P^P], 1136 (1) [CuP^P$_2$], 901 (14) [N^NCu$_2$P^P], 837 (11) [N^NCuP^P], 759 (1) [N^NCuP^P – Pyr], 707 (<1), 601 (100) [CuP^P]. – C$_{50}$H$_{39}$CuN$_4$OP$_2$ · 0.2 CH$_2$Cl$_2$ (837.36): calc. C 70.57, H 4.65, N 6.56; found C 70.86, H 4.66, N 6.43.

[(5-(Pyridin-2-yl)tetrazolate)Cu(PTEPhos)] (**7-Cu-18**). The complex **7-Cu-18** was synthesized following **GP3** from **7-H-Cu-18**. Yield: 33 mg, 42.5 μmol, 61%, white powder. – MS (FAB), m/z (%): 2181 (<1) [N^N$_2$Cu$_3$P^P$_3$], 1617 (<1) [N^N$_2$Cu$_3$P^P$_2$], 1554 (<1) [N^N$_2$Cu$_2$P^P$_2$], 1406 (2) [N^NCu$_2$P^P$_2$], 1197 (<1) [N^N$_3$Cu$_3$P^P], 840 (<1) [N^NCu$_2$P^P], 778 (3) [N^NCuP^P], 720 (2) [N^NCuP^P – N$_4$], 630 (55) [CuP^P]. – C$_{44}$H$_{36}$CuN$_5$OP$_2$ (776.28): calc. C 68.08, H 4.67, N 9.02; found C 67.88, H 4.74, N 8.81. – X-ray diffraction data of single crystals can be found in section 5.4.

[(2-(Tetrazol-5-yl)pyridine)Cu(PTEPhos)]BF$_4$ (**7-H-Cu-18**). The complex **7-H-Cu-18** was synthesized following **GP2** from ligand 2-(tetrazol-5-yl)pyridine (**7-H**) and PTEPhos (**18**). Yield: 100 mg, 116 μmol, 46%, yellow powder. – MS (FAB), m/z (%): 2071 (<1), 1616 (3) [N^N$_2$Cu$_3$P^P$_2$], 1505 (<1) [N^N$_2$Cu$_3$P^P$_2$ – 2 N$_4$], 1406 (10) [N^NCu$_2$P^P$_2$], 1297 (<1) [N^N$_3$Cu$_5$P^P – N$_2$], 1220 (<1) [N^N$_3$Cu$_2$P^PBF$_4$], 839 (8) [N^NCu$_2$P^P], 720 (2) [N^NCuP^P – N$_4$], 630 (100) [CuP^P]. – C$_{44}$H$_{37}$BCuF$_4$N$_5$OP$_2$ (864.10): calc. C 61.16, H 4.32, N 8.10; found C 61.10, H 4.35, N 7.84.

[(5-(Pyridin-2-yl)tetrazolate)Cu(XantPhos)] (**7-Cu-19**). a) The complex **7-Cu-19** was synthesized following **GP3** from **7-H-Cu-19**. Yield: 62 mg, 78.7 μmol, 37%, yellowish white powder. – MS (FAB), m/z (%): 1431 (<1) [N^NCu$_2$P^P$_2$], 1020 (<1), 886 (1) [N^N$_2$Cu$_2$P^P – 2N$_4$], 788 (1) [N^NCuP^P], 732 (1) [N^NCuP^P – N$_4$], 642 (6) [CuP^P], 604 (6), 578 (2) [P^P]. – C$_{45}$H$_{36}$CuN$_5$OP$_2$ (788.29): calc. C 68.56, H 4.60, N 8.88; found C 68.45, H 4.74, N 8.57. b) The complex **7-Cu-19** was synthesized following **GP4** from ligand 2-(tetrazol-5-yl)pyridine (**7-H**) and XantPhos (**19**). Yield: 297 mg, 0.377 mmol, 75%; yellowish powder. – C$_{45}$H$_{36}$CuN$_5$OP$_2$ · 0.4 CH$_2$Cl$_2$ (788.29): calc. C 66.31, H 4.51, N 8.52; found C 66.08, H 4.51, N 8.41. – X-ray diffraction data of single crystals can be found in section 5.4.

[(2-(Tetrazol-5-yl)pyridine)Cu(XantPhos)]BF$_4$ (**7-H-Cu-19**). The complex **7-H-Cu-19** was synthesized following **GP2** from ligand 2-(tetrazol-5-yl)pyridine (**7-H**) and XantPhos (**19**). Yield: 270 mg, 308 μmol, 77%, yellow powder. – MS (FAB), m/z (%): 2217 (<1) [N^N$_2$Cu$_3$P^P$_3$], 1641 (2) [N^N$_2$Cu$_3$P^P$_2$], 1430 (10) [N^NCu$_2$P^P$_2$], 1372 (<1) [N^NCuP^P$_2$], 958 (<1) [N^N$_2$Cu$_2$P^P – N$_3$], 851 (<1)

[N^NCu2P^P], 789 (1) [N^NCuP^P], 732 (2) [N^NCuP^P – N4], 642 (100) [CuP^P], 627 (7) [CuP^P – CH3]. – C45H37BCuF4N5OP2 (876.11): calc. C 61.69, H 4.26, N 7.99; found C 61.92, H 4.31, N 7.59. – The structure could be confirmed by X-ray diffraction, but no further analysis was possible due to very weak diffraction.

[(5-(Pyridin-2-yl)tetrazolate)Cu(DPTPhos)] (**7-Cu-21**). The complex was synthesized following **GP4**

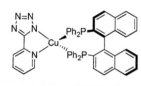

from starting materials ligand 2-(tetrazol-5-yl)pyridine (**7-H**) and DPTPhos (**21**). Yield: 213 mg, 279 µmol, 56%; yellowish powder. – MS (FAB), *m/z* (%): 1380 (12) [N^NCu2P^P2], 1171 (<1) [CuP^P2], 1121 (<1) [N^N3Cu2P^P], 973 (1) [N^N2Cu2P^P], 828 (5) [N^NCu2P^P], 802 (3) [N^NCu2P^P – N2], 763 (22) [N^NCuP^P], 706 (3)

[N^NCuP^P – N4], 663 (3), 617 (100) [CuP^P], 540 (3) [P^P]. – C42H32CuN5P2S · 0.5 CH2Cl2 (764.31): calc. C 63.27, H 4.12, N 8.68, S 3.97; found C 63.21, H 4.36, N 8.36, S 3.92.

[(5-(Pyridin-2-yl)tetrazolate)Cu(BINAP)] (**7-Cu-23**). The complex was synthesized following **GP4** from ligand 2-(tetrazol-5-yl)pyridine (**7-H**) and R-BINAP (**23**). Yield: 94 mg, 113 µmol, 28%; yellowish powder. – MS (FAB), *m/z* (%): 1373 (<1) [Cu2P^P2], 985 (<1) [N^N2Cu2P^P – N4], 965 (<1) [N^N2CuP^P – N], 896 (1) [N^NCu2P^P], 832 (6) [N^NCuP^P], 750 (<1) [Cu2P^P], 685 (10) [CuP^P], 613 (2), 437 (10) [P^P – PPh2].

– C50H36CuN5P2 · 0.2 CH2Cl2 (832.35): calc. C 70.99, H 4.32, N 8.25; found C 70.75, H 4.70, N 7.95. – X-ray diffraction data of single crystals can be found in section 5.4.

[(5-(Pyrazin-2-yl)tetrazolate)Cu(DPEPhos)] (**28-Cu-10**). The complex was synthesized according to **GP4** from 2-(tetrazol-5-yl)pyrazine (**28-H**) and DPEPhos (**10**). Yield: 319 mg, 426 µmol, 85%, yellow solid. – MS (FAB), *m/z* (%): 2310 (<1) [N^N3Cu3P^P3], 2099 (1) [N^N2Cu3P^P3], 1562 (2) [N^N2Cu3P^P2], 1351 (15) [N^NCu2P^P2], 1230 (<1) [N^N3Cu4P^P], 1140 (3) [CuP^P2], 813 (4) [N^NCu2P^P], 750 (2) [N^NCuP^P], 694 (6) [N^NCuP^P – N4],

602 (100) [CuP^P]. – C41H31CuN6OP2 · 0.2 CH2Cl2 (749.23): calc. C 64.58, H 4.13, N 10.97; found C 64.74, H 4.24, N 10.68. – X-ray diffraction data of single crystals can be found in section 5.4.

[(5-(Quinolin-2-yl)tetrazolate)Cu(DPEPhos)] (**29-Cu-10**). The complex was synthesized following

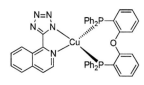

GP4 from tetrazole **29-H** and DPEPhos (**10**). Yield: 218 mg, 273 µmol, 55%; yellow powder. – MS (FAB), *m/z* (%): 1400 (17) [N^NCu$_2$P^P$_2$], 1322 (<1) [N^N$_3$Cu$_3$P^P], 1141 (1) [CuP^P$_2$], 1058 (2) [N^N$_2$Cu$_2$P^P], 954 (<1) [N^N$_2$CuP^P – N$_3$], 862 (3) [N^NCu$_2$P^P], 798 (8) [N^NCuP^P], 741 (9) [N^NCuP^P – N$_4$], 664 (1) [Cu$_2$P^P], 601 (100) [CuP^P]. – C$_{46}$H$_{34}$CuN$_5$OP$_2$ · 0.2 CH$_2$Cl$_2$ (798.29): calc. C 68.08, H 4.25, N 8.59; found C 68.33, H 4.53, N 8.12. – X-ray diffraction data of single crystals can be found in section 5.4.

[(5-(Isoquinolin-1-yl)tetrazolate)Cu(DPEPhos)] (**30-Cu-10**). Complex **30-Cu-10** was synthesized

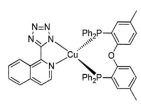

according to **GP4** from ligand 1-(tetrazol-5-yl)isoquinoline (**30-H**) and DPEPhos (**10**). Yield: 64.0 mg, 802 µmol, 23%; yellow solid. – MS (FAB), *m/z* (%): 1400 (13) [N^NCu$_2$P^P$_2$], 1139 (2) [CuP^P$_2$], 1058 (<1) [N^N$_2$Cu$_2$P^P], 951 (1) [N^N$_2$Cu$_2$P^P – N$_8$], 860 (4) [N^NCu$_2$P^P], 798 (27) [N^NCuP^P], 741 (9) [N^NCuP^P – N$_4$], 665 (3) [Cu$_2$P^P], 601 (100) [CuP^P]. – C$_{46}$H$_{34}$CuN$_5$OP$_2$ · 0.1 CH$_2$Cl$_2$ (798.29): calc. C 68.63, H 4.27, N 8.68; found C 68.52, H 4.48, N 8.33. – X-ray diffraction data of single crystals can be found in section 5.4.

[(5-(Isoquinolin-1-yl)tetrazolate)Cu(PTEPhos)] (**30-Cu-18**). Complex **30-Cu-18** was synthesized

according to **GP4** from ligand 1-(tetrazol-5-yl)isoquinoline (**30-H**) and PTEPhos (**18**). A yellow solid was obtained, which was further purified by repeated precipitation from its dichloromethane solution in diethyl ether. Yield: 86.0 mg, 104 µmol, 21%; yellow solid. – MS (FAB), *m/z* (%): 2282 (<1) [N^N$_2$Cu$_3$P^P$_3$], 1716 (2) [N^N$_2$Cu$_3$P^P$_2$], 1457 (17) [N^NCu$_2$P^P$_2$], 1285 (<1) [N^N$_3$Cu$_2$P^P], 1199 (<1) [CuP^P$_2$], 889 (6) [N^NCu$_2$P^P], 827 (1) [N^NCuP^P], 770 (4) [N^NCuP^P – N$_4$], 695 (2) [Cu$_2$P^P], 629 (100) [CuP^P]. – C$_{48}$H$_{38}$CuN$_5$OP$_2$ (826.36): calc. C 69.77, H 4.64, N 8.47; found C 69.51, H 4.57, N 8.38. – The structure could be confirmed by X-ray diffraction, but no further analysis was possible due to very weak diffraction.

[(5-([2](1,4)benzo[2](2,5)pyridinophan-13-yl)tetrazolate)Cu(DPEPhos)] (**31-Cu-10**). The complex was synthesized following **GP4** from (rac)-13-(tetrazol-5-yl)-[2](1,4)benzo[2](2,5)pyridinophane (**31-H**) and DPEPhos (**10**), and was already published by Dr. Joshua Kramer.[203] Due to its good solubility in diethyl ether it was purified by precipitation from a dichloromethane solution in cold diethyl ether/*n*-hexane (30 ml, 1:2). – Yield: 75.0 mg, 85.4 µmol, 34%; yellowish-white powder.

– MS (FAB), *m/z* (%): 2358 (<1) [N^N₂Cu₃P^P₃], 1820 (1) [N^N₂Cu₃P^P₂], 1480 (7) [N^NCu₂P^P₂], 1218 (1) [N^N₂Cu₂P^P], 1140 (<1) [CuP^P₂], 942 (6) [N^NCu₂P^P], 878 (8) [N^NCuP^P], 821 (4) [N^NCuP^P – N₄], 717 (2) [N^NCuP^P – (CH₂PhCH₂) – N₄], 601 (100) [CuP^P]. – C₅₂H₄₂CuN₅OP₂ · 0.3 CH₂Cl₂ (878.42): calc. C 69.49, H 4.75, N 7.75; found C 69.76, H 4.73, N 7.48. – X-ray diffraction data of single crystals can be found in section 5.4.

[(5-(*N*-Methylimidazol-2-yl)tetrazolate)Cu(DPEPhos)] (**34-Cu-10**). The complex was synthesized according to **GP4** from ligand **34-H** and DPEPhos (**10**), and purified by precipitation of its dichloromethane solution in *n*-hexane and further in diethyl ether/*n*-hexane (60 ml, 1:1). A white precipitate was obtained. Yield: 342 mg, 455 µmol, 68%. – MS (FAB), *m/z* (%): 1353 (9) [N^NCu₂P^P₂], 1139 (<1) [CuP^P₂], 964 (1) [N^N₂Cu₂P^P], 813 (4) [N^NCu₂P^P], 750 (5) [N^NCuP^P], 706 (1) [N^NCuP^P – Me – N₂], 694 (5) [N^NCuP^P – N₄], 601 (100) [CuP^P]. – C₄₁H₃₃CuN₆OP₂ · 0.15 CH₂Cl₂ (750.15): calc. C 64.61, H 4.52, N 10.99; found C 64.60, H 4.43, N 10.87. – X-ray diffraction data of single crystals can be found in section 5.4.

[(5-(*N*-Hexylimidazol-2-yl)tetrazolate)Cu(DPEPhos)] (**35-Cu-10**). The complex was obtained by following **GP4** with ligand **35-H** and DPEPhos (**10**), and precipitation in diethyl ether/*n*-hexane (35 ml, 3:4). Yield: 101 mg, 123 µmol, 49%; yellowish-white solid. – MS (FAB), *m/z* (%): 1423 (23) [N^NCu₂P^P₂], 1139 (1) [CuP^P₂], 1104 (3) [N^N₂Cu₂P^P], 885 (4) [N^NCu₂P^P], 820 (10) [N^NCuP^P], 764 (1) [N^NCuP^P – N₄], 735 (<1) [N^NCuP^P – C₂H₅ – N₄], 721 (1) [N^NCuP^P – C₃H₇ – N₄], 707 (1) [N^NCuP^P – C₄H₉ – N₄], 680 (1) [N^NCuP^P – C₆H₁₃ – N₄], 601 (100) [CuP^P]. – C₄₆H₄₃CuN₆OP₂ · 0.1 CH₂Cl₂ (821.37): calc. C 66.72, H 5.25, N 10.13; found C 66.64, H 5.30, N 10.01. – X-ray diffraction data of single crystals can be found in section 5.4.

[(5-(*N*-Methylbenzimidazol-2-yl)tetrazolate)Cu(DPEPhos)] (**36-Cu-10**). The complex was synthesized following **GP4** from ligand **36-H** and DPEPhos (**10**), and purified by precipitation in *n*-hexane and further in diethyl ether. A white solid was obtained. Yield: 214 mg, 267 µmol, 53%. – MS (FAB), *m/z* (%): 1403 (10) [N^NCu$_2$P^P$_2$], 1139 (1) [CuP^P$_2$], 1064 (1) [N^N$_2$Cu$_2$P^P], 865 (5) [N^NCu$_2$P^P], 801 (6) [N^NCuP^P], 744 (7) [N^NCuP^P – N$_4$], 706 (1), 668 (1) [CuP^P$_2$], 601 (100) [CuP^P]. – C$_{45}$H$_{35}$CuN$_6$OP$_2$· 0.1 CH$_2$Cl$_2$ (801.29): calc. C 66.89, H 4.38, N 10.38; found C 66.85, H 4.37, N 10.31.

[(5-(Benzoxazol-2-yl)tetrazolate)Cu$_2$(DPEPhos)$_2$]BF$_4$ (**c-(39-Cu-10)**). The complex **c-(39-Cu-10)** was obtained by adding a dichloromethane solution (3 ml) of [Cu(CH$_3$CN)$_4$]BF$_4$ (252 mg, 0.800 mmol, 1.00 equiv.) to a solution of 2-(tetrazol-5-yl)benzoxazole (**39-H**) (150 mg, 0.800 mg, 1.00 equiv.) with potassium hydroxide (67.3 mg, 1.20 mmol, 1.50 equiv.) in dichloromethane/ethanol (5 ml, 2:3) under nitrogen atmosphere after 3 h of stirring. DPEPhos (**10**) (431 mg, 0.800 mmol, 1.00 equiv.) in dichloromethane (3 ml) was added after 1 h, and the reaction mixture stirred at room temperature overnight. The solvents were evaporated from the reaction mixture, the residue extracted with few dichloromethane and the clear, yellow filtrate precipitated in diethyl ether (100 ml). A white solid precipitated slowly, which was collected, and precipitated again from dichloromethane/diethyl ether for further purification. Yield: 188 mg, 127 µmol, 32%, yellowish-white solid. – MS (FAB), *m/z* (%): 2424 (<1) [N^N$_3$Cu$_4$P^P$_3$], 2178 (1) [N^N$_2$Cu$_3$P^P$_3$], 2027 (<1) [N^N$_4$Cu$_4$P^P$_2$PF$_4$], 1639 (2) [N^N$_2$Cu$_3$P^P$_2$], 1489 (<1), 1390 (42) [N^NCu$_2$P^P$_2$], 1332 (1) [N^NCu$_2$P^P$_2$ – N$_4$], 1128 (<1) [N^N$_2$Cu$_2$P^PBF$_4$], 852 (5) [N^NCu$_2$P^P], 788 (1) [N^NCuP^P], 732 (2) [N^NCuP^P – N$_4$], 602 (100) [CuP^P]. – C$_{80}$H$_{60}$BCu$_2$F$_4$N$_5$O$_3$P$_4$ (1477.16): calc. C 64.76, H 4.08, N 4.71; found C 64.62, H 3.69, N 5.11.

[(5-(Benzothiazol-2-yl)tetrazolate)Cu(DPEPhos)] (**40-Cu-10**). The complex was synthesized from the tetrazole **40-H** (102 mg, 0.500 mmol, 1.00 equiv.) and DPEPhos (**10**). This was first deprotonated by potassium hydroxide (42.1 g, 0.750 mmol, 1.50 equiv.) in a dichloromethane/ethanol (5 ml, 2:3) mixture under nitrogen and stirred for 3 h, and subsequently the copper(I) salt [Cu(CH$_3$CN)$_4$]BF$_4$ (157 mg, 0.500 mmol, 1.00 equiv.) in 3 ml dry dichloromethane and after 1 h a dichloromethane solution of DPEPhos (269 mg in 3 ml, 0.500 mmol, 1.00 equiv.) were added. After stirring overnight, the solvents of the white suspension were evaporated and the residue extracted with dry acetone (6 ml). Precipitation of the filtrate in diethyl ether (60 ml) afforded 23 mg (28.6 µmol, 6%) of a white solid. – MS (FAB), m/z (%): 2448 (<1) [N^N$_3$Cu$_4$P^P$_3$ – N$_2$], 2210 (<1) [N^N$_2$Cu$_3$P^P$_3$], 1874 (<1) [N^N$_3$Cu$_3$P^P$_2$], 1671 (<1) [N^N$_2$Cu$_3$P^P$_2$], 1646 (1) [N^N$_2$Cu$_3$P^P$_2$ – N$_2$], 1608 (<1) [N^N$_2$Cu$_2$P^P$_2$], 1405 (6) [N^NCu$_2$P^P$_2$], 1140 (<1) [CuP^P$_2$], 868 (1) [N^NCu$_2$P^P], 844 (5) [N^NCu$_2$P^P – N$_2$], 804 (7) [N^NCuP^P], 747 (10) [N^NCuP^P – N$_4$], 600 (100) [CuP^P]. – X-ray diffraction data of single crystals can be found in section 5.4.

[(5-(4-Phenylpyridin-2-yl)tetrazolate)Cu(DPEPhos)] (**41-Cu-10**). The complex **41-Cu-10** was

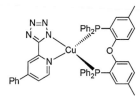

synthesized following **GP4** from 4-phenyl-2-(tetrazol-5-yl)-pyridine (**41-H**) and DPEPhos (**10**). The complex was precipitated twice from dichloromethane in diethyl ether (for further purification). Yield: 768 mg, 932 µmol, 62%; yellow solid. – MS (FAB), m/z (%): 2250 (1) [N^N$_2$Cu$_3$P^P$_3$], 1711 (3) [N^N$_2$Cu$_3$P^P$_2$], 1424 (12) [N^NCu$_2$P^P$_2$], 1140 (2) [CuP^P$_2$], 886 (5) [N^NCu$_2$P^P], 824 (5) [N^NCuP^P], 768 (9) [N^NCuP^P – N$_4$], 706 (1) [N^NCuP^P – Ph – N$_4$], 602 (100) [CuP^P]. – C$_{48}$H$_{36}$CuN$_5$OP$_2$ · 0.15 CH$_2$Cl$_2$ (824.33): calc. C 69.09, H 4.37, N 8.37; found C 69.00, H 4.42, N 8.19. – X-ray diffraction data of single crystals can be found in section 5.4.

[(5-(4-Phenylpyridin-2-yl)tetrazolate)Cu(PTEPhos)] (**41-Cu-18**). The complex **41-Cu-18** was obtained by following **GP4** from ligands 4-phenyl-2-(tetrazol-5-yl)-pyridine (**41-H**) and PTEPhos (**18**), and precipitation in diethyl ether/n-hexane (40 ml:30 ml). Yield: 61.0 mg, 71.6 µmol, 28%; yellow solid. – MS (FAB), m/z (%): 1482 (8) [N^NCu$_2$P^P$_2$], 1195 (1) [CuP^P$_2$], 1138 (<1) [CuP^P$_2$ – 4 Me], 914 (2) [N^NCu$_2$P^P], 852 (5) [N^NCuP^P], 795 (5) [N^NCuP^P – N$_4$], 736 (1) [N^NCuP^P – Ph – N$_4$ + O], 718 (2) [N^NCuP^P – Ph – N$_4$], 645 (6) [CuP^P + O], 629 (100) [CuP^P]. – C$_{50}$H$_{40}$CuN$_5$OP$_2$ · 0.2 CH$_2$Cl$_2$ (852.38): calc. C 69.35, H 4.68, N 8.06; found C 69.22, H 4.95, N 8.07.

[(5-(4-Phenylpyridin-2-yl)tetrazolate)Cu(XantPhos)] (**41-Cu-19**). a) The complex **41-Cu-19** was synthesized following **GP3** from [(4-phenyl-2-(tetrazol-5-yl)-pyridine)Cu(XantPhos)]PF$_6$ (202 mg, 200 µmol, 1.00 equiv.). Yield: 165 mg, 191 µmol, 95%; yellow powder. Crystals were obtained by layering diethyl ether upon a dichloromethane solution of **41-Cu-19**. – MS (FAB), m/z (%): 1790 (1) [N^N$_2$Cu$_3$P^P$_2$],

1505 (18) [N^NCu$_2$P^P$_2$], 1219 (<1) [CuP^P$_2$], 1149 (<1) [N^N$_2$Cu$_2$P^P], 957 (1) [N^N$_2$Cu$_2$P^P − 2 N$_4$ − Ph], 863 (2) [N^NCuP^P], 808 (7) [N^NCuP^P − N$_4$], 730 (2) [N^NCuP^P − N$_4$ − Ph], 641 (100) [CuP^P], 564 (7), 549 (6), 441 (5). − C$_{51}$H$_{40}$CuN$_5$OP$_2$ · 0.2 CH$_2$Cl$_2$ (864.39): calc. C 69.77, H 4.62, N 7.95; found C 69.59, H 4.77, N 7.61. b) The complex **41-Cu-19** was synthesized according to **GP4** from starting material 4-phenyl-2-(tetrazol-5-yl)pyridine (**41-H**) and XantPhos (**19**). The complex crystallized as yellow crystals from diethyl ether instead of precipitation. Yield: 287 mg, 332 μmol, 66%; yellow crystals. − MS (FAB), *m/z* (%): 1492 (<1) [N^NCu$_3$P^P$_2$ − Ph], 1429 (<1) [N^NCu$_2$P^P$_2$ − Ph], 1310 (<1) [N^NCuP^P$_2$ − Ph − N$_4$], 1219 (<1) [CuP^P$_2$], 1150 (<1) [N^N$_2$Cu$_2$P^P], 959 (<1) [N^N$_2$Cu$_2$P^P − 2 N$_4$ − Ph], 928 (1) [N^NCu$_2$P^P], 863 (1) [N^NCuP^P], 807 (6) [N^NCuP^P − N$_4$], 730 (2) [N^NCuP^P − N$_4$ − Ph], 641 (100) [CuP^P], 564 (5), 549 (3), 441 (3). − C$_{51}$H$_{40}$CuN$_5$OP$_2$ (864.39): calc. C 70.87, H 4.66, N 8.10; found C 70.58, H 4.68; N 7.97. − X-ray diffraction data of single crystals can be found in section 5.4.

[(5-(4-Phenylpyridin-2-yl)tetrazolate)Cu(DPTPhos)] (**41-Cu-21**). The complex was synthesized following **GP4** from 4-phenyl-2-(tetrazol-5-yl)pyridine (**41-H**) and DPTPhos (**21**). Yield: 254 mg, 302 μmol, 60%; yellow powder. − MS (FAB), *m/z* (%): 1458 (28) [N^NCu$_2$P^P$_2$], 1380 (<1) [N^NCu$_2$P^P$_2$ − Ph], 1276 (<1) [N^NCu$_2$P^P$_2$ − Ph − N$_3$], 1260 (<1) [N^NCu$_2$P^P$_2$ − Ph − N$_4$], 1171 (2) [CuP^P$_2$], 1127 (1) [N^N$_2$Cu$_2$P^P], 973 (1) [N^N$_2$Cu$_2$P^P − 2Ph], 904 (5) [N^NCu$_2$P^P], 840 (14) [N^NCuP^P], 617 (100) [CuP^P], 540 (3) [P^P]. − C$_{48}$H$_{36}$CuN$_5$P$_2$S · 0.25 CH$_2$Cl$_2$ (840.39): calc. C 67.26, H 4.27, N 8.13, S 3.72; found C 67.26, H 4.72, N 7.64, S 3.77. − X-ray diffraction data of single crystals can be found in section 5.4.

[(5-(4-Phenylpyridin-2-yl)tetrazolate)Cu(BINAP)] (**41-Cu-23**). The complex was synthesized following **GP4** from ligand 4-phenyl-2-(tetrazol-5-yl)-pyridine (**41-H**) and R-BINAP (**23**). Yield: 206 mg, 227 μmol, 57%; yellow powder. − MS (FAB), *m/z* (%): 1880 (4) [N^N$_2$Cu$_3$P^P$_2$], 1815 (<1) [N^N$_2$Cu$_2$P^P$_2$], 1594 (35) [N^NCu$_2$P^P$_2$], 1400 (<1) [N^NCuP^P$_2$ − Ph − N$_4$], 1308 (3) [CuP^P$_2$], 1257 (1) [N^N$_2$Cu$_3$P^P], 1193 (10) [N^N$_2$Cu$_2$P^P], 1062 (<1) [N^N$_2$Cu$_2$P^P − Ph − N$_4$], 972 (7) [N^NCu$_2$P^P], 909 (28) [N^NCuP^P], 852 (3) [N^NCuP^P − N$_4$], 792 (3) [N^NCuP^P − Ph − N$_3$], 776 (3) [N^NCuP^P − Ph − N$_4$], 685 (100) [CuP^P], 545 (6) [P^P − Ph], 500 (5), 437 (93)

[P^P – PPh₂]. – $C_{56}H_{40}CuN_5P_2 \cdot 0.15$ CH₂Cl₂ (908.44): calc. C 73.21, H 4.41, N 7.60; found C 73.10, H 4.55, N 7.30.

[(5-(4-Methylpyridin-2-yl)tetrazolate)Cu(DPEPhos)] (**42-Cu-10**). The complex **42-Cu-10** was obtained

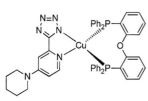

following **GP4** from ligand **42-H** and DPEPhos (**10**) as a white solid. Yield: 124 mg, 163 μmol, 27%. – MS (FAB), *m/z* (%): 1364 (6) [N^NCu₂P^P₂], 1139 (2) [CuP^P₂], 915 (<1) [N^N₂Cu₂P^P – N₄– Me], 885 (<1) [N^NCu₃P^P], 856 (<1) [N^NCu₃P^P – N₂], 824 (2) [N^NCu₂P^P], 762 (17) [N^NCuP^P], 708 (8) [N^NCuP^P – N₄], 663 (3), 601 (96) [CuP^P]. – $C_{43}H_{34}CuN_5OP_2$ (762.26): calc. 67.75, H 4.50, N 9.19; found C 67.83, H 4.54, N 8.72.

[(5-(4-Methoxypyridin-2-yl)tetrazolate)Cu(DPEPhos)] (**43-Cu-10**). The complex was synthesized according to **GP4** from ligand **43-H** and DPEPhos (**10**). Yield: 286 mg, 367 μmol, 74%. – MS (FAB), *m/z* (%): 1380 (9) [N^NCu₂P^P₂], 1303 (<1), 1231 (<1), 1139 (1) [CuP^P₂], 1018 (1) [N^N₂Cu₂P^P], 842 (4) [N^NCu₂P^P], 815 (1) [N^NCu₂P^P – N₂], 778 (6) [N^NCuP^P], 721 (7) [N^NCuP^P – N₄], 706 (1) [N^NCuP^P – N₄ – Me], 646 (2), 601 (100) [CuP^P]. – $C_{43}H_{34}CuN_5O_2P_2 \cdot 0.05$ CH₂Cl₂ (778.26): calc. C 66.08, H 4.39, N 8.95; found C 66.27, H 4.62, N 8.58. – X-ray diffraction data of single crystals can be found in section 5.4.

[(5-(4-(Piperidyl)pyridin-2-yl)tetrazolate)Cu(DPEPhos)] (**44-Cu-10**). The complex was synthesized following **GP4** from ligand **44-H** and DPEPhos (**10**), and purified twice by precipitation from a dichloromethane solution in diethyl ether. Yield: 237 mg, 285 μmol, 29%; greenish white powder. – MS (FAB), *m/z* (%): 1432 (9) [N^NCu₂P^P₂], 1229 (1) [N^NCuP^P₂ – C₅H₁₀N – N₄], 1154 (1), 1139 (2) [CuP^P₂], 1121 (2) [N^N₂Cu₂P^P], 984 (2) [N^N₂Cu₂P^P – C₅H₁₀N – N₄], 968 (1) [N^N₂Cu₂P^P – C₅H₁₀N – N₅], 938 (1) [N^N₂Cu₂P^P – C₅H₁₀N – N₇], 909 (1), 895 (5) [N^NCu₂P^P], 871 (3), 831 (24) [N^NCuP^P], 774 (9) [N^NCuP^P – N₄], 601 (100) [CuP^P]. – $C_{47}H_{41}CuN_6OP_2 \cdot 0.13$ CH₂Cl₂ (831.36): calc. C 67.04, H 5.16, N 9.95; found C 66.61, H 5.09, N 10.46. – X-ray diffraction data of single crystals can be found in section 5.4.

[(5-(2-Methylpyridin-6-yl)tetrazolate)Cu(DPEPhos)] (**45-Cu-10**). The complex was synthesized following **GP4** from ligand **45-H** and DPEPhos (**10**), and further purified by precipitation from a dichloromethane solution in diethyl ether under nitrogen atmosphere. Yield: 130 mg, 171 μmol, 28%; white powder. – MS (FAB), m/z (%): 1437 (<1) [N^N_2Cu_2P^P_2 – 2 Me – N_4], 1377 (<1) [N^N_2CuP^P_2 – 2 Me – N_4], 1363 (<1) [N^NCu_2P^P_2], 1323 (<1) [N^NCu_2P^P_2 – N_3], 1299 (<1) [N^NCuP^P_2], 1275 (<1), 1269 (<1) [N^NCuP^P_2 – Me – N], 1232 (<1) [N^NCuP^P_2 – Me – N_4], 1187 (<1), 1139 (<1) [CuP^P_2], 1054 (<1) [N^N_2Cu_3P^P], 930 (<1) [N^N_2Cu_2P^P – N_4], 918 (<1) [N^N_2Cu_2P^P – N_4 – Me], 762 (2) [N^NCuP^P], 705 (1) [N^NCuP^P – N_4], 601 (10) [CuP^P]. – C_43H_34CuN_5OP_2 · 0.2 CH_2Cl_2 (762.26): calc. C 67.75, H 4.50, N 9.19; found C 67.49, H 4.75, N 9.03. – X-ray diffraction data of single crystals can be found in section 5.4.

[(5-(2-Methoxypyridin-6-yl)tetrazolate)Cu(DPEPhos)] (**46-Cu-10**). The complex **46-Cu-10** was obtained following **GP4** from ligand **46-H** and DPEPhos (**10**) as a white solid. Yield: 99.0 mg, 127 μmol, 42%. – MS (FAB), m/z (%): 2158 (<1) [N^N_2Cu_3P^P_3], 1380 (<1) [N^NCu_2P^P_2], 1141 (<1) [CuP^P_2], 1025 (<1), 919 (<1), 841 (1) [N^NCu_2P^P], 779 (3) [N^NCuP^P], 722 (2) [N^NCuP^P – N_4], 602 (34) [CuP^P]. – C_43H_34CuN_5O_2P_2 (778.26): calc. C 66.36, H 4.40, N 9.00; found C 66.23, H 4.49, N 8.77. – X-ray diffraction data of single crystals can be found in section 5.4.

5.3 Supplemental Data

5.3.1 Photophysical Characterization

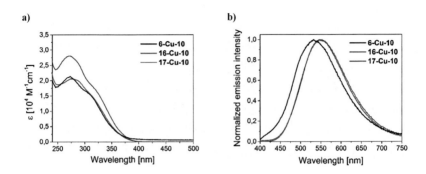

Figure 87. a) Absorption and b) emission spectra of the complexes [(N^N)Cu(DPEPhos)] **6-Cu-10**, **16-Cu-10** and **17-Cu-10** based on 5-(pyridine-2-yl)-triazolate ligands. The absorption of the complexes is recorded in CH_2Cl_2 at a concentration of 10^{-5} mol/l, while the emission is measured for amorphous powders at room temperature when excited at 350 nm.

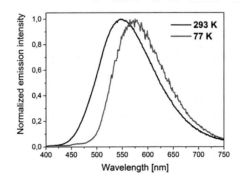

Figure 88. Emission spectra of the complex [(5-(pyridin-2-yl)-3-p-tolyl-1,2,4-triazolate)Cu(DPEPhos)] (**17-Cu-10**) as amorphous powder at 293 K and 77 K when excited at 350 nm.

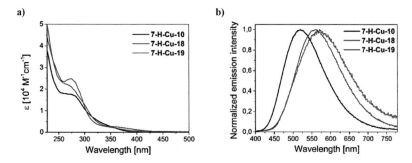

Figure 89. a) Absorption and b) emission spectra of the cationic complexes [(PyrTet)Cu(P^P)]BF$_4$ **7-H-Cu-10**, **7-H-Cu-18** and **7-H-Cu-19** with different bis(phosphine) ligands. The absorption of the complexes is recorded in CH$_2$Cl$_2$ at a concentration of 10^{-5} mol/l, while the emission is measured for amorphous powders at room temperature when excited at 350 nm.

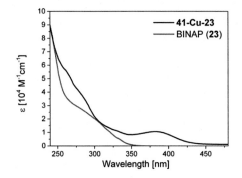

Figure 90. Absorption spectra of the mononuclear complex [(5-(4-phenylpyridin-2-yl)tetrazolate)Cu(BINAP)] (**41-Cu-23**) with the free ligand in CH$_2$Cl$_2$ at a concentration of 10^{-5} mol/l.

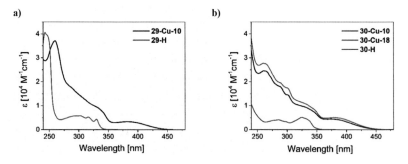

Figure 91. Absorption spectra of a) [(5-(quinolin-2-yl)tetrazolate)Cu(DPEPhos)] (**29-Cu-10**) in CH$_2$Cl$_2$ and the free ligand **29** in EtOH, and b) [(5-(isoquinolin-1-yl)tetrazolate)Cu(P^P)] **30-Cu-10** and **30-Cu-18** in CH$_2$Cl$_2$ and of the free ligand **30** in EtOH, at a concentration of 10^{-5} mol/l.

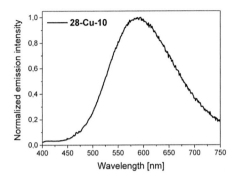

Figure 92. Emission spectrum of the mononuclear complexes [(5-(pyrazin-2-yl)tetrazolate)Cu(DPEPhos)] (**28-Cu-10**) as amorphous powder at room temperature when excited at 350 nm.

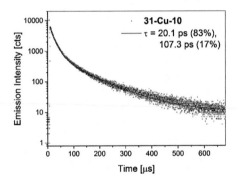

Figure 93. Microsecond luminescence decay dynamics of the complex [(N^N)Cu(DPEPhos)] **31-Cu-10** as powder detected at 549 nm (PL maximum) by the TCSPC technique. The red line represents a fit to a biexponential decay.

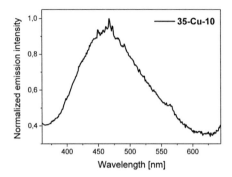

Figure 94. Emission spectrum of the complex [(5-(N-hexylimidazol-2-yl)tetrazolate)Cu(DPEPhos)] (**35-Cu-10**) as amorphous powder at room temperature when excited at 350 nm. Due to the weak photoluminescence, the signal-to-noise ratio is very low.

a)   b)

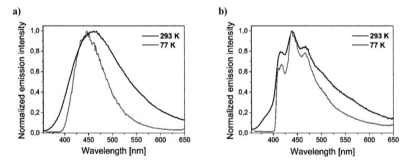

Figure 95. Emission spectra of a) [(5-(N-methylimidazol-2-yl)tetrazolate)Cu(DPEPhos)] (**34-Cu-10**) and b) [(5-(N-methylbenzimidazol-2-yl)tetrazolate)Cu(DPEPhos)] (**36-Cu-10**) at 293 K and 77 K as amorphous powder when excited at 350 nm.

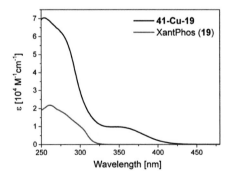

Figure 96. Absorption spectra of the complex [(5-(pyridin-2-yl)tetrazolate)Cu(XantPhos)] (**41-Cu-19**) and the free ligand in CH_2Cl_2 at a concentration of 10^{-5} mol/l.

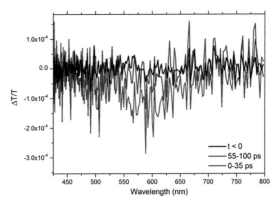

Figure 97. Broadband transient absorption kinetics of complex [(PyrTet)Cu(DPEPhos)] (**7-Cu-10**) in neat film in the wavelength region 400-800 nm with delay times of 35 and 100 ps after the pump pulse (400 nm) with long-lived background subtraction. Due to substantial spectral overlap of long-lived delayed fluorescence (positive signal) with excited state absorption (negative signal), transient absorption signals were averaged across multiple scans. Excited state absorption is observed from 530–700 nm.

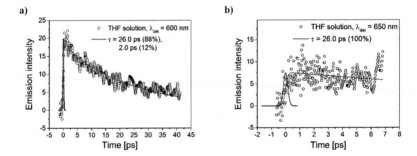

Figure 98. Picosecond luminescence decay dynamics of [(DPEPhos)Cu(PyrTet)] (**7-Cu-10**) in THF solution at a) 600 nm (PL maximum) and b) 650 nm (PL red-side), when excited at 400 nm. The solid red line represents fits to the a) biexponential and b) monoexponential decays. The instrument-response function is depicted as dotted line with 360 fs full-width half-maximum.

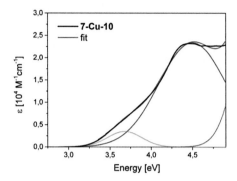

Figure 99. Absorption spectrum of model complex [(PyrTet)Cu(DPEPhos)] (**7-Cu-10**) at 250–430 nm, measured in CH₂Cl₂ at a concentration of 10⁻⁵ mol/l with long integration times to resolve the individual features (1 nm/min), and a fit of three Gaussians. The overall fit is given by the red line, with two Gaussians reflecting ligand-centered bands (blue and magenta line) and one Gaussian corresponding to the charge-transfer band (green line).

Table 51. Emission decay times of the complexes **7-Cu-10**, **16-Cu-10** and **17-Cu-10** at 293 K obtained from fitting of the experimental data with mono- or biexponentials. A weighted-average lifetime (τ_{ave}) was calculated by the equation $\tau_{ave} = \Sigma A_i \tau_i / \Sigma A_i$ with A_i as the pre-exponential factors for the lifetime. A SpectraLED ($\lambda = 314/355$ nm, pulse FWHM = 493 ns) was used as excitation source, if not otherwise noted.

Complex	τ_1 [µs]	A_1	τ_2 [µs]	A_2	τ_{ave} [µs]
7-Cu-10[a]	22.00	1			22.0
16-Cu-10	6.03	0.3068	11.66	0.6932	9.9
17-Cu-10[a]	12.66	1			12.7

[a] A flashlight ($\lambda = 250$–2000 nm, 3 µs pulse) was used as excitation source.

Table 52. Emission decay times of the complexes **7-Cu-10**, **16-Cu-10** and **17-Cu-10** at 77 K obtained from fitting of the experimental data with mono- or biexponentials. A weighted-average lifetime (τ_{ave}) was calculated by the equation $\tau_{ave} = \Sigma A_i \tau_i / \Sigma A_i$ with A_i as the pre-exponential factors for the lifetime. A SpectraLED ($\lambda = 314/355$ nm, pulse FWHM = 493 ns) was used as excitation source, if not otherwise noted.

Complex	τ_1 [µs]	A_1	τ_2 [µs]	A_2	τ_{ave} [µs]
7-Cu-10[a]	111.49	1			111.5
16-Cu-10	95.68	0.5655	161.35	0.4345	124.2
17-Cu-10	124.08	0.6327	259.06	0.3673	173.7

[a] A flashlight ($\lambda = 250$–2000 nm, 3 µs pulse) was used as excitation source.

Table 53. Emission decay times of the complexes **7-Cu-18** and **7-Cu-19** at 293 K obtained from fitting of the experimental data with a monoexponentials. A weighted-average lifetime (τ_{ave}) was calculated by the equation $\tau_{ave} = \Sigma A_i \tau_i / \Sigma A_i$ with A_i as the pre-exponential factors for the lifetime. A SpectraLED ($\lambda = 314/355$ nm, pulse FWHM = 493 ns) was used as excitation source, if not otherwise noted.

Complex	τ_1 [µs]	A_1	τ_{ave} [µs]
7-Cu-18[a]	21.12	1	21.1
7-Cu-19[a]	27.21	1	27.2

[a] A flashlight ($\lambda = 250–2000$ nm, 3 µs pulse) was used as excitation source.

Table 54. Emission decay times of the complexes **28-Cu-10**, **29-Cu-10**, **30-Cu-10**, **30-Cu-18**, **31-Cu-10**, **40-Cu-10** at 293 K obtained from fitting of the experimental data with mono- or biexponentials. A weighted-average lifetime (τ_{ave}) was calculated by the equation $\tau_{ave} = \Sigma A_i \tau_i / \Sigma A_i$ with A_i as the pre-exponential factors for the lifetime. A SpectraLED ($\lambda = 314/355$ nm, pulse FWHM = 493 ns) was used as excitation source, if not otherwise noted.

Complex	τ_1 [µs]	A_1	τ_2 [µs]	A_2	τ_{ave} [µs]
28-Cu-10	1.59	0.5826	8.10	0.4174	4.3
29-Cu-10	12.99	0.7186	21.07	0.2814	15.2
30-Cu-10	25.17	0.6520	67.83	0.3480	40.0
30-Cu-18[a]	60.38	0.3419	103.21	0.6581	88.6
31-Cu-10[a]	20.05	0.8268	107.31	0.1732	35.2
40-Cu-10[a]	19.56	1			19.6

[a] A flashlight ($\lambda = 250–2000$ nm, 3 µs pulse) was used as excitation source.

Table 55. Emission decay times of the complexes **29-Cu-10**, **30-Cu-10**, **30-Cu-18**, **31-Cu-10**, **40-Cu-10** at 77 K obtained from fitting of the experimental data with mono-, bi- or triexponentials. A weighted-average lifetime (τ_{ave}) was calculated by the equation $\tau_{ave} = \Sigma A_i \tau_i / \Sigma A_i$ with A_i as the pre-exponential factors for the lifetime. A SpectraLED ($\lambda = 314/355$ nm, pulse FWHM = 493 ns) was used as excitation source, if not otherwise noted.

Complex	τ_1 [µs]	A_1	τ_2 [µs]	A_2	τ_3 [µs]	A_3	τ_{ave} [µs]
29-Cu-10	169.75	1					169.8
30-Cu-10	552.47	0.3247	2426.38	0.0428	172.08	0.6325	392.1
30-Cu-18	616.85	0.1401	273.63	0.8463	2060.95	0.0136	346.0
31-Cu-10[a]	2173.38	0.01814	9286.03	0.1467	215.25	0.6719	1546.3

[a] A flashlight ($\lambda = 250–2000$ nm, 3 µs pulse) was used as excitation source.

Table 56. Emission decay times of the complexes **41-Cu-10**, **41-Cu-18**, **41-Cu-19** and **42-Cu-10**, **43-Cu-10**, **45-Cu-10**, **47-Cu-10**, **48-Cu-10** at 293 K obtained from fitting of the experimental data with mono-, bi- or triexponentials. A weighted-average lifetime (τ_{ave}) was calculated by the equation $\tau_{ave} = \Sigma A_i \tau_i / \Sigma A_i$ with A_i as the pre-exponential factors for the lifetime. A SpectraLED (λ = 314/355 nm, pulse FWHM = 493 ns) was used as excitation source, if not otherwise noted.

Complex	τ_1 [μs]	A_1	τ_2 [μs]	A_2	τ_3 [μs]	A_3	τ_{ave} [μs]
41-Cu-10	18.62	1					18.6
41-Cu-10 crystal	9.78	0.4754	21.45	0.5246			15.9
41-Cu-18	7.28	0.5062	9.78	0.4938			8.5
41-Cu-19	12.67	1					12.7
42-Cu-10	17.43	1					17.4
43-Cu-10	4.99	0.4258	9.79	0.5742			7.8
45-Cu-10	4.68	0.8386	10.55	0.1567	67.23	0.0037	5.8
47-Cu-10	14.53	1					14.5
48-Cu-10	14.28	1					14.3

Table 57. Emission decay times of the complexes **41-Cu-18**, **41-Cu-19** and **42-Cu-10**, **43-Cu-10**, **45-Cu-10**, **47-Cu-10**, **48-Cu-10** at 77 K obtained from fitting of the experimental data with mono-, bi- or triexponentials. A weighted-average lifetime (τ_{ave}) was calculated by the equation $\tau_{ave} = \Sigma A_i \tau_i / \Sigma A_i$ with A_i as the pre-exponential factors for the lifetime. A SpectraLED (λ = 314/355 nm, pulse FWHM = 493 ns) was used as excitation source, if not otherwise noted.

Complex	τ_1 [μs]	A_1	τ_2 [μs]	A_2	τ_3 [μs]	A_3	τ_{ave} [μs]
41-Cu-18	87.94		0.7912	135.16	0.2088		97.8
41-Cu-19	105.35		1				105.4
42-Cu-10	129.12		1				129.1
43-Cu-10	176.27		1				176.3
45-Cu-10	140.58		0.5839	423.09	0.4019	1360.87	0.0142	271.4
47-Cu-10	81.24		0.9204	128.15	0.0796		85.0
48-Cu-10	144.89	1					144.9

5.3.2 Theoretical Calculations[V]

a)   b)

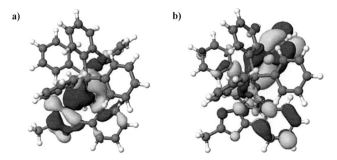

Figure 100. Frontier orbitals a) HOMO and b) LUMO of the complex [(5-(pyridin-2-yl)-3-methyl-1,2,4-triazolate)Cu(DPEPhos)] (**16-Cu-10**), as calculated by DFT using the BP86 functional with def2-SV(P) basis set. The initial structure was generated by adding a exchanging the N(3) against C(Me) in **7-Cu-10**.

a)   b)

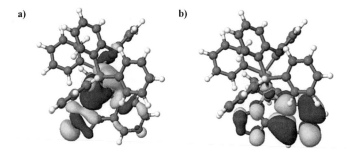

Figure 101. Frontier orbitals a) HOMO and b) LUMO of the complex [(5-(pyrazin-2-yl)-tetrazolate)Cu(DPEPhos)] (**28-Cu-10**), as calculated by DFT using the BP86 functional with def2-SV(P) basis set. The crystal structure is used as initial geometry, and is optimized in the ground and triplet state.

a)   b)

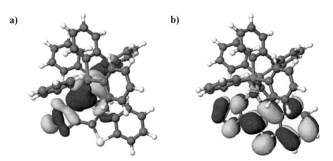

Figure 102. Frontier orbitals a) HOMO and b) LUMO of the complex [(5-(benzothiazol-2-yl)-tetrazolate)Cu(DPEPhos)] (**40-Cu-10**), as calculated by DFT using the BP86 functional with def2-SV(P) basis set. The crystal structure is used as initial geometry, and is optimized in the ground and triplet state.

a)   b)

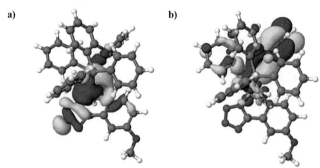

Figure 103. Frontier orbitals a) HOMO and b) LUMO of the complex [(5-(4-methoxypyridin-2-yl)-tetrazolate)Cu(DPEPhos)] (**43-Cu-10**), as calculated by DFT using the BP86 functional with def2-SV(P) basis set. The crystal structure is used as initial geometry, and is optimized in the ground and triplet state.

a)   b)

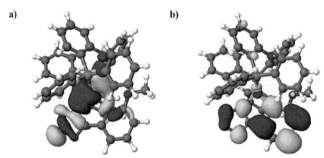

Figure 104. Frontier orbitals a) HOMO and b) LUMO of the complex [(5-(2-methoxypyridin-6-yl)-tetrazolate)Cu(DPEPhos)] (**48-Cu-10**), as calculated by DFT using the BP86 functional with def2-SV(P) basis set. The crystal structure is used as initial geometry, and is optimized in the ground and triplet state.

a)   b)

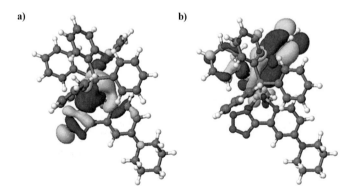

Figure 105. Frontier orbitals a) HOMO and b) LUMO of the complex [(5-(4-piperidylpyridin-2-yl)-tetrazolate)Cu(DPEPhos)] (**45-Cu-10**), as calculated by DFT using the BP86 functional with def2-SV(P) basis set. The crystal structure is used as initial geometry, and is optimized in the ground and triplet state.

Table 58. Calculated HOMO and LUMO energies of the complexes [(R-PyrTet)Cu(DPEPhos)] **42-Cu-10**, **43-Cu-10**, **45-Cu-10**, **46-Cu-10** in comparison to the model complex **7-Cu-10**, using the BP86 functional.

Complex	HOMO [eV]	LUMO [eV]	$\Delta E_{HOMO-LUMO}$ [eV]
7-Cu-10	–4.58	–2.23	2.35
42-Cu-10	–4.54	–2.15	2.39
43-Cu-10	–4.50	–2.02	2.35
45-Cu-10	–4.57	–2.16	2.41
46-Cu-10	–4.50	–2.13	2.37

5.3.3 Electrochemical Data[VI]

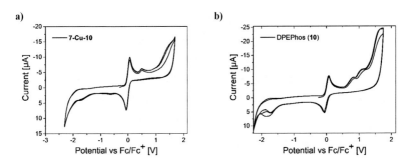

a)   b)

Figure 106. Cyclovoltammetric scan of a) complex [(PyrTet)Cu(DPEPhos)] (**7-Cu-10**) and b) the free ligand DPEPhos (**10**) in CH_2Cl_2. Irreversible oxidation peaks are found at +0.50 V for complex **7-Cu-10**, and at +0.83 eV and +1.08 V for the free ligand. The peaks at –1.72 V in both scans are attributed to solvent.

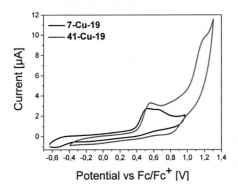

Figure 107. Cyclovoltammetric scans of the copper(I) complexes [(5-(pyridin-2-yl)tetrazolate)Cu(XantPhos)] (**7-Cu-19**) and [(5-(4-phenylpyridin-2-yl)tetrazolate)Cu(XantPhos)] (**41-Cu-19**) in CH_2Cl_2.

5.3.4 Thermogravimetric Data[VII]

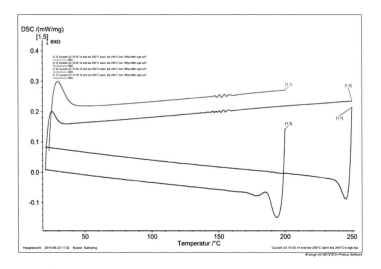

Figure 108. DSC measurement of the copper(I) complex [(PyrTet)Cu(DPEPhos)] (**7-Cu-10**) as amorphous powder. The sample was first heated up to 200 °C (heating circle: 1.1, green line, cooling circle: 1.3, blue line), and in a second circle to 240 °C (heating circle: 1.4, purple line, cooling circle: 1.3, brown line). No phase transition or melting point is observed in the temperature range of 20–240 °C.

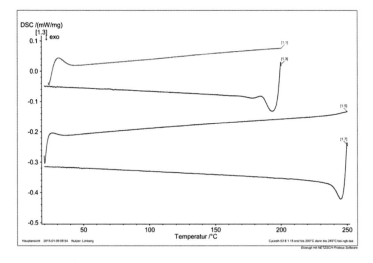

Figure 109. DSC measurement of the copper(I) complex [(PyrTet)Cu(DPEPhos)] (**7-Cu-10**) as neat film. The sample was first heated up to 200 °C (heating circle: 1.1, green line, cooling circle: 1.3, blue line), and in a second circle to 240 °C (heating circle: 1.4, purple line, cooling circle: 1.3, brown line). No phase transition or melting point is observed in the temperature range of 20–240 °C.

5.4 Crystallographic Data

Crystallographic data (excluding structure factors) for selected structures reported in this work have been deposited with the Cambridge Crystallographic Data Centre as supplementary publication no. CCDC 922230 (**7-Cu-8**), CCDC 922231 (**7-H-Cu-8**), CCDC 922232 (**7-Cu-10**), CCDC 922233 (**7-H-Cu-10**), CCDC 922234 (**7-Cu-19**), and CCDC 922235 (**7-Cu-18**), CCDC 1047149 (**5-H-Cu-8**), CCDC 1047150 (**6-H-Cu-8-PF$_6$**), CCDC 1047151 (**c-(5-H-Cu-9-PF$_6$)**), CCDC 1047152 (**5-H-Cu-10**), CCDC 1047153 (**6-H-Cu-10**), CCDC 1047154 (**c-(5-Cu-8-PF$_6$)**), CCDC 1047155 (**c-(6-Cu-8-PF$_6$)**), CCDC 1047156 (**c-(6-Cu-9)**), CCDC 1047157 (**c-(6-Cu-9-PF$_6$)**), CCDC 1047158 (**c-(7-Cu-9)**), CCDC 1047159 (**c-(7-Cu-9-PF$_6$)**), and CCDC 1047160 (**5-Cu-10**).

[(2-(Pyridin-2-yl)benzimidazole)Cu(PPh₃)₂]BF₄ (5-H-Cu-8) – CCDC 1047149

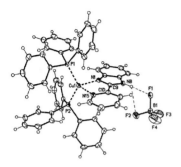

Crystal Data

$C_{48}H_{39}CuN_3P_2 \cdot BF_4$	$Z = 2$
$M_r = 870.11$	$F(000) = 896$
Triclinic, $P\text{-}1$ (no.2)	$D_x = 1.401$ Mg m^{-3}
$a = 11.278$ (1) Å	Mo $K\alpha$ radiation, $\lambda = 0.71073$ Å
$b = 12.684$ (1) Å	Cell parameters from 167 reflections
$c = 16.171$ (1) Å	$\theta = 2.5–25.0°$
$\alpha = 83.52$ (1)°	$\mu = 0.66$ mm^{-1}
$\beta = 76.41$ (1)°	$T = 123$ K
$\gamma = 66.60$ (1)°	Blocks, yellow
$V = 2063.1$ (3) Å^3	$0.50 \times 0.45 \times 0.40$ mm

Data collection

Bruker-Nonius KappaCCD diffractometer	7867 reflections with $I > 2\sigma(I)$
Radiation source: fine-focus sealed tube	$R_{int} = 0.021$
rotation in ϕ and ω, n° scans	$\theta_{max} = 27.5°$, $\theta_{min} = 2.6°$
Absorption correction: multi-scan *SADABS* (Sheldrick, 2008)	$h = -14{\rightarrow}14$
$T_{min} = 0.720$, $T_{max} = 0.790$	$k = -15{\rightarrow}16$
26082 measured reflections	$l = -20{\rightarrow}20$
9426 independent reflections	

Refinement

Refinement on F^2	Primary atom site location: heavy-atom method
Least-squares matrix: full	Secondary atom site location: difference Fourier map
$R[F^2 > 2\sigma(F^2)] = 0.033$	Hydrogen site location: difference Fourier map
$wR(F^2) = 0.084$	H atoms treated by a mixture of independent and constrained refinement
$S = 1.02$	$w = 1/[\sigma^2(F_o^2) + (0.0357P)^2 + 1.5498P]$ where $P = (F_o^2 + 2F_c^2)/3$
9426 reflections	$(\Delta/\sigma)_{max} = 0.001$
535 parameters	$\Delta\rangle_{max} = 0.83$ e Å^{-3}
1 restraint	$\Delta\rangle_{min} = -0.55$ e Å^{-3}

[(2-(1,2,3-Triazol-5-yl)pyridine)Cu(PPh₃)₂]PF₆ (6-H-Cu-8-PF₆) – CCDC 1047150

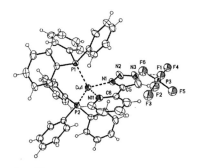

Crystal Data

C₄₃H₃₆CuN₄P₂·F₆P	$Z = 2$
$M_r = 879.21$	$F(000) = 900$
Triclinic, P-1 (no.2)	$D_x = 1.446$ Mg m⁻³
$a = 12.0277$ (5) Å	Mo $K\alpha$ radiation, $\lambda = 0.71073$ Å
$b = 12.9010$ (5) Å	Cell parameters from 7032 reflections
$c = 14.6324$ (5) Å	$\theta = 1.0–27.5°$
$\alpha = 100.140$ (2)°	$\mu = 0.72$ mm⁻¹
$\beta = 113.803$ (2)°	$T = 123$ K
$\gamma = 94.065$ (2)°	Blocks, colourless
$V = 2019.36$ (14) Å³	0.50 × 0.30 × 0.25 mm

Data collection

Bruker APEXII diffractometer	8193 reflections with $I > 2\sigma(I)$
Radiation source: fine-focus sealed tube	$R_{int} = 0.031$
rotation in φ and ω, 2° scans	$\theta_{max} = 27.5°$, $\theta_{min} = 3.6°$
Absorption correction: multi-scan *MULABS* (*PLATON* Spek 2008, cf. Blessing 1995)	$h = -15 \rightarrow 15$
$T_{min} = 0.729$, $T_{max} = 0.837$	$k = -16 \rightarrow 16$
22436 measured reflections	$l = -19 \rightarrow 18$
9143 independent reflections	

Refinement

Refinement on F^2	Primary atom site location: heavy-atom method
Least-squares matrix: full	Secondary atom site location: difference Fourier map
$R[F^2 > 2\sigma(F^2)] = 0.064$	Hydrogen site location: mixed
$wR(F^2) = 0.173$	H atoms treated by a mixture of independent and constrained refinement
$S = 1.06$	$w = 1/[\sigma^2(F_o^2) + (0.0777P)^2 + 5.553P]$ where $P = (F_o^2 + 2F_c^2)/3$
9143 reflections	$(\Delta/\sigma)_{max} = 0.001$
505 parameters	$\Delta\rangle_{max} = 1.28$ e Å⁻³
681 restraints	$\Delta\rangle_{min} = -1.48$ e Å⁻³

[(2-(Tetrazol-5-yl)pyridine)Cu(PPh₃)₂]BF₄ (7-H-Cu-8) – CCDC 922231

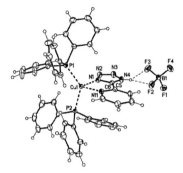

Empirical formula	C42 H35 B Cu F4 N5 P2
	C42 H35 Cu N5 P2 - B F4
Formula weight	822.04
Temperature	123(2) K
Wavelength	0.71073 A
Crystal system, space group	Triclinic, P-1 (no.2)
Unit cell dimensions	a = 10.889(1) A alpha = 108.70(1) deg.
	b = 11.950(1) A beta = 96.41(1) deg.
	c = 16.851(2) A gamma = 107.01(1) deg.
Volume	1934.3(3) A^3
Z, Calculated density	2, 1.411 Mg/m^3
Absorption coefficient	0.705 mm^-1
F(000)	844
Crystal size	0.24 x 0.18 x 0.06 mm
Theta range for data collection	3.01 to 27.48 deg.
Limiting indices	-14<=h<=14, -15<=k<=15, -21<=l<=21
Reflections collected / unique	43246 / 8862 [R(int) = 0.0314]
Completeness to theta = 27.48	99.8 %
Absorption correction	Semi-empirical from equivalents
Max. and min. transmission	0.9562 and 0.8627
Refinement method	Full-matrix least-squares on F^2
Data / restraints / parameters	8862 / 0 / 500
Goodness-of-fit on F^2	1.032
Final R indices [I>2sigma(I)]	R1 = 0.0306, wR2 = 0.0670
R indices (all data)	R1 = 0.0440, wR2 = 0.0727
Largest diff. peak and hole	0.412 and -0.338 e.A^-3

[(2-(Pyridin-2-yl)benzimidazole)$_2$Cu$_2$(dppm)$_2$](PF$_6$)$_2$ (c-(5-H-Cu-9-PF$_6$)) – CCDC 1047151

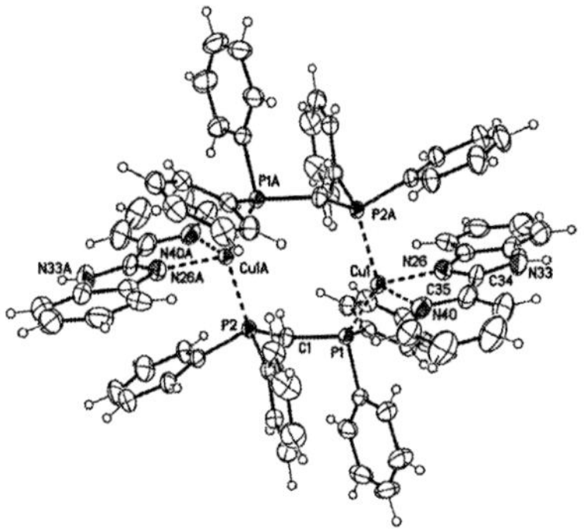

Empirical formula	C82 H82 Cu2 F12 N6 O2 P6 C74 H62 Cu2 N6 P4 - 2 PF6 - 2 Et2O
Formula weight	1724.44
Temperature	123(2) K
Wavelength	0.71073 A
Crystal system, space group	Monoclinic, P2(1)/n (No.14)
Unit cell dimensions	a = 15.024(1) A alpha = 90 deg. b = 14.129(2) A beta = 108.60(1) deg. c = 19.206(2) A gamma = 90 deg.
Volume	3864.0(7) A^3
Z, Calculated density	2, 1.482 Mg/m^3
Absorption coefficient	0.756 mm^-1
F(000)	1776
Crystal size	0.35 x 0.15 x 0.10 mm
Theta range for data collection	3.02 to 27.48 deg.
Limiting indices	-19<=h<=19, -18<=k<=18, -24<=l<=24
Reflections collected / unique	53397 / 8841 [R(int) = 0.0406]
Completeness to theta = 27.48	99.8 %
Absorption correction	Semi-empirical from equivalents
Max. and min. transmission	0.9281 and 0.6397
Refinement method	Full-matrix least-squares on F^2
Data / restraints / parameters	8841 / 100 / 499
Goodness-of-fit on F^2	1.026
Final R indices [I>2sigma(I)] R indices (all data)	R1 = 0.0519, wR2 = 0.1323 R1 = 0.0686, wR2 = 0.1450
Largest diff. peak and hole	1.239 and -0.686 e.A^-3

[(2-(Pyridin-2-yl)benzimidazole)Cu(DPEPhos)]BF₄ (5-H-Cu-10) – CCDC 1047152

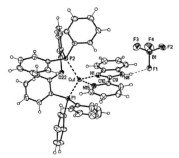

Crystal Data

C₄₈H₃₇CuN₃OP₂·BF₄·0.5(CH₂Cl₂)	$F(000) = 1900$
$M_r = 926.56$	$D_x = 1.420$ Mg m⁻³
Monoclinic, $P2_1/n$ *(no.14)*	Mo $K\alpha$ radiation, $\lambda = 0.71073$ Å
$a = 11.423$ (1) Å	Cell parameters from 244 reflections
$b = 14.922$ (2) Å	$\theta = 2.5–25.0°$
$c = 25.552$ (3) Å	$\mu = 0.70$ mm⁻¹
$\beta = 95.86$ (1)°	$T = 123$ K
$V = 4332.7$ (9) Å³	Blocks, yellow
$Z = 4$	0.60 × 0.40 × 0.35 mm

Data collection

Bruker-Nonius KappaCCD diffractometer	8922 reflections with $I > 2\sigma(I)$
Radiation source: fine-focus sealed tube	$R_{int} = 0.020$
rotation in ϕ and ω, 1° scans	$\theta_{max} = 27.5°$, $\theta_{min} = 2.7°$
Absorption correction: multi-scan *SADABS* (Sheldrick, 2008)	$h = -14 \to 14$
$T_{min} = 0.720$, $T_{max} = 0.790$	$k = -19 \to 19$
62632 measured reflections	$l = -33 \to 33$
9942 independent reflections	

Refinement

Refinement on F^2	Primary atom site location: structure-invariant direct methods
Least-squares matrix: full	Secondary atom site location: difference Fourier map
$R[F^2 > 2\sigma(F^2)] = 0.037$	Hydrogen site location: mixed
$wR(F^2) = 0.089$	H atoms treated by a mixture of independent and constrained refinement
$S = 1.06$	$w = 1/[\sigma^2(F_o^2) + (0.029P)^2 + 5.845P]$ where $P = (F_o^2 + 2F_c^2)/3$
9942 reflections	$(\Delta/\sigma)_{max} = 0.003$
566 parameters	$\Delta\rangle_{max} = 0.84$ e Å⁻³
22 restraints	$\Delta\rangle_{min} = -0.73$ e Å⁻³

[(2-(1,2,3-Triazol-5-yl)pyridine)Cu(DPEPhos)]BF₄ (6-H-Cu-10) – CCDC 1047153

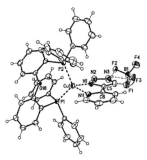

Crystal Data

C₄₃H₃₄CuN₄OP₂·BF₄·0.5(C₄H₁₀O)	$F(000) = 3592$
$M_r = 872.09$	$D_x = 1.394$ Mg m⁻³
Monoclinic, *C2/c (no.15)*	Cu $K\alpha$ radiation, $\lambda = 1.54178$ Å
$a = 23.3084\ (8)$ Å	Cell parameters from 9821 reflections
$b = 15.1103\ (8)$ Å	$\theta = 3.5–72.1°$
$c = 23.8529\ (10)$ Å	$\mu = 1.97$ mm⁻¹
$\beta = 98.530\ (2)°$	$T = 123$ K
$V = 8308.0\ (6)$ Å³	Blocks, yellow
$Z = 8$	$0.16 \times 0.10 \times 0.06$ mm

Data collection

Bruker D8 Venture diffractometer with Photon100 detector	7290 reflections with $I > 2\sigma(I)$
Radiation source: IμS microfocus	$R_{int} = 0.037$
rotation in ϕ and ω, 0.5°, shutterless scans	$\theta_{max} = 72.1°$, $\theta_{min} = 3.5°$
Absorption correction: multi-scan *SADABS* (Sheldrick, 2015)	$h = -28 \rightarrow 28$
$T_{min} = 0.795$, $T_{max} = 0.864$	$k = -18 \rightarrow 17$
56053 measured reflections	$l = -29 \rightarrow 29$
8177 independent reflections	

Refinement

Refinement on F^2	Primary atom site location: structure-invariant direct methods
Least-squares matrix: full	Secondary atom site location: difference Fourier map
$R[F^2 > 2\sigma(F^2)] = 0.031$	Hydrogen site location: difference Fourier map
$wR(F^2) = 0.084$	H atoms treated by a mixture of independent and constrained refinement
$S = 1.03$	$w = 1/[\sigma^2(F_o^2) + (0.042P)^2 + 10.490P]$ where $P = (F_o^2 + 2F_c^2)/3$
8177 reflections	$(\Delta/\sigma)_{max} = 0.001$
553 parameters	$\Delta\rangle_{max} = 0.66$ e Å⁻³
62 restraints	$\Delta\rangle_{min} = -0.34$ e Å⁻³

[(2-(Tetrazol-5-yl)pyridine)Cu(DPEPhos)]|(5-(pyridin-2-yl)tetrazolate)Cu (DPEPhos)]PF₆ (7-H-Cu-10-PF₆) – CCDC 922233

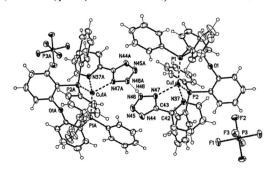

Empirical formula	C86 H69 Cl4 Cu2 F6 N10 O2 P5
	[(C42 H32 Cu N5 O P2)(C42 H33 Cu N5 O P2)]+ [P F6]- - 2 CH2Cl2
Formula weight	1812.24
Temperature	123(2) K
Wavelength	0.71073 A
Crystal system, space group	Triclinic, P-1 (No.2)
Unit cell dimensions	a = 12.294(1) A alpha = 71.00(1) deg.
	b = 13.170(1) A beta = 85.67(1) deg.
	c = 14.471(1) A gamma = 69.54(1) deg.
Volume	2073.7(3) A^3
Z, Calculated density	1, 1.451 Mg/m^3
Absorption coefficient	0.806 mm^-1
F(000)	926
Crystal size	0.45 x 0.25 x 0.10 mm
Theta range for data collection	2.93 to 27.48 deg.
Limiting indices	-15<=h<=15, -17<=k<=17, -18<=l<=18
Reflections collected / unique	42424 / 9464 [R(int) = 0.0271]
Completeness to theta = 27.48	99.7 %
Absorption correction	Semi-empirical from equivalents
Max. and min. transmission	0.9281 and 0.7847
Refinement method	Full-matrix least-squares on F^2
Data / restraints / parameters	9464 / 22 / 523
Goodness-of-fit on F^2	1.029
Final R indices [I>2sigma(I)]	R1 = 0.0565, wR2 = 0.1509
R indices (all data)	R1 = 0.0675, wR2 = 0.1615
Largest diff. peak and hole	1.441 and -1.853 e.A^-3

[(2-(Pyridin-2-yl)benzimidazolat)$_3$Cu$_4$(PPh$_3$)$_6$]PF$_6$ (c-(5-Cu-8-PF$_6$))   –   CCDC 1047154

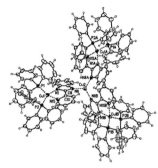

Empirical formula	C146 H117 Cu4 F6 N10 P7
	C144 H114 Cu4 N10 P6 - PF6 - CH3CN
Formula weight	2596.45
Temperature	123(2) K
Wavelength	0.71073 A
Crystal system, space group	Trigonal, R-3 (no.148)
Unit cell dimensions	a = 25.7574(8) A alpha = 90 deg.
	b = 25.7574(8) A beta = 90 deg.
	c = 33.5059(13) A gamma = 120 deg.
Volume	19251.1(11) A^3
Z, Calculated density	6, 1.344 Mg/m^3
Absorption coefficient	0.806 mm^-1
F(000)	8028
Crystal size	0.32 x 0.20 x 0.08 mm
Theta range for data collection	3.04 to 25.03 deg.
Limiting indices	-27<=h<=24, -28<=k<=30, -39<=l<=21
Reflections collected / unique	20137 / 7301 [R(int) = 0.0787]
Completeness to theta = 25.03	96.5 %
Absorption correction	Semi-empirical from equivalents
Max. and min. transmission	0.93321 and 0.78174
Refinement method	Full-matrix least-squares on F^2
Data / restraints / parameters	7301 / 21 / 546
Goodness-of-fit on F^2	1.066
Final R indices [I>2sigma(I)]	R1 = 0.0866, wR2 = 0.1955
R indices (all data)	R1 = 0.1215, wR2 = 0.2258
Largest diff. peak and hole	1.112 and -0.720 e.A^-3

[(5-(Pyridin-2-yl)1,2,3-triazolate)$_2$Cu$_3$(PPh$_3$)$_6$]PF$_6$ (c-(6-Cu-8-PF$_6$))   –   CCDC 1047155

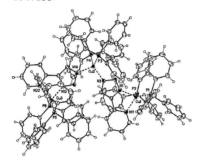

Crystal Data

C$_{122}$H$_{100}$Cu$_3$N$_8$P$_6$·F$_6$P·C$_4$H$_{10}$O	$Z = 2$
$M_r = 2273.62$	$F(000) = 2352$
Triclinic, P-1 (no.2)	$D_x = 1.358$ Mg m^{-3}
$a = 15.4306$ (2) Å	Mo $K\alpha$ radiation, $\lambda = 0.71073$ Å
$b = 19.2247$ (2) Å	Cell parameters from 23847 reflections
$c = 19.3598$ (2) Å	$\theta = 1.0$–27.5°
$\alpha = 84.874$ (1)°	$\mu = 0.74$ mm^{-1}
$\beta = 87.647$ (1)°	$T = 123$ K
$\gamma = 76.399$ (1)°	Plates, colourless
$V = 5558.49$ (11) Å^3	$0.40 \times 0.32 \times 0.08$ mm

Data collection

Bruker APEXII diffractometer	21791 reflections with $I > 2\sigma(I)$
Radiation source: fine-focus sealed tube	$R_{int} = 0.032$
rotation in ϕ and ω, 1° scans	$\theta_{max} = 27.5°$, $\theta_{min} = 1.5°$
Absorption correction: multi-scan *MULABS* (*PLATON* Spek 2008, cf. Blessing 1995)	$h = -19 \rightarrow 20$
$T_{min} = 0.813$, $T_{max} = 0.900$	$k = -24 \rightarrow 24$
81116 measured reflections	$l = -25 \rightarrow 25$
25272 independent reflections	

Refinement

Refinement on F^2	Primary atom site location: heavy-atom method
Least-squares matrix: full	Secondary atom site location: difference Fourier map
$R[F^2 > 2\sigma(F^2)] = 0.051$	Hydrogen site location: inferred from neighbouring sites
$wR(F^2) = 0.135$	H-atom parameters constrained
$S = 1.03$	$w = 1/[\sigma^2(F_o^2) + (0.0556P)^2 + 13.0748P]$ where $P = (F_o^2 + 2F_c^2)/3$
25272 reflections	$(\Delta/\sigma)_{max} = 0.001$
1340 parameters	$\Delta\rangle_{max} = 1.79$ e Å^{-3}
434 restraints	$\Delta\rangle_{min} = -1.18$ e Å^{-3}

[(5-(Pyridin-2-yl)tetrazolate)Cu(PPh₃)₂]BF₄ (7-Cu-8) – CCDC 922230

$$[(5\text{-(Pyridin-2-yl)tetrazolate)}Cu(PPh_3)_2]BF_4 \ (7\text{-Cu-8}) - CCDC\ 922230$$

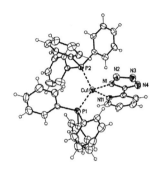

Empirical formula	C42 H34 Cu N5 P2
Formula weight	734.22
Temperature	123(2) K
Wavelength	0.71073 A
Crystal system, space group	Monoclinic, P2(1)/n (no.14)
Unit cell dimensions	a = 9.0816(2) A alpha = 90 deg.
	b = 18.2546(6) A beta = 91.356(2) deg.
	c = 22.2364(6) A gamma = 90 deg.
Volume	3685.34(18) A^3
Z, Calculated density	4, 1.323 Mg/m^3
Absorption coefficient	0.717 mm^-1
F(000)	1520
Crystal size	0.24 x 0.10 x 0.06 mm
Theta range for data collection	2.97 to 25.03 deg.
Limiting indices	-10<=h<=10, -21<=k<=20, -26<=l<=21
Reflections collected / unique	22790 / 6500 [R(int) = 0.1038]
Completeness to theta = 25.03	99.7 %
Absorption correction	Semi-empirical from equivalents
Max. and min. transmission	0.94806 and 0.84718
Refinement method	Full-matrix least-squares on F^2
Data / restraints / parameters	6500 / 8 / 420
Goodness-of-fit on F^2	1.139
Final R indices [I>2sigma(I)]	R1 = 0.0765, wR2 = 0.1337
R indices (all data)	R1 = 0.1137, wR2 = 0.1489
Largest diff. peak and hole	0.457 and -0.385 e.A^-3

[(5-(Pyridin-2-yl)1,2,3-triazolate)₂Cu₄(dppm)₄](BF₄)₂ (c-(6-Cu-9)) – CCDC 1047156

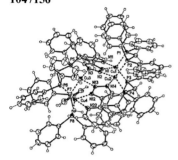

Crystal Data

$C_{114}H_{98}Cu_4N_8P_8 \cdot 2(BF_4)$	$Z = 2$
$M_r = 2255.54$	$F(000) = 2312$
Triclinic, $P\text{-}1$ (no.2)	$D_x = 1.378$ Mg m^{-3}
$a = 12.7716$ (6) Å	Cu $K\alpha$ radiation, $\lambda = 1.54178$ Å
$b = 15.7488$ (7) Å	Cell parameters from 9975 reflections
$c = 28.7966$ (13) Å	$\theta = 3.1–68.3°$
$\alpha = 87.544$ (2)°	$\mu = 2.53$ mm^{-1}
$\beta = 80.793$ (2)°	$T = 123$ K
$\gamma = 71.984$ (2)°	Plates, colourless
$V = 5437.0$ (4) Å^3	$0.28 \times 0.20 \times 0.08$ mm

Data collection

Bruker D8 Venture diffractometer with PHOTON 100 detector	17026 reflections with $I > 2\sigma(I)$
Radiation source: IμS microfocus	$R_{int} = 0.036$
rotation in ϕ and ω, 0.5°, shutterless scans	$\theta_{max} = 68.3°$, $\theta_{min} = 3.0°$
Absorption correction: multi-scan *SADABS* (Sheldrick, 2014)	$h = -15\rightarrow15$
$T_{min} = 0.592$, $T_{max} = 0.753$	$k = -18\rightarrow17$
68531 measured reflections	$l = -34\rightarrow34$
19857 independent reflections	

Refinement

Refinement on F^2	Primary atom site location: structure-invariant direct methods
Least-squares matrix: full	Secondary atom site location: difference Fourier map
$R[F^2 > 2\sigma(F^2)] = 0.045$	Hydrogen site location: inferred from neighbouring sites
$wR(F^2) = 0.116$	H-atom parameters constrained
$S = 1.02$	$w = 1/[\sigma^2(F_o^2) + (0.053P)^2 + 9.770P]$ where $P = (F_o^2 + 2F_c^2)/3$
19857 reflections	$(\Delta/\sigma)_{max} = 0.003$
1287 parameters	$\Delta\rangle_{max} = 1.40$ e Å^{-3}
396 restraints	$\Delta\rangle_{min} = -1.04$ e Å^{-3}

[(5-(Pyridin-2-yl)1,2,3-triazolate)₂Cu₄(dppm)₄](PF₆)₂ (c-(6-Cu-9-PF₆)) – CCDC 1047157

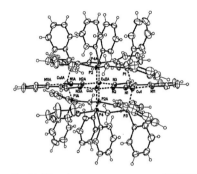

Empirical formula	C116 H102 Cl4 Cu4 F12 N8 P10
	C114 H98 Cu4 N8 P8 - 2 PF6 - 2 CH2Cl2
Formula weight	2541.72
Temperature	173(2) K
Wavelength	0.71073 A
Crystal system, space group	Triclinic, P-1 (no.2)
Unit cell dimensions	a = 13.082(2) A alpha = 88.87(1) deg.
	b = 13.714(2) A beta = 78.22(1) deg.
	c = 16.919(2) A gamma = 72.43(1) deg.
Volume	2830.0(7) A^3
Z, Calculated density	1, 1.491 Mg/m^3
Absorption coefficient	1.050 mm^-1
F(000)	1296
Crystal size	0.40 x 0.24 x 0.08 mm
Theta range for data collection	2.97 to 27.48 deg.
Limiting indices	-16<=h<=16, -17<=k<=17, -21<=l<=21
Reflections collected / unique	53126 / 12938 [R(int) = 0.0348]
Completeness to theta = 27.48	99.8 %
Absorption correction	Semi-empirical from equivalents
Max. and min. transmission	0.9208 and 0.7232
Refinement method	Full-matrix least-squares on F^2
Data / restraints / parameters	12938 / 148 / 681
Goodness-of-fit on F^2	1.033
Final R indices [I>2sigma(I)]	R1 = 0.0506, wR2 = 0.1272
R indices (all data)	R1 = 0.0703, wR2 = 0.1392
Largest diff. peak and hole	1.244 and -1.149 e.A^-3

[(5-(Pyridin-2-yl)tetrazolate)$_2$Cu$_4$(dppm)$_4$](BF$_4$)$_2$ (c-(7-Cu-9)) – CCDC 1047158

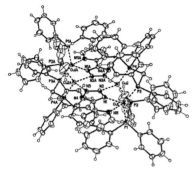

Crystal Data

C$_{112}$H$_{96}$Cu$_4$N$_{10}$P$_8$·2(BF$_4$)·2(CH$_2$Cl$_2$)	$F(000) = 2480$
$M_r = 2427.37$	$D_x = 1.449$ Mg m^{-3}
Monoclinic, $P2_1/c$ *(no.14)*	Mo $K\alpha$ radiation, $\lambda = 0.71073$ Å
$a = 15.264$ (2) Å	Cell parameters from 337 reflections
$b = 22.970$ (3) Å	$\theta = 2.5–25.0°$
$c = 17.299$ (3) Å	$\mu = 1.03$ mm^{-1}
$\beta = 113.43$ (1)°	$T = 123$ K
$V = 5565.2$ (15) Å^3	Plates, pale green
$Z = 2$	$0.32 \times 0.16 \times 0.08$ mm

Data collection

Bruker-Nonius KappaCCD diffractometer	10071 reflections with $I > 2\sigma(I)$
Radiation source: fine-focus sealed tube	$R_{int} = 0.038$
rotation in ϕ and ω, 1° scans	$\theta_{max} = 27.5°$, $\theta_{min} = 2.6°$
Absorption correction: multi-scan *SADABS* (Sheldrcik, 2008)	$h = -19\rightarrow19$
$T_{min} = 0.798$, $T_{max} = 0.914$	$k = -27\rightarrow29$
56450 measured reflections	$l = -22\rightarrow22$
12758 independent reflections	

Refinement

Refinement on F^2	Primary atom site location: heavy-atom method
Least-squares matrix: full	Secondary atom site location: difference Fourier map
$R[F^2 > 2\sigma(F^2)] = 0.045$	Hydrogen site location: inferred from neighbouring sites
$wR(F^2) = 0.112$	H-atom parameters constrained
$S = 1.03$	$w = 1/[\sigma^2(F_o^2) + (0.0467P)^2 + 8.8275P]$ where $P = (F_o^2 + 2F_c^2)/3$
12758 reflections	$(\Delta/\sigma)_{max} = 0.001$
693 parameters	$\Delta\rangle_{max} = 1.24$ e Å^{-3}
49 restraints	$\Delta\rangle_{min} = -0.95$ e Å^{-3}

[(5-(Pyridin-2-yl)tetrazolate)₂Cu₄(dppm)₄](BF₄)₂   (c-(7-Cu-9-PF₆))   –   CCDC 1047159

[(5-(Pyridin-2-yl)tetrazolate)$_2$Cu$_4$(dppm)$_4$](BF$_4$)$_2$ (c-(7-Cu-9-PF$_6$)) – CCDC 1047159

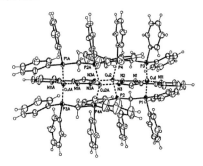

Empirical formula	C114 H100 Cl4 Cu4 F12 N10 P10
	C112 H96 Cu4 N10 P8 -
	2 PF6 - 2 CH2Cl2
Formula weight	2543.70
Temperature	123(2) K
Wavelength	0.71073 A
Crystal system, space group	Monoclinic, P2(1)/c (No.14)
Unit cell dimensions	a = 15.335(2) A alpha = 90 deg.
	b = 23.104(3) A beta = 112.48(2) deg.
	c = 17.089(2) A gamma = 90 deg.
Volume	5594.6(12) A^3
Z, Calculated density	2, 1.510 Mg/m^3
Absorption coefficient	1.062 mm^-1
F(000)	2592
Crystal size	0.30 x 0.30 x 0.06 mm
Theta range for data collection	2.94 to 27.48 deg.
Limiting indices	-19<=h<=19, -27<=k<=30, -22<=l<=22
Reflections collected / unique	47084 / 12773 [R(int) = 0.0484]
Completeness to theta = 27.48	99.7 %
Absorption correction	Semi-empirical from equivalents
Max. and min. transmission	0.9422 and 0.6798
Refinement method	Full-matrix least-squares on F^2
Data / restraints / parameters	12773 / 21 / 694
Goodness-of-fit on F^2	1.027
Final R indices [I>2sigma(I)]	R1 = 0.0588, wR2 = 0.1431
R indices (all data)	R1 = 0.0899, wR2 = 0.1592
Largest diff. peak and hole	1.634 and -1.427 e.A^-3

[(2-(Pyridin-2-yl)benzimidazolat)Cu(DPEPhos)] (5-Cu-10) – CCDC 1047160

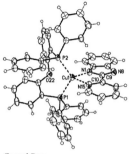

Crystal Data

$C_{48}H_{36}CuN_3OP_2$	$F(000) = 1648$
$M_r = 796.28$	$D_x = 1.376$ Mg m^{-3}
Monoclinic, $P2_1/n$ *(no.14)*	Mo $K\alpha$ radiation, $\lambda = 0.71073$ Å
$a = 9.144$ (1) Å	Cell parameters from 619 reflections
$b = 18.331$ (2) Å	$\theta = 2.5–25.0°$
$c = 23.049$ (2) Å	$\mu = 0.69$ mm^{-1}
$\beta = 95.93$ (1)°	$T = 123$ K
$V = 3842.8$ (7) Å^3	Plates, yellow
$Z = 4$	$0.30 \times 0.08 \times 0.04$ mm

Data collection

Bruker-nobnius KappaCCD diffractometer	4986 reflections with $I > 2\sigma(I)$
Radiation source: fine-focus sealed tube	$R_{int} = 0.116$
rotation in ϕ and ω, 1° scans	$\theta_{max} = 27.5°$, $\theta_{min} = 2.6°$
Absorption correction: multi-scan SADABS (Sheldrick, 2008)	$h = -11 \rightarrow 11$
$T_{min} = 0.596$, $T_{max} = 0.982$	$k = -23 \rightarrow 23$
32257 measured reflections	$l = -29 \rightarrow 29$
8809 independent reflections	

Refinement

Refinement on F^2	Primary atom site location: structure-invariant direct methods
Least-squares matrix: full	Secondary atom site location: difference Fourier map
$R[F^2 > 2\sigma(F^2)] = 0.110$	Hydrogen site location: inferred from neighbouring sites
$wR(F^2) = 0.309$	H-atom parameters constrained
$S = 1.05$	$w = 1/[\sigma^2(F_o^2) + (0.186P)^2 + 1.1P]$ where $P = (F_o^2 + 2F_c^2)/3$
8809 reflections	$(\Delta/\sigma)_{max} < 0.001$
496 parameters	$\Delta\rangle_{max} = 2.97$ e Å^{-3}
777 restraints	$\Delta\rangle_{min} = -1.69$ e Å^{-3}

[(5-(Pyridin-2-yl)tetrazolate)Cu(DPEPhos)] (7-Cu-10) – CCDC 922232

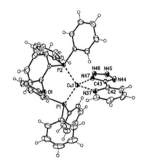

Empirical formula	C42 H32 Cu N5 O P2
Formula weight	748.21
Temperature	123(2) K
Wavelength	0.71073 A
Crystal system, space group	Monoclinic, P2(1)/n (No.14)
Unit cell dimensions	a = 9.0162(4) A alpha = 90 deg.
	b = 17.8097(12) A beta = 90.648(6) deg.
	c = 22.3584(19) A gamma = 90 deg.
Volume	3590.0(4) A^3
Z, Calculated density	4, 1.384 Mg/m^3
Absorption coefficient	0.739 mm^-1
F(000)	1544
Crystal size	0.45 x 0.20 x 0.15 mm
Theta range for data collection	2.96 to 27.48 deg.
Limiting indices	-11<=h<=11, -23<=k<=23, -28<=l<=29
Reflections collected / unique	30988 / 8194 [R(int) = 0.0338]
Completeness to theta = 27.48	99.7 %
Absorption correction	Semi-empirical from equivalents
Max. and min. transmission	0.9010 and 0.7510
Refinement method	Full-matrix least-squares on F^2
Data / restraints / parameters	8194 / 0 / 460
Goodness-of-fit on F^2	1.051
Final R indices [I>2sigma(I)]	R1 = 0.0343, wR2 = 0.0721
R indices (all data)	R1 = 0.0494, wR2 = 0.0779
Largest diff. peak and hole	0.388 and -0.358 e.A^-3

[(5-(Pyridin-2-yl)tetrazolate)Cu(PTEPhos)] (7-Cu-18) – CCDC 922235

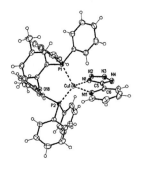

Empirical formula	C44 H36 Cu N5 O P2
Formula weight	776.26
Temperature	123(2) K
Wavelength	0.71073 A
Crystal system, space group	Monoclinic, P2(1)/n
Unit cell dimensions	a = 9.3177(7) A alpha = 90 deg.
	b = 18.3378(9) A beta = 90.654(7) deg.
	c = 22.3498(18) A gamma = 90 deg.
Volume	3818.6(5) A^3
Z, Calculated density	4, 1.350 Mg/m^3
Absorption coefficient	0.698 mm^-1
F(000)	1608
Crystal size	0.24 x 0.12 x 0.06 mm
Theta range for data collection	2.95 to 27.48 deg.
Limiting indices	-12<=h<=12, -23<=k<=23, -29<=l<=29
Reflections collected / unique	67074 / 8734 [R(int) = 0.0328]
Completeness to theta = 27.48	99.9 %
Absorption correction	Semi-empirical from equivalents
Max. and min. transmission	0.9562 and 0.8603
Refinement method	Full-matrix least-squares on F^2
Data / restraints / parameters	8734 / 0 / 480
Goodness-of-fit on F^2	1.075
Final R indices [I>2sigma(I)]	R1 = 0.0307, wR2 = 0.0691
R indices (all data)	R1 = 0.0418, wR2 = 0.0743
Largest diff. peak and hole	0.393 and -0.366 e.A^-3

[(5-(Pyridin-2-yl)tetrazolate)Cu(Xantphos)] (7-Cu-19) – CCDC 922234

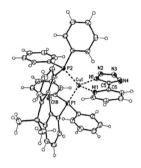

Empirical formula	C45 H36 Cu N5 O P2
Formula weight	788.27
Temperature	123(2) K
Wavelength	0.71073 A
Crystal system, space group	Monoclinic, P2(1)/n (no.14)
Unit cell dimensions	a = 10.0720(6) A alpha = 90 deg. b = 20.3844(18) A beta = 94.184(5) deg. c = 17.9511(18) A gamma = 90 deg.
Volume	3675.7(5) A^3
Z, Calculated density	4, 1.424 Mg/m^3
Absorption coefficient	0.726 mm^-1
F(000)	1632
Crystal size	0.50 x 0.20 x 0.10 mm
Theta range for data collection	3.01 to 27.48 deg.
Limiting indices	-13<=h<=12, -26<=k<=26, -20<=l<=23
Reflections collected / unique	28344 / 8393 [R(int) = 0.0259]
Completeness to theta = 27.48	99.6 %
Absorption correction	Semi-empirical from equivalents
Max. and min. transmission	0.9281 and 0.7120
Refinement method	Full-matrix least-squares on F^2
Data / restraints / parameters	8393 / 0 / 489
Goodness-of-fit on F^2	1.035
Final R indices [I>2sigma(I)] R indices (all data)	R1 = 0.0270, wR2 = 0.0661 R1 = 0.0334, wR2 = 0.0696
Largest diff. peak and hole	0.359 and -0.435 e.A^-3

[(5-(Pyridin-2-yl)tetrazolate)Cu(BINAP)] (7-Cu-23)

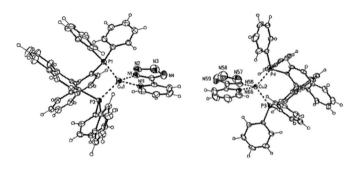

Empirical formula	C50 H36 Cu N5 P2
Formula weight	832.32
Temperature	123(2) K
Wavelength	0.71073 A
Crystal system, space group	Orthorhombic, P2(1)2(1)2(1) (no.19)
Unit cell dimensions	a = 12.652(2) A alpha = 90 deg. b = 15.812(2) A beta = 90 deg. c = 41.293(3) A gamma = 90 deg.
Volume	8260.8(18) A^3
Z, Calculated density	8, 1.338 Mg/m^3
Absorption coefficient	0.648 mm^-1
F(000)	3440
Crystal size	0.20 x 0.10 x 0.05 mm
Theta range for data collection	2.95 to 25.03 deg.
Limiting indices	-12<=h<=15, -18<=k<=18, -44<=l<=49
Reflections collected / unique	41173 / 14510 [R(int) = 0.0622]
Completeness to theta = 25.03	99.8 %
Absorption correction	Semi-empirical from equivalents
Max. and min. transmission	0.9703 and 0.7458
Refinement method	Full-matrix least-squares on F^2
Data / restraints / parameters	14510 / 0 / 1045
Goodness-of-fit on F^2	1.048
Final R indices [I>2sigma(I)] R indices (all data)	R1 = 0.0530, wR2 = 0.1062 R1 = 0.0753, wR2 = 0.1149
Absolute structure parameter	0.000(12)
Largest diff. peak and hole	0.929 and -0.447 e.A^-3

[(5-(Pyrazin-2-yl)tetrazolate)Cu(DPEPhos)] (28-Cu-10)

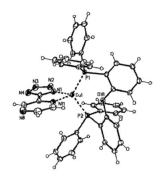

Empirical formula	C41 H31 Cu N6 O P2
Formula weight	749.20
Temperature	123(2) K
Wavelength	0.71073 A
Crystal system, space group	Monoclinic, P2(1)/n (no.14)
Unit cell dimensions	a = 11.157(1) A alpha = 90 deg. b = 22.589(2) A beta = 93.35(1) deg. c = 13.841(1) A gamma = 90 deg.
Volume	3482.3(5) A^3
Z, Calculated density	4, 1.429 Mg/m^3
Absorption coefficient	0.763 mm^-1
F(000)	1544
Crystal size	0.40 x 0.30 x 0.20 mm
Theta range for data collection	2.57 to 27.48 deg.
Limiting indices	-14<=h<=14, -29<=k<=29, -17<=l<=17
Reflections collected / unique	59759 / 7983 [R(int) = 0.0220]
Completeness to theta = 27.48	99.8 %
Absorption correction	Semi-empirical from equivalents
Max. and min. transmission	0.8621 and 0.7868
Refinement method	Full-matrix least-squares on F^2
Data / restraints / parameters	7983 / 0 / 460
Goodness-of-fit on F^2	1.046
Final R indices [I>2sigma(I)] R indices (all data)	R1 = 0.0244, wR2 = 0.0606 R1 = 0.0284, wR2 = 0.0631
Largest diff. peak and hole	0.325 and -0.366 e.A^-3

[(5-(Quinolin-2-yl)tetrazolate)Cu(DPEPhos)] (29-Cu-10)

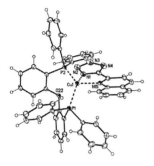

Crystal Data

$C_{46}H_{34}CuN_5OP_2$	$F(000) = 1648$
$M_r = 798.26$	$D_x = 1.420$ Mg m^{-3}
Monoclinic, $P2_1/n$ *(no.14)*	Mo $K\alpha$ radiation, $\lambda = 0.71073$ Å
$a = 11.7872$ (4) Å	Cell parameters from 9744 reflections
$b = 22.5563$ (8) Å	$\theta = 2.5$–$27.5°$
$c = 14.0828$ (5) Å	$\mu = 0.72$ mm^{-1}
$\beta = 94.030$ (1)°	$T = 123$ K
$V = 3735.0$ (2) Å^3	Plates, yellow
$Z = 4$	$0.20 \times 0.12 \times 0.06$ mm

Data collection

Bruker D8 Venture diffractometer with Photon 100 detector	7710 reflections with $I > 2\sigma(I)$
Radiation source: IµS microfocus	$R_{int} = 0.025$
rotation in ϕ and ω, 05°, shutterless scans	$\theta_{max} = 27.6°$, $\theta_{min} = 2.3°$
Absorption correction: multi-scan *SADABS* (Sheldrick, 2014)	$h = -15\rightarrow15$
$T_{min} = 0.838$, $T_{max} = 0.963$	$k = -29\rightarrow28$
46062 measured reflections	$l = -18\rightarrow18$
8617 independent reflections	

Refinement

Refinement on F^2	Primary atom site location: structure-invariant direct methods
Least-squares matrix: full	Secondary atom site location: difference Fourier map
$R[F^2 > 2\sigma(F^2)] = 0.027$	Hydrogen site location: difference Fourier map
$wR(F^2) = 0.071$	H-atom parameters constrained
$S = 1.04$	$w = 1/[\sigma^2(F_o^2) + (0.0341P)^2 + 2.0738P]$ where $P = (F_o^2 + 2F_c^2)/3$
8617 reflections	$(\Delta/\sigma)_{max} = 0.002$
496 parameters	$\Delta\rangle_{max} = 0.31$ e Å^{-3}
0 restraints	$\Delta\rangle_{min} = -0.49$ e Å^{-3}

[(5-(Isoquinolin-1-yl)tetrazolate)Cu(DPEPhos)] (30-Cu-10)

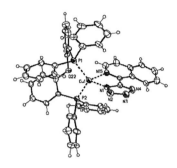

Empirical formula	C46 H34 Cu N5 O P2
Formula weight	798.26
Temperature	123(2) K
Wavelength	0.71073 A
Crystal system, space group	Triclinic, P-1 (no.2)
Unit cell dimensions	a = 10.2892(7) A alpha = 96.152(4) deg. b = 12.5867(6) A beta = 105.389(3) deg. c = 15.0591(9) A gamma = 91.692(3) deg.
Volume	1866.09(19) A^3
Z, Calculated density	2, 1.421 Mg/m^3
Absorption coefficient	0.716 mm^-1
F(000)	824
Crystal size	0.20 x 0.06 x 0.02 mm
Theta range for data collection	3.01 to 25.02 deg.
Limiting indices	-12<=h<=12, -14<=k<=14, -17<=l<=17
Reflections collected / unique	15471 / 6361 [R(int) = 0.1061]
Completeness to theta = 25.02	96.6 %
Absorption correction	Semi-empirical from equivalents
Max. and min. transmission	0.90295 and 0.78897
Refinement method	Full-matrix least-squares on F^2
Data / restraints / parameters	6361 / 0 / 496
Goodness-of-fit on F^2	1.201
Final R indices [I>2sigma(I)]	R1 = 0.0964, wR2 = 0.1667
R indices (all data)	R1 = 0.1570, wR2 = 0.1944
Largest diff. peak and hole	0.665 and -0.500 e.A^-3

[(5-([2](1,4)benzo[2](2,5)pyridinophan-13-yl)tetrazolate)Cu(DPEPhos)]   (31-Cu-10)

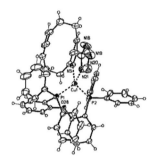

Empirical formula	C53 H44 Cl2 Cu N5 O P2
	C52 H42 Cu N5 O P2 - C H2 Cl2 solvent 'squeezed out'
Formula weight	963.31
Temperature	123(2) K
Wavelength	0.71073 A
Crystal system, space group	Monoclinic, P2(1)/c (no.14)
Unit cell dimensions	a = 11.768(1) A alpha = 90 deg.
	b = 13.648(1) A beta = 95.02(1) deg.
	c = 28.942(2) A gamma = 90 deg.
Volume	4630.5(6) A^3
Z, Calculated density	4, 1.382 Mg/m^3
Absorption coefficient	0.702 mm^-1
F(000)	1992
Crystal size	0.16 x 0.08 x 0.04 mm
Theta range for data collection	2.59 to 26.00 deg.
Limiting indices	-14<=h<=14, 0<=k<=16, 0<=l<=35
Reflections collected / unique	9058 / 9058 [R(int) = 0.0000]
Completeness to theta = 25.00	99.6 %
Absorption correction	Semi-empirical from equivalents
Max. and min. transmission	0.9703 and 0.7416
Refinement method	Full-matrix least-squares on F^2
Data / restraints / parameters	9058 / 0 / 550
Goodness-of-fit on F^2	1.050
Final R indices [I>2sigma(I)]	R1 = 0.0641, wR2 = 0.1502
R indices (all data)	R1 = 0.1069, wR2 = 0.1644
Largest diff. peak and hole	0.647 and -0.629 e.A^-3

[(5-(N-Methylimidazol-2-yl)tetrazolate)Cu(DPEPHos)] (34-Cu-10)

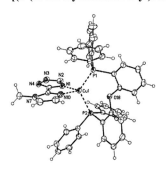

Crystal Data

$C_{41}H_{33}CuN_6OP_2 \cdot 2(H_2O)$	$F(000) = 1632$
$M_r = 787.24$	$D_x = 1.418$ Mg m^{-3}
Monoclinic, $P2_1/n$ *(no.14)*	Mo $K\alpha$ radiation, $\lambda = 0.71073$ Å
$a = 11.315$ (1) Å	Cell parameters from 211 reflections
$b = 23.135$ (2) Å	$\theta = 2.5$–$25.0°$
$c = 14.167$ (1) Å	$\mu = 0.73$ mm^{-1}
$\beta = 96.24$ (1)°	$T = 123$ K
$V = 3686.6$ (5) Å^3	Blocks, yellow
$Z = 4$	$0.55 \times 0.35 \times 0.20$ mm

Data collection

Bruker-Nonius KappaCCD diffractometer	6050 reflections with $I > 2\sigma(I)$
Radiation source: fine-focus sealed tube	$R_{int} = 0.038$
rotation in ϕ and ω, 1° scans	$\theta_{max} = 25.3°$, $\theta_{min} = 2.5°$
Absorption correction: multi-scan *SADABS* (Sheldrick, 2008)	$h = -13 \rightarrow 13$
$T_{min} = 0.689$, $T_{max} = 0.865$	$k = -27 \rightarrow 27$
47161 measured reflections	$l = -16 \rightarrow 16$
6660 independent reflections	

Refinement

Refinement on F^2	Primary atom site location: structure-invariant direct methods
Least-squares matrix: full	Secondary atom site location: difference Fourier map
$R[F^2 > 2\sigma(F^2)] = 0.042$	Hydrogen site location: difference Fourier map
$wR(F^2) = 0.111$	H atoms treated by a mixture of independent and constrained refinement
$S = 1.06$	$w = 1/[\sigma^2(F_o^2) + (0.046P)^2 + 6.720P]$ where $P = (F_o^2 + 2F_c^2)/3$
6660 reflections	$(\Delta/\sigma)_{max} < 0.001$
491 parameters	$\Delta\rangle_{max} = 1.70$ e Å^{-3}
6 restraints	$\Delta\rangle_{min} = -0.76$ e Å^{-3}

[(5-(N-Hexylimidazol-2-yl)tetrazolate)Cu(DPEPHos)] (35-Cu-10)

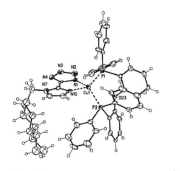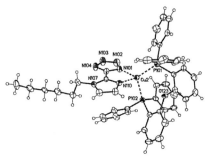

Crystal Data

$C_{46}H_{43}CuN_6OP_2$	$Z = 4$
$M_r = 821.34$	$F(000) = 1712$
Triclinic, $P\text{-}1$ (no.2)	$D_x = 1.349$ Mg m^{-3}
$a = 9.263$ (1) Å	Mo $K\alpha$ radiation, $\lambda = 0.71073$ Å
$b = 19.584$ (2) Å	Cell parameters from 215 reflections
$c = 22.610$ (1) Å	$\theta = 2.5$–$25.0°$
$\alpha = 87.62$ (1)°	$\mu = 0.66$ mm^{-1}
$\beta = 83.11$ (1)°	$T = 123$ K
$\gamma = 83.58$ (1)°	Plates, yellow
$V = 4044.8$ (6) Å^3	$0.20 \times 0.08 \times 0.04$ mm

Data collection

Bruker-Nonius KappaCCD diffractometer	8699 reflections with $I > 2\sigma(I)$
Radiation source: fine-focus sealed tube	$R_{int} = 0.118$
rotation in ϕ and ω, 1° scans	$\theta_{max} = 26.0°$, $\theta_{min} = 2.5°$
Absorption correction: multi-scan *SADABS* (Sheldrick, 2008)	$h = -11 \rightarrow 11$
$T_{min} = 0.846$, $T_{max} = 0.982$	$k = -24 \rightarrow 24$
51360 measured reflections	$l = -27 \rightarrow 27$
15844 independent reflections	

Refinement

Refinement on F^2	Primary atom site location: structure-invariant direct methods
Least-squares matrix: full	Secondary atom site location: difference Fourier map
$R[F^2 > 2\sigma(F^2)] = 0.094$	Hydrogen site location: inferred from neighbouring sites
$wR(F^2) = 0.240$	H-atom parameters constrained
$S = 1.06$	$w = 1/[\sigma^2(F_o^2) + (0.080P)^2 + 23.P]$ where $P = (F_o^2 + 2F_c^2)/3$
15844 reflections	$(\Delta/\sigma)_{max} < 0.001$
1000 parameters	$\Delta\rangle_{max} = 1.09$ e Å^{-3}
723 restraints	$\Delta\rangle_{min} = -0.69$ e Å^{-3}

[(5-(Benzothiazol-2-yl)tetrazolate)Cu(DPEPhos)] (40-Cu-10)

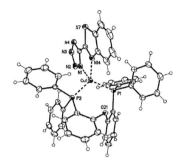

Crystal Data

$C_{44}H_{32}CuN_5OP_2S$	$F(000) = 1656$
$M_r = 804.28$	$D_x = 1.445$ Mg m^{-3}
Monoclinic, $P2_1/n$ *(no.14)*	Mo $K\alpha$ radiation, $\lambda = 0.71073$ Å
$a = 11.439$ (1) Å	Cell parameters from 320 reflections
$b = 22.988$ (2) Å	$\theta = 2.5–25.0°$
$c = 14.077$ (1) Å	$\mu = 0.78$ mm^{-1}
$\beta = 92.49$ (1)°	$T = 123$ K
$V = 3698.2$ (5) Å^3	Blocks, colourless
$Z = 4$	$0.20 \times 0.10 \times 0.10$ mm

Data collection

Bruker-Nonius KappaCCD diffractometer	7134 reflections with $I > 2\sigma(I)$
Radiation source: fine-focus sealed tube	$R_{int} = 0.035$
rotation in ϕ and ω, 1° scans	$\theta_{max} = 27.5°$, $\theta_{min} = 2.5°$
Absorption correction: multi-scan *SADABS* (Sheldrick, 2008)	$h = -14 \rightarrow 14$
$T_{min} = 0.782$, $T_{max} = 0.928$	$k = -29 \rightarrow 29$
61736 measured reflections	$l = -18 \rightarrow 18$
8459 independent reflections	

Refinement

Refinement on F^2	Primary atom site location: structure-invariant direct methods
Least-squares matrix: full	Secondary atom site location: difference Fourier map
$R[F^2 > 2\sigma(F^2)] = 0.031$	Hydrogen site location: difference Fourier map
$wR(F^2) = 0.073$	H-atom parameters constrained
$S = 1.05$	$w = 1/[\sigma^2(F_o^2) + (0.0287P)^2 + 2.546P]$ where $P = (F_o^2 + 2F_c^2)/3$
8459 reflections	$(\Delta/\sigma)_{max} = 0.002$
487 parameters	$\Delta\rangle_{max} = 0.33$ e Å^{-3}
0 restraints	$\Delta\rangle_{min} = -0.33$ e Å^{-3}

[(5-(4-Phenylpyridin-2-yl)tetrazolate)Cu(DPEPhos)] (41-Cu-10)

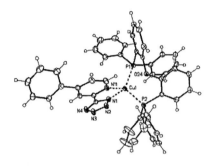

Empirical formula	C48 H36 Cu N5 O P2
Formula weight	824.30
Temperature	123(2) K
Wavelength	0.71073 A
Crystal system, space group	Monoclinic, P2(1)/n (no.14)
Unit cell dimensions	a = 11.284(1) A alpha = 90 deg. b = 23.909(2) A beta = 91.46(1) deg. c = 14.602(2) A gamma = 90 deg.
Volume	3938.2(7) A^3
Z, Calculated density	4, 1.390 Mg/m^3
Absorption coefficient	0.681 mm^-1
F(000)	1704
Crystal size	0.40 x 0.20 x 0.08 mm
Theta range for data collection	2.91 to 27.48 deg.
Limiting indices	-14<=h<=14, -31<=k<=31, -18<=l<=18
Reflections collected / unique	66986 / 9016 [R(int) = 0.0318]
Completeness to theta = 27.48	99.7 %
Absorption correction	Semi-empirical from equivalents
Max. and min. transmission	0.9422 and 0.8204
Refinement method	Full-matrix least-squares on F^2
Data / restraints / parameters	9016 / 0 / 514
Goodness-of-fit on F^2	1.066
Final R indices [I>2sigma(I)] R indices (all data)	R1 = 0.0308, wR2 = 0.0708 R1 = 0.0392, wR2 = 0.0746
Largest diff. peak and hole	0.399 and -0.345 e.A^-3

[(5-(4-Phenylpyridin-2-yl)tetrazolate)Cu(XantPhos)] (41-Cu-19)

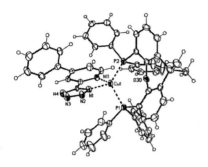

Empirical formula	C51 H40 Cu N5 O P2
Formula weight	864.36
Temperature	123(2) K
Wavelength	0.71073 A
Crystal system, space group	Monoclinic, P2(1)/c (no.14)
Unit cell dimensions	a = 17.828(3) A alpha = 90 deg.
	b = 11.828(2) A beta = 108.75(2) deg.
	c = 21.652(4) A gamma = 90 deg.
Volume	4323.4(13) A^3
Z, Calculated density	4, 1.328 Mg/m^3
Absorption coefficient	0.624 mm^-1
F(000)	1792
Crystal size	0.15 x 0.09 x 0.02 mm
Theta range for data collection	2.96 to 25.00 deg.
Limiting indices	-21<=h<=21, -14<=k<=14, -25<=l<=25
Reflections collected / unique	46447 / 7580 [R(int) = 0.1783]
Completeness to theta = 25.00	99.6 %
Absorption correction	Semi-empirical from equivalents
Max. and min. transmission	0.9965 and 0.8752
Refinement method	Full-matrix least-squares on F^2
Data / restraints / parameters	7580 / 0 / 541
Goodness-of-fit on F^2	1.043
Final R indices [I>2sigma(I)]	R1 = 0.0710, wR2 = 0.1065
R indices (all data)	R1 = 0.1549, wR2 = 0.1313
Largest diff. peak and hole	0.657 and -0.442 e.A^-3

[(5-(4-Phenylpyridin-2-yl)tetrazolate)Cu(DPTPhos)] (41-Cu-21)

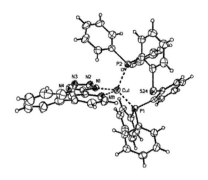

Empirical formula	C53 H48 Cl2 Cu N5 O P2 S
	C48 H36 Cu N5 P2 S - Et2O - CH2Cl2
Formula weight	999.40
Temperature	123(2) K
Wavelength	0.71073 A
Crystal system, space group	Triclinic, P-1 (no.2)
Unit cell dimensions	a = 12.474(1) A alpha = 100.54(1) deg.
	b = 14.101(1) A beta = 96.61(1) deg.
	c = 14.283(2) A gamma = 94.70(1) deg.
Volume	2439.8(4) A^3
Z, Calculated density	2, 1.360 Mg/m^3
Absorption coefficient	0.710 mm^-1
F(000)	1036
Crystal size	0.24 x 0.12 x 0.06 mm
Theta range for data collection	2.93 to 25.03 deg.
Limiting indices	-14<=h<=14, -16<=k<=16, -17<=l<=17
Reflections collected / unique	27939 / 8553 [R(int) = 0.0867]
Completeness to theta = 25.03	99.3 %
Absorption correction	Semi-empirical from equivalents
Max. and min. transmission	0.9562 and 0.7251
Refinement method	Full-matrix least-squares on F^2
Data / restraints / parameters	8553 / 82 / 586
Goodness-of-fit on F^2	1.024
Final R indices [I>2sigma(I)]	R1 = 0.0712, wR2 = 0.1683
R indices (all data)	R1 = 0.1316, wR2 = 0.1980
Largest diff. peak and hole	0.675 and -0.645 e.A^-3

[(5-(4-Methoxypyridin-2-yl)tetrazolate)Cu(DPEPhos)] (43-Cu-10)

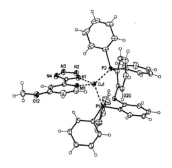

Crystal Data

$C_{43}H_{34}CuN_5O_2P_2$	$Z = 2$
$M_r = 778.23$	$F(000) = 804$
Triclinic, $P\text{-}1$	$D_x = 1.410$ Mg m^{-3}
$a = 10.2136$ (6) Å	Cu $K\alpha$ radiation, $\lambda = 1.54178$ Å
$b = 12.5768$ (7) Å	Cell parameters from 9977 reflections
$c = 14.8874$ (9) Å	$\theta = 3.1–72.1°$
$\alpha = 95.760$ (2)°	$\mu = 2.03$ mm^{-1}
$\beta = 105.248$ (2)°	$T = 123$ K
$\gamma = 91.76°$	Blocks, colourless
$V = 1832.46$ (19) Å^3	$0.10 \times 0.08 \times 0.06$ mm

Data collection

Bruker D8 Venture diffractometer with Photon100 detector	6440 reflections with $I > 2\sigma(I)$
Radiation source: IµS microfocus	$R_{int} = 0.036$
rotation in ϕ and ω, 1°, shutterless scans	$\theta_{max} = 72.2°$, $\theta_{min} = 3.1°$
Absorption correction: multi-scan *SADABS* (Sheldrick, 2015)	$h = -12 \rightarrow 12$
$T_{min} = 0.818$, $T_{max} = 0.915$	$k = -15 \rightarrow 15$
34807 measured reflections	$l = -18 \rightarrow 18$
7193 independent reflections	

Refinement

Refinement on F^2	Primary atom site location: heavy-atom method
Least-squares matrix: full	Secondary atom site location: difference Fourier map
$R[F^2 > 2\sigma(F^2)] = 0.028$	Hydrogen site location: difference Fourier map
$wR(F^2) = 0.073$	H-atom parameters constrained
$S = 1.01$	$w = 1/[\sigma^2(F_o^2) + (0.0284P)^2 + 1.2748P]$ where $P = (F_o^2 + 2F_c^2)/3$
7193 reflections	$(\Delta/\sigma)_{max} = 0.002$
479 parameters	$\Delta\rangle_{max} = 0.49$ e Å^{-3}
0 restraints	$\Delta\rangle_{min} = -0.41$ e Å^{-3}

[(5-(4-(Piperidyl)pyridin-2-yl)tetrazolate)Cu(DPEPhos)] (44-Cu-10)

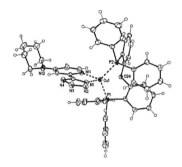

Crystal Data

$C_{47}H_{41}CuN_6OP_2$	$F(000) = 1728$
$M_r = 831.34$	$D_x = 1.354$ Mg m^{-3}
Monoclinic, $P2_1/c$ *(no.14)*	Mo $K\alpha$ radiation, $\lambda = 0.71073$ Å
$a = 10.962$ (1) Å	Cell parameters from 255 reflections
$b = 28.243$ (3) Å	$\theta = 2.5–25.0°$
$c = 13.171$ (1) Å	$\mu = 0.66$ mm^{-1}
$\beta = 91.06$ (1)°	$T = 123$ K
$V = 4077.0$ (6) Å^3	Blocks, colourless
$Z = 4$	$0.30 \times 0.25 \times 0.15$ mm

Data collection

Bruker-Nonius KappacCD diffractometer	7537 reflections with $I > 2\sigma(I)$
Radiation source: fine-focus sealed tube	$R_{int} = 0.036$
rotation in ϕ and ω, 1° scans	$\theta_{max} = 27.5°$, $\theta_{min} = 2.5°$
Absorption correction: multi-scan *SADABS* (Sheldrick, 2008)	$h = -14\rightarrow14$
$T_{min} = 0.826$, $T_{max} = 0.901$	$k = -36\rightarrow36$
48443 measured reflections	$l = -17\rightarrow16$
9346 independent reflections	

Refinement

Refinement on F^2	Primary atom site location: structure-invariant direct methods
Least-squares matrix: full	Secondary atom site location: difference Fourier map
$R[F^2 > 2\sigma(F^2)] = 0.037$	Hydrogen site location: difference Fourier map
$wR(F^2) = 0.080$	H-atom parameters constrained
$S = 1.07$	$w = 1/[\sigma^2(F_o^2) + (0.0249P)^2 + 3.442P]$ where $P = (F_o^2 + 2F_c^2)/3$
9346 reflections	$(\Delta/\sigma)_{max} = 0.001$
514 parameters	$\Delta\rangle_{max} = 0.38$ e Å^{-3}
0 restraints	$\Delta\rangle_{min} = -0.38$ e Å^{-3}

[(5-(2-Methylpyridin-6-yl)tetrazolate)Cu(DPEPhos)] (45-Cu-10)

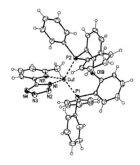

Crystal Data

$C_{43}H_{34}CuN_5OP_2$	$F(000) = 1576$
$M_r = 762.23$	$D_x = 1.354$ Mg m^{-3}
Monoclinic, $P2_1/n$ *(no.14)*	Mo $K\alpha$ radiation, $\lambda = 0.71073$ Å
$a = 9.2151 (7)$ Å	Cell parameters from 247 reflections
$b = 17.9878 (16)$ Å	$\theta = 2.5–25.0°$
$c = 22.5563 (14)$ Å	$\mu = 0.71$ mm^{-1}
$\beta = 91.147 (5)°$	$T = 123$ K
$V = 3738.2 (5)$ Å^3	Rods, yellow
$Z = 4$	$0.40 \times 0.12 \times 0.06$ mm

Data collection

Bruker-Nonius KappaCCD diffractometer	6666 reflections with $I > 2\sigma(I)$
Radiation source: fine-focus sealed tube	$R_{int} = 0.048$
rotation in ϕ and ω, 1° scans	$\theta_{max} = 27.5°$, $\theta_{min} = 2.6°$
Absorption correction: multi-scan *SADABS* (Sheldrick, 2008)	$h = -11 \rightarrow 11$
$T_{min} = 0.860$, $T_{max} = 0.942$	$k = -23 \rightarrow 22$
39543 measured reflections	$l = -28 \rightarrow 29$
8572 independent reflections	

Refinement

Refinement on F^2	Primary atom site location: structure-invariant direct methods
Least-squares matrix: full	Secondary atom site location: difference Fourier map
$R[F^2 > 2\sigma(F^2)] = 0.044$	Hydrogen site location: mixed
$wR(F^2) = 0.093$	H-atom parameters constrained
$S = 1.04$	$w = 1/[\sigma^2(F_o^2) + (0.0305P)^2 + 3.597P]$ where $P = (F_o^2 + 2F_c^2)/3$
8572 reflections	$(\Delta/\sigma)_{max} = 0.001$
461 parameters	$\Delta\rangle_{max} = 0.52$ e Å^{-3}
138 restraints	$\Delta\rangle_{min} = -0.36$ e Å^{-3}

[(5-(2-Methoxypyridin-6-yl)tetrazolate)Cu(DPEPhos)] (46-Cu-10)

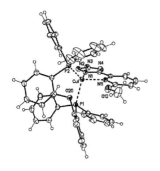

Crystal Data

$C_{43}H_{34}CuN_5O_2P_2$	$F(000) = 1608$
$M_r = 778.23$	$D_x = 1.368$ Mg m^{-3}
Monoclinic, $P2_1/c$	Mo $K\alpha$ radiation, $\lambda = 0.71073$ Å
$a = 11.723$ (1) Å	Cell parameters from 279 reflections
$b = 18.497$ (2) Å	$\theta = 2.5–25.0°$
$c = 17.881$ (2) Å	$\mu = 0.71$ mm^{-1}
$\beta = 103.02$ (1)°	$T = 123$ K
$V = 3777.6$ (7) Å^3	Blocks, colourless
$Z = 4$	$0.30 \times 0.20 \times 0.10$ mm

Data collection

Bruker-Nonius KappaCCD diffractometer	7370 reflections with $I > 2\sigma(I)$
Radiation source: fine-focus sealed tube	$R_{int} = 0.035$
rotation in ϕ and ω, 1° scans	$\theta_{max} = 27.5°$, $\theta_{min} = 2.6°$
Absorption correction: multi-scan *SADABS* (Sheldrick, 2008)	$h = -14 \rightarrow 15$
$T_{min} = 0.819$, $T_{max} = 0.928$	$k = -24 \rightarrow 24$
63599 measured reflections	$l = -23 \rightarrow 23$
8654 independent reflections	

Refinement

Refinement on F^2	Primary atom site location: structure-invariant direct methods
Least-squares matrix: full	Secondary atom site location: difference Fourier map
$R[F^2 > 2\sigma(F^2)] = 0.032$	Hydrogen site location: difference Fourier map
$wR(F^2) = 0.078$	H-atom parameters constrained
$S = 1.06$	$w = 1/[\sigma^2(F_o^2) + (0.0305P)^2 + 2.970P]$ where $P = (F_o^2 + 2F_c^2)/3$
8654 reflections	$(\Delta/\sigma)_{max} = 0.001$
479 parameters	$\Delta\rangle_{max} = 0.42$ e Å^{-3}
0 restraints	$\Delta\rangle_{min} = -0.33$ e Å^{-3}

6. Abbreviations

A	Ampere	DIOP	2,3-*O*-isopropylidene-2,3-dihydroxy-1,4-bis(diphenylphosphinobutane)
Å	Angstrom		
Ar	aromatic	DMF	*N,N*-dimethylformamide
ATR	attenuated total reflection	DMSO	dimethylsulfoxide
BINAP	2,2'-bis(diphenylphosphino)-1,1'-binaphthyl	DPEPhos	bis(2-(diphenylphosphino)-phenyl)ether
bs	broad singlet	dppm	bis(diphenylphosphino)-methane
Bu	butyl		
c	centi-	DPTPhos	bis(2-diphenylphosphino-phenyl)thioether
Cd	Candela		
C_{eff}	current efficiency	DRIFT	diffused reflectance infrared fourier transform spectroscopy
cH	cyclohexane		
CIE	Commission internationale de l'éclairage	DSC	differential scanning calorimetry
		DSSC	dye-sensitized solar cell
CT	charge transfer	e.g.	exempli gratia
CV	cyclovoltammetry	EA	elemental analysis
CVD	chemical vapor deposition	EE	ethyl acetate
d	doublet	EI	electron ionization
DBFPhos	4,6-bis(diphenylphosphino)-dibenzofuran	EML	emission layer
DEPT	Distortionless Enhancement by Polarization Transfer	EQE	external quantum yield
		equiv.	equivalent(s)
DFT	density functional theory	Et	ethyl

etc.	et cetera		J	coupling constant
ETL	electron transport layer		K	kilo
EtOH	ethanol		KIT	Karlsruhe Institute of Technology
eV	electron volt			
f	femto-		l	liter
FAB	fast atom bombardment		L	luminance
FOG	fluorescence optical gating		LED	light-emitting diode
FWHM	full-width at half maximum		LEEC	light-emitting electrochemical cell
g	gramm			
GC	gas chromatography		lm	lumen
GP	general procedure		LUMO	lowest unoccupied molecular orbital
h	hour(s)			
HOMO	highest occupied molecular orbital		m	medium
			M	mega
			M	molar
HTL	hole transport layer		m	multiplet, meter, mili-
Hz	Hertz		mCPBA	$meta$-chloroperoxybenzoic acid
I	current			
i.e.	id est		Me	methyl
IR	infra red		MeOH	methanol
IRF	instrument response function		min	minute(s)
ISC	intersystem crossing		MLCT	metal-to-ligand charge transfer
ITO	"indium tin oxide": nonstochiometric composite of SnO_2 and In_2O_3		(ML+IL)CT	metal- and ligand-to-ligand charge transfer

(ML+XL)CT	metal- and halide-to-ligand charge transfer	PMMA	poly(methyl methacrylate)
		ppm	parts per million
MS	mass spectrometry	PPV	poly(1,4-phenylene vinylene)
MTBE	methyl-*tert*-butyl ether	PTEPhos	bis(2-diphenylphosphino-*p*-tolyl)ether
n	nano-		
N^N	bidentate ligand coordinating via N-atoms	PVK	poly(n-vinylcarbazole)
		PyrBimH	2-(pyridin-2-yl)benzimidazole
*n*BuLi	*n*-butyllithium	PyrTetH	2-(tetrazol-5-yl)pyridine
NMR	nuclear magnetic resonance	PyrTriH	2-(1,2,3-triazol-5-yl)pyridine
OLED	organic light-emitting diode	q	quartet
p	pico-	quin	quintet
P^P	bidentate ligand coordinating via P-atoms	quind	doublet of quintet
PEDOT:PSS	polyethylene dioxythiophene doped with polystyrene sulfonic acid	R_f	retardation factor
		RISC	reverse intersystem crossing
P_{eff}	power efficiency	rt	room temperature
PFO	polyfluorene	s	singlet, seconds, strong
Ph	phenyl	SMOLED	small molecule organic light-emitting diode
PhanePhos	4,12-bis(diphenylphosphino)-[2.2]-paracyclophane	SOC	spin-orbit coupling
PL	photoluminescence	t	triplet
PLED	polymer light-emitting diode	TA	transient absorption
PLQY	photoluminescence quantum yield	TADF	thermally activated delayed fluorescence

TCSPC	time-correlated single photon counting	WOLED	white organic light-emitting diode
TCTA	tris(4-carbazoyl-9-ylphenyl)-amine	XantPhos	4,5-bis(diphenylphosphino)-9,9-dimethylxanthene
td	doublet of triple	XLCT	halide-to-ligand charge transfer
tert	tertiary		
TGA	thermogravimetric analysis	Z	atomic number
THF	tetrahydrofuran	α-NPD	N,N'-di(1-naphthyl)-N,N'-diphenyl-(1,1'-biphenyl)-4,4'-diamine
TLC	thin-layer chromatography		
TMEDA	N,N,N',N'-tetra-methylenediamine	μ	mikro-
TMS	trimethylsilane		
TPBi	1,3,5-tris(1-phenyl-1H-benzimidazol-2-yl)benzene		
TPQ	triplet-polaron quenching		
tt	triplet of triplet		
TTA	triplet-triplet annihilation		
U_{turnon}	turn-on voltage		
UV	ultraviolet		
V	Volt(age)		
Vis	visible		
vw	very weak		
w	weak		

7. References

[1] Photonic Industry Report 2013 (accessed from http://www.photonics21.org on 09.12.14)

[2] Press release of The Royal Swedish Academy of Sciences, http://www.nobelprize.org/nobel_prizes/physics/laureates/2014/press.html, 09.12.14

[3] F. So, J. Kido, P. Burrows, *MRS Bull.* **2011**, *33*, 663–669.

[4] http://fineartamerica.com/featured/1800s-edison-incandescent-electric-lamp-patent-print-lightbulb-blueprint-art-sara-harris.html (accessed on 27.12.14)

[5] http://home.howstuffworks.com/question151.htm (accessed on 27.12.14)

[6] http://www.ledtubes-light.com/sale-1595414-1630lm-epistar-t8-t10-led-fluorescent-tube-light-bulb-with-warm-natural-cool-white.html (accessed on 27.12.14)

[7] http://www.darbylighting.com/led-lights.html (accessed on 27.12.14)

[8] http://www.tesa.de/imedia/253/11881253,dim:460x353,mode:exact.jpg (accessed on 27.12.14)

[9] M. C. Gather, A. Köhnen, K. Meerholz, *Adv. Mater.* **2011**, *23*, 233–248.

[10] A. Bernanose, M. Comte, P. Vouaux, *J. Chim. Phys. Phys. Chim. Biol.* **1953**, *50*, 64–68.

[11] C. W. Tang, S. A. Van Slyke, *Appl. Phys. Lett.* **1987**, *51*, 913–915.

[12] J. H. Burroughes, D. D. C. Bradley, A. R. Brown, R. N. Marks, K. Mackay, R. H. Friend, P. L. Burns, A. B. Holmes, *Nature* **1990**, 539–541.

[13] D. D. C. Bradley, *Synth. Met.* **1993**, *54*, 401–415.

[14] Press release of The Royal Swedish Academy of Sciences, http://www.nobelprize.org/nobel_prizes/chemistry/laureates/2000/press.html, 12.12.14

[15] S. Lamansky, P. Djurovich, D. Murphy, F. Abdel-Razzaq, H.-E. Lee, C. Adachi, P. E. Burrows, S. R. Forrest, M. E. Thompson, *J. Am. Chem. Soc.* **2001**, *123*, 4304–4312.

[16] A. Tsuboyama, H. Iwawaki, M. Furugori, T. Mukaide, J. Kamatani, S. Igawa, T. Moriyama, S. Miura, T. Takiguchi, S. Okada, M. Hoshino, K. Ueno, *J. Am. Chem. Soc.* **2003**, *125*, 12971–12979.

[17] J. P. Duan, P. P. Sun, C. H. Cheng, *Adv. Mater.* **2003**, *15*, 224–228.

[18] G. J. Hedley, A. Ruseckas, I. D. W. Samuel, *Chem. Phys. Lett.* **2008**, *450*, 292–296.

[19] A. F. Rausch, M. E. Thompson, H. Yersin, *J. Phys. Chem. A* **2009**, *113*, 5927–5932.

[20] C.-H. Yang, M. Mauro, F. Polo, S. Watanabe, I. Muenster, R. Fröhlich, L. De Cola, *Chem. Mater.* **2012**, *24*, 3684–3695.

[21]   K. S. Yook, J. Y. Lee, *Adv. Mater.* **2012**, *24*, 3169–3190.

[22]   R. C. Kwong, S. Sibley, T. Dubovoy, M. Baldo, S. R. Forrest, M. E. Thompson, *Chem. Mater.* **1999**, *11*, 3709–3713.

[23]   W. Lu, B. Mi, M. C. W. Chan, Z. Hui, C. Che, N. Zhu, S. Lee, *J. Am. Chem. Soc.* **2004**, *126*, 4958–4971.

[24]   J. G. Williams, S. Develay, D. Rochester, L. Murphy, *Coord. Chem. Rev.* **2008**, *252*, 2596–2611.

[25]   C.-M. Che, C.-C. Kwok, S.-W. Lai, A. F. Rausch, W. J. Finkenzeller, N. Zhu, H. Yersin, *Chemistry* **2010**, *16*, 233–247.

[26]   Z. M. Hudson, C. Sun, M. G. Helander, Y.-L. Chang, Z.-H. Lu, S. Wang, *J. Am. Chem. Soc.* **2012**, *134*, 13930–3.

[27]   M. Mydlak, M. Mauro, F. Polo, M. Felicetti, J. Leonhardt, G. Diener, L. De Cola, C. A. Strassert, *Chem. Mater.* **2011**, *23*, 3659–3667.

[28]   M. A. Baldo, D. F. O'Brien, Y. You, A. Shoustikov, S. Sibley, M. E. Thompson, S. R. Forrest, *Nature* **1998**, *395*, 151–154.

[29]   M. A. Baldo, M. E. Thompson, S. R. Forrest, *Nature* **2000**, *403*, 750–753.

[30]   H. Yersin, A. F. Rausch, R. Czerwieniec, in *Phys. Org. Semicond.*, Second Edition, Eds. W. Brütting, C. Adachi, Wiley VCH: Weinheim, **2012**, 371–426.

[31]   Y. Tao, K. Yuan, T. Chen, P. Xu, H. Li, R. Chen, C. Zheng, L. Zhang, W. Huang, *Adv. Mater.* **2014**, *26*, 7931–7958.

[32]   H. Uoyama, K. Goushi, K. Shizu, H. Nomura, C. Adachi, *Nature* **2012**, *492*, 234–238

[33]   Q. Zhang, B. Li, S. Huang, H. Nomura, H. Tanaka, C. Adachi, *Nat. Photonics* **2014**, *8*, 326–332.

[34]   BMBF: „Organische Elektronik – Hightech aus Kunststoff", http://www.bmbf.de/de/16267.php (accessed on 23.12.2014).

[35]   Press release: Brillant leuchtend, biegsam und transparent OLED - Displays machen Zukunftsvisionen wahr - Technik auf organischer Basis (accessed on 10.12.14)

[36]   Y.-L. Loo, I. McCulloch, *MRS Bull.* **2008**, *33*, 653–662.

[37]   L. S. Hung, C. W. Tang, M. G. Mason, *Appl. Phys. Lett.* **1997**, *70*, 152–154.

[38]   D. Hertel, C. D. Müller, K. Meerholz, *Chem. Unserer Zeit* **2005**, *39*, 336–347.

[39]   M. Deußen, H. Bässler, *Chem. Unserer Zeit* **1997**, *31*, 76–86.

[40]   H. Bässler, A. Köhler, *Top. Curr. Chem.* **2012**, *312*, 1–65.

[41]   Q. Pei, *Material Matters* **2007**, *2.3*, 26–34.

42   N. J. Turro, V. Ramamurthy, J. C. Scaiano, *Principles of Molecular Photochemistry – An Introduction*, University Science Books: Sausalito, **2009**.

43   Dissertation M. Rothmann, *Designing novel host materials for blue phosphorescent organic light-emitting diodes*, 2009.

44   J. R. Lakowicz, *Principles of Fluorescence Spectroscopy*, Third edition, Springer: New York, **2006**.

45   B. Valeur, *Molecular Fluorescence: Principles and Applications*, Wiley-VCH: Weinheim, **2001**.

46   L. Yang, F. Okuda, K. Kobayashi, K. Nozaki, Y. Tanabe, Y. Ishii, M.-A. Haga, *Inorg. Chem.* **2008**, *47*, 7154–7165.

47   H. Yersin, A. F. Rausch, R. Czerwieniec, T. Hofbeck, T. Fischer, *Coord. Chem. Rev.* **2011**, *255*, 2622–2652.

48   H. Yersin, W. J. Finkenzeller, in *Highly Efficient OLEDs with Phosphorescent Materials*, Ed. H. Yersin, Wiley-VCH: Weinheim, **2008**, 1–98.

49   OLED association White paper, B. Young, *Continued Advances in OLED Technologies and Materials Accelerate the Emergence of New High Performance Displays and Lighting Products*, **2010** (accessed from http://www.oled-a.org/techpaper_details.cfm?ID=3 on 23.12.14).

50   Q. Zhang, T. Komino, S. Huang, S. Matsunami, K. Goushi, C. Adachi, *Adv. Funct. Mater.* **2012**, *22*, 2327–2336.

51   K. S. Yook, J. Y. Lee, *Adv. Mater.* **2012**, *24*, 3169–3190.

52   M. A. Baldo, S. R. Forrest, *Phys. Rev. B* **2000**, *62*, 10958–10966.

53   E. Krausz, J. Higgins, H. Riesen, *Inorg. Chem.* **1993**, *32*, 4053–4056.

54   J. S. Kim, P. K. H. Ho, C. E. Murphy, N. Baynes, R. H. Friend, *Adv. Mater.* **2002**, *14*, 206–209.

55   F. So, D. Kondakov, *Adv. Mater.* **2010**, *22*, 3762–3777.

56   A. Fleissner, K. Stegmaier, C. Melzer, H. von Seggern, T. Schwalm, M. Rehahn, *Chem. Mater.* **2009**, *21*, 4288–4298.

57   V. Sivasubramaniam, F. Brodkorb, S. Hanning, H. P. Loebl, V. van Elsbergen, H. Boerner, U. Scherf, M. Kreyenschmidt, *J. Fluor. Chem.* **2009**, *130*, 640–649.

58   R. Seifert, I. Rabelo De Moraes, S. Scholz, M. C. Gather, B. Lüssem, K. Leo, *Org. Electron.* **2013**, *14*, 115–123.

59   W. Bijnens, J. Manca, T. Wu, M. D'Olicslaeger, D. Vanderzande, J. Gelan, W. De Ceuninck, L. De Schepper, L. M. Stals, *Synth. Met.* **1996**, *83*, 261–265.

60   S. F. Lim, L. Ke, W. Wang, S. J. Chua, *Appl. Phys. Lett.* **2001**, *78*, 2116–2118.

61   D. Y. Kondakov, R. H. Young, *J. Appl. Phys.* **2010**, *108*, 074513.

[62] Y. Noguchi, H.-J. Kim, R. Ishino, K. Goushi, C. Adachi, Y. Nakayama, H. Ishii, *Org. Electron.* **2015**, *17*, 184–191.

[63] S. Möller, S. R. Forrest, *J. Appl. Phys.* **2002**, *91*, 3324–3327.

[64] T. Bocksrocker, J. Hoffmann, C. Eschenbaum, A. Pargner, J. Preinfalk, F. Maier-Flaig, U. Lemmer, *Org. Electron.* **2013**, *14*, 396–401.

[65] Y. Sun, S. R. Forrest, *J. Appl. Phys.* **2006**, *100*, 073106.

[66] Y. Sun, S. R. Forrest, *Nat. Photonics* **2008**, *2*, 483–487.

[67] S.-Y. Kim, W.-I. Jeong, C. Mayr, Y.-S. Park, K.-H. Kim, J.-H. Lee, C.-K. Moon, W. Brütting, J.-J. Kim, *Adv. Funct. Mater.* **2013**, *23*, 3896–3900.

[68] C. Murawski, P. Liehm, K. Leo, M. C. Gather, *Adv. Funct. Mater.* **2014**, *24*, 1117–1124.

[69] W. Brütting, J. Frischeisen, T. D. Schmidt, B. J. Scholz, C. Mayr, *Phys. status solidi* **2013**, *210*, 44–65.

[70] L. Penninck, F. Steinbacher, R. Krause, K. Neyts, *Org. Electron.* **2012**, *13*, 3079–3084.

[71] D.C. Müller, A. Falcou, N. Reckefuss, M. Rohjan, V. Wiederhirn, P. Rudati, H. Frohne, O. Nuyken, H. Becker, K. Meerholz, *Nature* **2003**, *421*, 829–833.

[72] D. Volz, T. Baumann, H. Flügge, M. Mydlak, T. Grab, M. Bächle, C. Barner-Kowollik, S. Bräse, *J. Mater. Chem.* **2012**, *22*, 20786–20790.

[73] E. Bellmann, S. E. Shaheen, S. Thayumanavan, S. Barlow, R. H. Grubbs, S. R. Marder, B. Kippelen, N. Peyghambarian, *Chem. Mater.* **1998**, *10*, 1668–1676.

[74] A. Bacher, C. H. Erdelen, W. Paulus, H. Ringsdorf, H. Schmidt, P. Schuhmacher, *Macromolecules* **1999**, *32*, 4551–4557.

[75] S. Inaoka, D. B. Roitman, R. C. Advincula, *Chem. Mater.* **2005**, *17*, 6781–6789.

[76] A. Teichler, J. Perelaer, U. S. Schubert, *J. Mater. Chem. C* **2013**, *1*, 1910–1925.

[77] G. Hernandez-Sosa, N. Bornemann, I. Ringle, M. Agari, E. Dörsam, N. Mechau, U. Lemmer, *Adv. Funct. Mater.* **2013**, *23*, 3164–3171.

[78] D. Tanaka, H. Sasabe, Y.-J. Li, S.-J. Su, T. Takeda, J. Kido, *Jpn. J. Appl. Phys.* **2007**, *46*, L10–L12.

[79] S. Reineke, F. Lindner, G. Schwartz, N. Seidler, K. Walzer, B. Lüssem, K. Leo, *Nature* **2009**, *459*, 234–238.

[80] http://www.oled-info.com/lg/udc_white_oled_technology_exceeds_100_lm_w (accessed on 27.12.14)

[81] G. Vicentini, L. Zinner, J. Zukerman-Schpector, K. Zinner, *Coord. Chem. Rev.* **2000**, *196*, 353–382.

[82]   Y. Ma, Y. Wang, *Coord. Chem. Rev.* **2010**, *254*, 972–990.

[83]   A. de Bettencourt-Dias, *Dalton Trans.* **2007**, 2229–22241.

[84]   H. Wang, P. He, H. Yan, J. Shi, M. Gong, *Inorg. Chem. Commun.* **2011**, *14*, 1183–1185.

[85]   A. N. Gusev, V. F. Shul, S. B. Meshkova, P. G. Doga, M. Hasegawa, G. G. Aleksandrov, I. L. Eremenko, W. Linert, *Inorg. Chim. Acta* **2012**, *387*, 321–326.

[86]   D. Ma, D. Wang, B. Li, Z. Hong, S. Lu, L. Wang, X. Zhao, N. Miriam, N. Takadab, Y. Ichino, K. Yase, H. Zhang, X. Jing, F. Wang, *Synth. Met.* **1999**, *102*, 1136–1137.

[87]   C. J. Liang, D. Zhao, Z. R. Hong, D. X. Zhao, X. Y. Liu, W. L. Li, J. B. Peng, J. Q. Yu, C. S. Lee, S. T. Lee, *Appl. Phys. Lett.* **2000**, *76*, 67–69.

[88]   D. Kourkoulos, C. Karakus, D. Hertel, R. Alle, S. Schmeding, J. Hummel, N. Risch, E. Holder, K. Meerholz, *Dalton Trans.* **2013**, 13612–13621.

[89]   G. Li, T. Fleetham, J. Li, *Adv. Mater.* **2014**, *26*, 2931–2936.

[90]   M. Cocchi, J. Kalinowski, V. Fattori, J. a. G. Williams, L. Murphy, *Appl. Phys. Lett.* **2009**, *94*, 073309.

[91]   C. Adachi, M. A. Baldo, M. E. Thompson, S. R. Forrest, *J. Appl. Phys.* **2001**, *90*, 5048–5051.

[92]   R. L. Rudnick, S. Gao, in *Treatise on Geochemistry*, Ed. H. D. H. K. Turekian, Pergamon: Oxford, **2003**, 1–64.

[93]   R. Czerwieniec, J. Yu, H. Yersin, *Inorg. Chem.* **2011**, *50*, 8293–8301.

[94]   D. M. Zink, D. Volz, T. Baumann, M. Mydlak, H. Flügge, J. Friedrichs, M. Nieger, S. Bräse, *Chem. Mater.* **2013**, *25*, 4471–4486.

[95]   http://www.finanzen.net/rohstoffe/kupferpreis/euro (accessed on 01.02.15)

[96]   http://pm-prices.heraeus.com/Heraeus_CurrentPrices.aspx?Lang=DE&Minor=False#__utma= 257634834.1549887489.1422822824.1422822824.1422822824.1&__utmb=257634834.2.10.14 22822824&__utmc=257634834&__utmx=-&__utmz=257634834.1422822824.1.1.utmcsr= google|utmccn=%28organic%29|utmcmd=organic|utmctr=%28not%20provided%29&__utmv=- &__utmk=149413237 (accessed on 01.02.15)

[97]   D. Volz, M. Wallesch, C. Fléchon, M. Danz, A. Verma, J. M. Navarro, S. Bräse, T. Baumann, *Green Chem.* **2015**, *17*, 1988–2011.

[98]   M. Montalti, A. Credi, L. Prodi, M. T. Gandolfi, *Handbook of Photochemistry*, Third Edition. CRC Press, Taylor and Francis Group: Boca Raton, **2006**.

[99]   J.C. Deaton, S. C. Switalski, D. Y. Kondakov, R. H. Young, T. D. Pawlik, D. J. Giesen, S. B. Harkins, A. J. M. Miller, S. F. Mickenberg, J. C. Peters, *J. Am. Chem. Soc.* **2010**, *132*, 9499–9508.

[100] D. Volz, Y. Chen, M. Wallesch, R. Liu, C. Fléchon, D. M. Zink, J. Friedrichs, H. Flügge, R. Steininger, J. Göttlicher, C. Heske, L. Weinhardt, S. Bräse, F. So, T. Baumann, *Funct. Mater.* **2015**, *27*, 2538–2543.

[101] C. W. Hsu, C. C. Lin, M. W. Chung, Y. Chi, G. H. Lee, P.T. Chou, C. H. Chang, P. Y. Chen, *J. Am. Chem. Soc.* **2011**, *133*, 12085–12099.

[102] L. Bergmann, M. Nieger, J. Friedrichs, T. Baumann, S. Bräse, in preparation, **2016**. *The Coordination and Photo-Chemistry of Copper(I) Complexes: Variation of N^N Ligands from Imidazole to Tetrazole and bridging or non-bridging Phosphines.*

[103] R. Peng, M. Li, D. Li, *Coord. Chem. Rev.* **2010**, *254*, 1–18.

[104] M. Wallesch, D. Volz, D. M. Zink, U. Schepers, M. Nieger, T. Baumann, S. Bräse, *Chem. Eur. J.* **2014**, *20*, 6578–6590.

[105] L. Maini, D. Braga, P. P. Mazzeo, B. Ventura, *Dalton Trans.* **2012**, *41*, 531–539.

[106] D. M. Zink, T. Grab, T. Baumann, M. Nieger, E. C. Barnes, W. Klopper, S. Bräse, *Organometallics* **2011**, *30*, 3275–3283.

[107] D. Volz, A. F. Hirschbiel, D. M. Zink, J. Friedrichs, M. Nieger, T. Baumann, S. Bräse, C. Barner-Kowollik, *J. Mater. Chem. C* **2014**, *2*, 1457–1462.

[108] M. J. Leitl, F. R. Küchle, H. A. Mayer, L. Wesemann, H. Yersin, *J. Phys. Chem. A* **2013**, *117*, 11823–11836.

[109] H. Kitagawa, Y. Ozawa, K. Toriumi, *Chem. Commun.* **2010**, *46*, 6302–6304.

[110] P. C. Ford, E. Cariati, J. Bourassa, *J. Chem. Rev.* **1999**, *99*, 3625–3648.

[111] P. P. Mazzeo, L. Maini, A. Petrolati, V. Fattori, K. Shankland, D. Braga, *Dalton Trans.* **2014**, *43*, 9448–55.

[112] Z. Liu, P. I. Djurovich, M. T. Whited, M. E. Thompson, *Inorg. Chem.* **2012**, *51*, 230–236.

[113] H. Ohara, A. Kobayashi, M. Kato, *Dalton Trans.* **2014**, *43*, 17317–17323.

[114] T. Gneuß, M. J. Leitl, L. H. Finger, N. Rau, H. Yersin, J. Sundermeyer, *Dalton Trans.* **2015**, doi: 10.1039/c4dt02631d.

[115] A. Tsuboyama, K. Kuge, M. Furugori, S. Okada, M. Hoshino, *Inorg. Chem.* **2007**, *46*, 1992–2001.

[116] J.-L. Chen, X. Cao, J. Wang, L. He, Z. Liu, H. Wen, Z.-N. Chen, *Inorg. Chem.* **2013**, *52*, 9727–9740.

[117] M. Osawa, M. Hoshino, M. Hashimoto, I. Kawata, S. Igawa, M. Yashima, *Dalton Trans.* **2015**, doi: 10.1039/c4dt02853h.

[118] W. M. Mara, K. A. Fransted, L. X. Chen, *Coord. Chem. Rev.* **2015**, doi: 10.1016/j.ccr.2014.06.013.

[119] T. J. Penfold, S. Karlsson, G. Capano, F. A. Lima, J. Rittmann, M. Reinhard, M. H. Rittmann-Frank, O. Braem, E. Barano, R. Abela, I. Tavernelli, U. Rothlisberger, C. J. Milne, M. Chergui, *Phys. Chem. A* **2013**, *117*, 4591–4601.

[120] M. W. Mara, N. E. Jackson, J. Huang, A. B. Stickrath, X. Zhang, N. A. Gothard, M. A. Ratner, L. X. Chen, *J. Phys. Chem. B* **2013**, *117*, 1921–1931.

[121] N. Armaroli, G. Accorsi, M. Holler, O. Moudam, J.-F. Nierengarten, Z. Zhou, R. T. Wegh, R. Welter, *Adv. Mater.* **2006**, *18*, 1313–1316.

[122] S. Keller, E. C. Constable, C. E. Housecroft, M. Neuburger, A. Prescimone, G. Longo, A. Pertegás, M. Sessolo, H. J. Bolink, *Dalton. Trans.* **2014**, *43*, 16593–16596.

[123] D. G. Cuttell, S.-M. Kuang, P. E. Fanwick, D. R. McMillin, R. A. Walton, *J. Am. Chem. Soc.* **2002**, *124*, 6–7.

[124] Z. A. Siddique, Y. Yamamoto, T. Ohno, K. Nozaki, K. *Inorg. Chem.* **2003**, *42*, 6366–6378.

[125] Miller, A. J. M.; Dempsey, J. L.; Peters, J. C. *Inorg. Chem.* **2007**, *46*, 7244–7246.

[126] J. Min, Q. Zhang, W. Sun, Y. Cheng, Wang, L. *Dalton Trans.* **2011**, *40*, 686–693.

[127] V. A. Krylova, P. I. Djurovich, J. W. Aronson, R. Haiges, M. T. Whited, M. E. Thompson, *Organometallics* **2012**, *31*, 7983–7993.

[128] V. A. Krylova, P. I. Djurovich, B. L. Conley, R. Haiges, M. T. Whited, T. J. Williams, M. E. Thompson, *Chem. Commun.* **2014**, *50*, 7176–7179.

[129] M. Osawa, I. Kawata, R. Ishii, S. Igawa, M. Hashimoto, M. Hoshino, *J. Mater. Chem. C* **2013**, *1*, 4375–4383.

[130] M. Osawa, *Chem. Commun.* **2014**, *50*, 1801–1803.

[131] A. Kaeser, M. Mohankumar, J. Mohanraj, F. Monti, M. Holler, J.-J. Cid, O. Moudam, I. Nierengarten, L. Karmazin-Brelot, C. Duhayon, B. Delavaux-Nicot, N. Armaroli, J.-F. Nierengarten, *Inorg. Chem.* **2013**, *52*, 12140–12151.

[132] C. Femoni, S. Muzzioli, A. Palazzi, S. Stagni, S. Zacchini, F. Monti, G. Accorsi, M. Bolognesi, N. Armaroli, M. Massi, G. Valenti, M. Marcaccio, *Dalton Trans.* **2013**, *42*, 997–1010.

[133] D. J. Fife, W. M. Moore, K. W. Morse, *Inorg. Chem.* **1984**, *23*, 1684–1691.

[134] S. J. Lippard, J. J. Mayerle, *Inorg. Chem.* **1972**, *11*, 753–759.

[135] R. Ahuja, M. Nethaji, A. G. Samuelson, *Inorg. Chim. Acta* **2011**, *372*, 220–226.

[136] D. M. Zink, T. Baumann, J. Friedrichs, M. Nieger, S. Bräse, *Inorg. Chem.* **2013**, *52*, 13509–13520.

[137] L. Maini, P. P. Mazzeo, F. Farinella, V. Fattori, D. Braga, *Faraday Discuss.* **2014**, *170*, 93–107.

[138]   L. Bergmann, G. J. Hedley, T. Baumann, S. Bräse, I. D. W. Samuel, *Direct observation of intersystem crossing in a thermally activated delayed fluorescence copper complex in the solid state, Sci. Adv.* **2015**, *2*, e150088.

[139]   J. R. Kirchhoff, R. E. Gamache, M. W. Blaskie, A. A. Del Paggio, R. K. Lengel, D. R. McMillin, *Inorg. Chem.* **1983**, *22*, 2380–2384.

[140]   X.-L. Chen, R. Yu, Q.-K. Zhang, L.-J. Zhou, X. Wu, Q. Zhang, C.-Z. Lu, *Chem. Mater.* **2013**, *25*, 3910–3920.

[141]   H. Yersin, R. Czerwieniec, A. Hupfer, *Proc. SPIE* **2012**, *8435*, 843508.

[142]   H. Wang, L. Xie, Q. Peng, L. Meng, Y. Wang, Y. Yi, P. Wang, *Adv. Mater.* **2014**, *26*, 5198–5204.

[143]   H. Tanaka, K. Shizu, H. Nakanotani, C. Adachi, *Chem. Mater.* **2013**, *25*, 3766–3771.

[144]   T. Hofbeck, U. Monkowius, H. Yersin, *J. Am. Chem. Soc.* **2015**, *137*, 399–404.

[145]   R. Czerwieniec, K. Kowalski, H. Yersin, *Dalton Trans.* **2013**, *42*, 9826–9830.

[146]   C. L. Linfoot, M. J. Leitl, P. Richardson, A. F. Rausch, O. Chepelin, F. J. White, H. Yersin, N. Robertson, *Inorg. Chem.* **2014**, *53*, 10854–10861.

[147]   R. Czerwieniec, H. Yersin, *Diversity of Cu(I) Complexes Showing Thermally Activated Delayed Fluorescence – Profound Analysis, J. Am. Chem. Soc.* **2015**, accepted.

[148]   M. J. Leitl, V. a Krylova, P. I. Djurovich, M. E. Thompson, H. Yersin, *J. Am. Chem. Soc.* **2014**, *136*, 16032–16038.

[149]   G. D. Hager, G. A. Crosby, *J. Am. Chem. Soc.* **1975**, *97*, 7031–7037.

[150]   W. J. Finkenzeller, H. Yersin, *Chem. Phys. Lett.* **2003**, *377*, 299–305.

[151]   M. Iwamura, S. Takeuchi, T. Tahara, *J. Am. Chem. Soc.* **2007**, *129*, 5248–5256.

[152]   M. Iwamura, H. Watanabe, K. Ishii, S. Takeuchi, T. Tahara, *J. Am. Chem. Soc.* **2011**, *133*, 7728–7736.

[153]   Y. Leydet, D. M. Bassani, G. Jonusauskas, N. D. McClenaghan, *J. Am. Chem. Soc.* **2007**, *129*, 8688–8689.

[154]   G. B. Shaw, C. D. Grant, H. Shirota, E. W. C. Jr, G. J. Meyer, L. X. Chen, *J. Am. Chem. Soc.* **2007**, *129*, 2147–60.

[155]   J. Huang, M. W. Mara, a B. Stickrath, O. Kokhan, M. R. Harpham, K. Haldrup, M. L. Shelby, X. Zhang, R. Ruppert, J.-P. Sauvage, L. X. Chen, *Dalton Trans.* **2014**, *43*, 17615–17623.

[156]   G. Capano, M. Chergui, U. Rothlisberger, I. Tavernelli, T. J. Penfold, *J. Phys. Chem. A* **2014**, *118*, 9861–9869.

[157]   A. Laviecambot, M. Cantuel, Y. Leydet, G. Jonusauskas, D. Bassani, N. McClenaghan, *Coord. Chem. Rev.* **2008**, *252*, 2572–2584.

[158] M. N. Berberan-Santos, J. M. M. Garcia, *J. Am. Chem. Soc.* **1996**, *118*, 9391–9394.

[159] A. Penzkofer, A. Tyagi, E. Slyusareva, A. Sizykh, *Chem. Phys.* **2010**, *378*, 58–65.

[160] G. Méhes, H. Nomura, Q. Zhang, T. Nakagawa, C. Adachi, *Angew. Chemie* **2012**, *124*, 11473–11477.

[161] Q. Zhang, Q. Zhou, Y. Cheng, L. Wang, D. Ma, X. Jing, F. Wang, *Adv. Mater.* **2004**, *16*, 432–436.

[162] K. J. Lotito, J. C. Peters, *Chem. Commun.* **2010**, *46*, 3690–3692.

[163] T. McCormick, W.-L. Jia, S. Wang, *Inorg. Chem.* **2006**, *45*, 147–55.

[164] X. Liu, W. Sun, L. Zou, Z. Xie, X. Li, C. Lu, L. Wang, Y. Cheng, *Dalton Trans.* **2012**, *41*, 1312–1319.

[165] A. Barbieri, G. Accorsi, N. Armaroli, *Chem. Commun.* **2008**, 2185–93.

[166] H. Araki, K. Tsuge, Y. Sasaki, S. Ishizaka, N. Kitamura, *Inorg. Chem.* **2005**, *44*, 9667–9675.

[167] L. Bergmann, J. Friedrichs, M. Mydlak, T. Baumann, M. Nieger, S. Bräse, *Chem. Commun.* **2013**, *49*, 6501–6503.

[168] C. A. Tolman, *J. Am. Chem. Soc.* **1970**, *92*, 2956–2965.

[169] P. Dierkes, P. W. N. M. van Leeuwen, *Dalton Trans.* **1999**, 1519–1530.

[170] J. K. Bera, M. Nethaji, A. G. Samuelson, *Inorg. Chem.* **1999**, *38*, 218–228.

[171] P. D. Harvey, M. Drouin, T. Zhang, *Inorg. Chem.* **1997**, *36*, 4998–5005.

[172] M. Kranenburg, Y. E. M. van der Burgt, P. C. J. Kramer, P. W. N. M. van Leeuwen, *Organometallics* **1995**, *14*, 3081–3089.

[173] L. S. Kallander, Q. Lu, W. Chen, T. Tomaszek, G. Yang, D. Tew, T. D. Meek, G. A. Hofmann, C. K. Schulz-Pritchard, W. W. Smith, C. A. Janson, M. D. Ryan, G-F. Zhang, K. O. Johanson, R. B. Kirkpatrick, T. F. Ho, P. W. Fisher, M. R. Mattern, R. K. Johnson, M. J. Hansbury, J. D. Winkler, K. W. Ward, D. F. Veber, S. K. Thompson, *J. Med. Chem.* **2005**, *48*, 5644–5647.

[174] W. G. Finnegan, R. A. Henry, R. Lofquist, *J. Am. Chem. Soc.* **1958**, *80*, 3908–3911.

[175] H. Detert, D. Schollmeier, *Synthesis* **1999**, *6*, 999–1004.

[176] T. Hou, J. Bian, X. Yue, S. Yue, J. Ma, *Inorg. Chim. Acta* **2013**, *394*, 15–20.

[177] Z. Si, J. Li, B. Li, S. Liu, W. Li, *J. Lumin.* **2009**, *129*, 181–186.

[178] A. Kaeser, O. Moudam, G. Accorsi, I. Séguy, J. Navarro, A. Belbakra, C. Duhayon, N. Armaroli, B. Delavaux-Nicot, J.-F. Nierengarten, *Eur. J. Inorg. Chem.* **2014**, *2014*, 1345–1355.

[179] P. Coppens, J. Sokolow, E. Trzop, A. Makal, Y. Chen, *J. Phys. Chem. Lett.* **2013**, *4*, 579–582.

[180] D. Volz, M. Nieger, J. Friedrichs, T. Baumann, S. Bräse, *Langmuir* **2013**, *29*, 3034–3044.

[181] L. Zhang, B. Li, Z. Su, *Langmuir* **2009**, *25*, 2068–2074.

[182] L. Zhang, B. Li, Z. Su, *J. Phys. Chem. C* **2009**, *113*, 13968–13973.

[183] A. Y. Kovalevsky, M. Gembicky, I. V Novozhilova, P. Coppens, *Inorg. Chem.* **2003**, *42*, 8794–802.

[184] K. F. Freed, J. Jortner, *J. Chem. Phys.* **1970**, *52*, 6272–6291.

[185] D. M. Zink, M. Bächle, T. Baumann, M. Nieger, M. Kühn, C. Wang, W. Klopper, U. Monkowius, T. Hofbeck, H. Yersin, S. Bräse, *Inorg. Chem.* **2013**, *52*, 2292–2305.

[186] S. Sakaki, H. Mizutani, Y. Kase, K. Inokuchi, T. Arai, T. Hamada, *J. Chem. Soc., Dalt. Trans.* **1996**, 1909–1914.

[187] S. Kuang, D. G. Cuttell, D. R. McMillin, P. E. Fanwick, R. A. Walton, *Inorg. Chem.* **2002**, *41*, 3313–3322.

[188] A. Wada, Q. Zhang, T. Yasuda, I. Takasu, S. Enomoto, C. Adachi, *Chem. Commun.* **2012**, *48*, 5340–5342.

[189] E. Orselli, G. S. Kottas, A. E. Konradsson, P. Coppo, R. Fröhlich, L. de Cola, A. van Dijken, M. Büchel, H. Börner, *Inorg. Chem.* **2007**, *46*, 11082–11093.

[190] R. Hage, R. Prins, J. G. Haasnoot, J. Reedijk, J. G. Vos, *Dalton Trans.* **1987**, 1389–1395.

[191] L. Qin, Q. Zhang, W. Sun, J. Wang, C. Lu, Y. Cheng, L. Wang, *Dalton Trans.* **2009**, 9388–9391.

[192] V. a Krylova, P. I. Djurovich, M. T. Whited, M. E. Thompson, *Chem. Commun.* **2010**, *46*, 6696–6698.

[193] X. Zhang, L. Song, M. Hong, H. Shi, K. Xu, Q. Lin, Y. Zhao, Y. Tian, J. Sun, K. Shu, W. Chai, *Polyhedron* **2014**, *81*, 687–694.

[194] C. Sarcher, A. Lühl, F. C. Falk, S. Lebedkin, M. Kühn, C. Wang, J. Paradies, M. M. Kappes, W. Klopper, P. W. Roesky, *Eur. J. Inorg. Chem.* **2012**, *2012*, 5033–5042.

[195] E. Zuidema, P. E. Goudriaan, B. H. G. Swennenhuis, P. C. J. Kamer, P. W. N. M. van Leeuwen, M. Lutz, A. L. Spek, *Organometallics* **2010**, *29*, 1210–1221.

[196] Z. Freixa, P. W. N. M. van Leeuwen, P. Van Leeuwen, *Dalton Trans.* **2003**, 1890–1901.

[197] Y. A. Veits, E. V. Mutsenek, E. G. Neganova, I. P. Beletskaya, *Russ. J. Org. Chem.* **2001**, *37*, 1583–1586.

[198] R. Ragni, E. A. Plummer, K. Brunner, J. W. Hofstraat, F. Babudri, G. M. Farinola, F. Naso, L. De Cola, *J. Mater. Chem.* **2006**, *16*, 1161–1170.

[199] S. Bettington, M. Tavasli, M. R. Bryce, A. Beeby, H. Al-Attar, A. P. Monkman, *Chem. Eur. J.* **2007**, *13*, 1423–1431.

[200]   Z. P. Demko, K. B. Sharpless, *J. Org. Chem.* **2001**, *66*, 7945–7950.

[201]   J. M. McManus, R. M. Herbst, J. Org. Chem. **1959**, *24*, 1462–1464.

[202]   Y. Zhou, C. Yao, R. Ni, G. Yang, *Synth. Commun.* **2010**, *40*, 2624–2632.

[203]   Dissertation J. Kramer, *Neue Funktionalisierungs- und Anwendungsmöglichkeiten von [2.2]Paracyclophanen und [2](1,4)Benzo[2](2,5)pyridinophanen*, **2013**.

[204]   O. Daugulis, H.-Q. Do, *Org. Lett.* **2010**, *12*, 2517–2519.

[205]   Patent M. R. Collins, V. Natarajan, *Inhibitors of Hepatitis C Virus Protease, and Compositions and Treatments Using the Same*, WO2006043145A1, patent application.

[206]   K. Dickore, K. Susse, K. Bode, *Liebigs Ann. Chem.* **1970**, *733*, 70–87.

[207]   H. P. Kokatla, P. F. Thomson, S. Bae, V. R. Doddi, M. K. Lakshman, *J. Org. Chem.* **2011**, *76*, 7842–7848.

[208]   W. K. Fife, *J. Org. Chem.* **1983**, *48*, 1375–1377.

[209]   R. T. Shuman, P. L. Ornstein, J. W. Paschal, P. D. Gesellchen, *J. Org. Chem.* **1990**, *55*, 738–741.

[210]   E. Busto, V. Gotor-Fernández, V. Gotor, *Tetrahedron Asymmetry* **2006**, *17*, 1007–1016.

[211]   P. C. Kamer, P. W. van Leeuwen, J. N. Reek, *Acc. Chem. Res.* **2001**, *34*, 895–904.

[212]   D. Volz, M. Wallesch, S. L. Grage, J. Göttlicher, R. Steininger, D. Batchelor, T. Vitova, A. S. Ulrich, C. Heske, L. Weinhardt, T. Baumann, S. Bräse, *Inorg. Chem.* **2014**, *53*, 7837–7847.

[213]   H.-B. Zhu, S.-Y. Zhang, X. Lu, W.-N. Yang, S.-H. Gou, J. Chen, *Z. anorg. allg. Chem.* **2011**, *637*, 1423–1426.

[214]   M. Knorr, F. Guyon, A. Khatyr, C. Strohmann, M. Allain, S. M. Aly, A. Lapprand, D. Fortin, P. D. Harvey, *Inorg. Chem.* **2012**, *51*, 9917–9934.

[215]   K. M. Henline, C. Wang, R. D. Pike, J. C. Ahern, B. Sousa, H. H. Patterson, A. T. Kerr, C. L. Cahill, *Cryst. Growth Des.* **2014**, *14*, 1449–1458.

[216]   I.-H. Park, S. S. Lee, *CrystEngComm.* **2011**, *13*, 6520–6525.

[217]   A. Gallego, O. Castillo, C. J. Gómez-García, F. Zamora, S. Delgado, *Inorg. Chem.* **2012**, *51*, 718–727.

[218]   I. Papazoglou, P. J. Cox, A. G. Papadopoulos, M. P. Sigalas, P. Aslanidis, *Dalton Trans.* **2013**, *42*, 2755–2764.

[219]   M. A. Tsiaggali, E. G. Andreadou, a G. Hatzidimitriou, a a Pantazaki, P. Aslanidis, *J. Inorg. Biochem.* **2013**, *121*, 121–128.

[220]   N. Armaroli, G. Accorsi, F. Cardinali, A. Listorti, in *Photochem. Photophysics Coord. Compd. I*, Eds. V. Balzani, S. Campagna, Springer: Berlin, Heidelberg, **2007**, 69–115.

[221]   Private communication with Prof. Dr. Hartmut Yersin, University of Regensburg.

[222]   D. Lai, J. I. Zink, *Inorg. Chem.* **1993**, *32*, 2594–2596.

[223]   S. Igawa, M. Hashimoto, I. Kawata, M. Yashima, M. Hoshino, M. Osawa, *J. Mater. Chem. C*
       2013, *1*, 542–551.

[224]   M. S. Masar, C. a. Mirkin, C. L. Stern, L. N. Zakharov, A. L. Rheingold, *Inorg. Chem.* **2004**, *43*,
       4693–4701.

[225]   E. Breitmaier, G. Jung, *Organische Chemie*, Fourth Edition,Thieme: Stuttgart, **2001**.

[226]   M. W. Haenel, D. Jakubik, E. Rothenberger, G. Schroth, *Chem. Ber.* **1991**, 1705–1710.

[227]   A. Makal, J. Benedict, E. Trzop, J. Sokolow, B. Fournier, Y. Chen, J. A. Kalinowski, T. Graber,
       R. Henning, P. Coppens, *J. Phys. Chem. A* **2012**, *116*, 3359–3365.

[228]   I. I. Vorontsov, T. Graber, A. Y. Kovalevsky, I. V Novozhilova, M. Gembicky, Y.-S. Chen,
       P. Coppens, *J. Am. Chem. Soc.* **2009**, *131*, 6566–6573.

[229]   K. P. C. Vollhardt, N. E. Schore, *Organische Chemie*, Fifth Edition, Wiley-VCH: Weinheim,
       2011.

[230]   M. Nishikawa, Y. Takara, Y. Hattori, K. Nomoto, T. Kusamoto, S. Kume, H. Nishihara, *Inorg.
       Chem.* **2013**, *52*, 8962–8970.

[231]   K. Hisada, S. Ito, M. Yamamoto, *J. Phys. Chem. B* **1997**, *101*, 6827–6833.

[232]   S. C. J. Meskersa, J. Hübner, M. Oestreich, H. Bässle, *Chem. Phys. Lett.* **2001**, *339*, 223–228.

[233]   M. Iwamura, S. Takeuchi, T. Tahara, *Phys. Chem. Chem. Phys.* **2014**, *16*, 4143–4154.

[234]   S.J. Strickler, R.A. Berg, *J. Chem. Phys.* **1962**, *37*, 814–822.

[235]   R.S. Knox, H. van Amerongen, *J. Phys. Chem. B* **2002**, *106*, 5289–5293.

[236]   S. Y. Lee, T. Yasuda, Y. S. Yang, Q. Zhang, C. Adachi, *Angew. Chem. Int. Ed.* **2014**, *53*, 1–7.

[237]   S. Youn Lee, T. Yasuda, H. Nomura, C. Adachi, *Appl. Phys. Lett.* **2012**, *101*, 093306.

[238]   K. Sato, K. Shizu, K. Yoshimura, A. Kawada, H. Miyazaki, C. Adachi, *Phys. Rev. Lett.* **2013**,
       110, 247401.

[239]   K. Masui, H. Nakanotani, C. Adachi, *Org. Electron.* **2013**, *14*, 2721–2726.

[240]   P. Chen, Z. Xiong, Q. Peng, J. Bai, S. Zhang, F. Li, *Adv. Opt. Mater.* **2014**, *2*, 142–148.

[241]   S. Y. Lee, T. Yasuda, Y. S. Yang, Q. Zhang, C. Adachi, *Angew. Chem., Int. Ed.* **2014**, *53*,
       6402–6406.

[242]   K. A. Fransted, N. E. Jackson, R. Zong, M. W. Mara, J. Huang, M. R. Harpham, M. L. Shelby,
       R. P. Thummel, L. X. Chen, *J. Phys. Chem. A* **2014**, *118*, 10497–10506.

[243] X. L. Xin, M. Chen, Y. B. Ai, F. L. Yang, X. L. Li, F. Li, *Inorg. Chem.* **2014**, *53*, 2922–2931.

[244] Q. Zhang, Q. Zhou, Y. Cheng, L. Wang, D. Ma, X. Jing, F. Wang, *Adv. Funct. Mater.* **2006**, *16*, 1203–1208.

[245] J. C. Ribierre, A. Ruseckas, K. Knights, S. V. Staton, N. Cumpstey, P. L. Burn, I. D. W. Samuel, *Phys. Rev. Lett.* **2008**, *100*, 017402.

[246] G. M. Sheldrick, *Acta Crystallogr.* **2008**, *A64*, 112–122.

[247] G. M. Sheldrick, *Acta Crystallogr.* **2015**, *C71*, 3–8.

[248] N. C. Greenham, I. D. W. Samuel, G. R. Hayes, R. T. Phillips, Y. A. R. R. Kessener, S. C. Moratti, A. B. Holmes, R. H. Friend, *Chem. Phys. Lett.* **1995**, *241*, 89–96.

[249] T. J. Smith, K. J. Stevenson, in *Handbook of Electrochemistry*, Ed. C. G. Zoski, Elsevier Science, **2007**, 73–110.

[250] A. D. Becke, *Phys. Rev. A* **1988**, *38*, 3098–3100.

[251] J. P. Perdew, *Phys. Rev. B* **1986**, *33*, 8822–8827.

[252] M. Häser, R. Ahlrichs, *J. Comput. Chem.* **1989**, *10*, 104–111.

[253] F. Weigend, M. Häser, M. *Theor. Chem. Acc.* **1997**, *97*, 331–340.

[254] M. Sierka, A. Hogekamp, R. Ahlrichs, *J. Chem. Phys.* **2003**, *118*, 9136–9148.

[255] A. D. Becke, *J. Chem. Phys.* **1993**, *98*, 5648–5652.

[256] C. Lee, W. Yang, R. Parr, G. *Phys. Rev. B* **1988**, 37, 785–789.

[257] S. H. Vosko, L. Wilk, M. Nusair, *Can. J. Phys.* **1980**, *58*, 1200–1211.

[258] P. J. Stephens, F. J. Devlin, C. F. Chabalowski, M. J. Frisch, *J. Phys. Chem.* **1994**, *98*, 11623–11627.

[259] F. Weigend, R. Ahlrichs, *Phys. Chem. Chem. Phys.* **2005**, *7*, 3297–3305.

[260] D. Rappoport, J. Furche, *J. Chem. Phys.* **2010**, *133*, 134105.

[261] TURBOMOLE V6.4 2012, a development of University of Karlsruhe and Forschungszentrum Karlsruhe GmbH, 1989–2007, TURBOMOLE GmbH, since 2007; available from http://www. turbomole.com

[262] W. C. Still, M. Kahn, A. Mitra, *J. Organ. Chem.* **1978**, *43*, 2923–2925.

[263] H. G. O. Becker et al., *Organikum*, 18th Edition, Deutscher Verlag der Wissenschaften: Berlin, **1990**.

[264] R. N. Butler, in *Comprehensive Heterocyclic Chemistry II*, Eds. A. R. Katritzky, C. W. Rees, E. F. V. Scriven, Pergamon Press: Oxford, **1996**, *4*, 621–678.

[265] T. Marimuthu, M. D. Bala, H. B. Friedrich, *J. Chem. Crystallogr.* **2011**, *42*, 251–257.

[266] G. L. Moxham, H. Randell-Sly, S. K. Brayshaw, A. S. Weller, M. C. Willis, *Chem. Eur. J.* **2008**, *14*, 8383–8397.

[267] A. Xie, M. Cao, Y. Liu, L. Feng, X. Hu, W. Dong, *Eur. J. Org. Chem.* **2014**, *2014*, 436–441.

[268] N. M. Lân, R. Burgard, C. Wentrup, *J. Org. Chem.* **2004**, *69*, 2033–2036.

[269] B. Akhlaghinia, S. Rezazadeh, *J. Braz. Chem. Soc.* **2012**, *23*, 2197–2203.

[270] C. B. Reese, Z. Pei-zhuo, *J. Chem. Soc., Perkin Trans.* **1993**, *1*, 2291–2301.

[271] W. Crow, A. Khan, M. Paddon-Row, *Aust. J. Chem.* **1975**, *28*, 1741–1754.

[272] C. B. Reese, Z. Pei-Zhuo, *J. Chem. Soc., Perkin Trans.* **1993**, *1*, 2291–2301.

[273] D. Zhang, C. Zhou, R. Wang, *Catal. Commun.* **2012**, *22*, 83–88.

[274] R. H. Good, G. Jones, J. R. Phipps, *J. Chem. Soc. Perkin Trans. 1* **1972**, 2441–2445.

[275] E. Gómez, C. Avendaño, A. McKillop, *Tetrahedron* **1986**, *42*, 2625–2634.

[276] G. Liso, G. Trapani, A. Latrofa, *J. Heterocyclic Chem.* **1987**, *24*, 1683–1684.

[277] S. Frère, V. Thiéry, T. Besson, *Synth. Commun.* **2003**, *33*, 3795–3804.

[278] T. Sakamoto, S. Kaneda, S. Nishimura, H. Yamanaka, *Chem. Pharm. Bull.* **1985**, *33*, 565–571.

[279] H. J. Park, J. N. Kim, H. Yoo, K. Wee, S. O. Kang, D. W. Cho, U. C. Yoon, *J. Org. Chem.* **2013**, *78*, 8054–8064.

[280] J. I. Murray, R. Woscholski, A. C. Spivey, *Chem. Commun.* **2014**, *50*, 13608–13611.

[281] W. Zhao, X. Wang, C. Yang, *Synth. Commun.* **2014**, *44*, 150–160.

[282] I. Tsukamoto, H. Koshio, T. Kuramochi, C. Saitoh, H. Yanai-Inamura, C. Kitada-Nozawa, E. Yamamoto, T. Yatsu, Y. Shimada, S. Sakamoto, S.-I. Tsukamoto, *Bioorg. Med. Chem.* **2009**, *17*, 3130–3141.

[283] S. Duric, C. C. Tzschucke, *Org. Lett.* **2011**, *13*, 2310–2313.

[284] Y. Oguro, N. Miyamoto, T. Takagi, K. Okada, Y. Awazu, H. Miki, A. Hori, K. Kamiyama, S. Imamura, *Bioorg. Med. Chem.* **2010**, *18*, 7150–7163.

[285] Y. Kobayashi, I. Kumadaki, *Chem. Pharm. Bull.* **1969**, *17*, 510–514.

8. Appendix

8.1 Curriculum Vitae

LARISSA BERGMANN

*02/22/1986, Stühlingen, Germany

Education

11/2011 – 03/2015	**Doctoral thesis** at CYNORA GmbH, in cooperation with Prof. Dr. Stefan Bräse, Institute of Organic Chemistry, Karlsruhe Institute of Technology (KIT)
06 – 12/2013	**Research project** at Organic Semiconductor Optoelectronics group, University of St. Andrews, UK, in collaboration with Prof. Dr. Ifor D.W. Samuel, School of Physics and Astronomy
10/2006 – 09/2011	Studies in Chemistry at Karlsruhe Institute of Technology (KIT), **Diploma degree** (09/2011)
02 – 08/2011	**Diploma thesis** at CYNORA GmbH, in cooperation with Prof. Dr. Stefan Bräse, Institute of Organic Chemistry, Karlsruhe Institute of Technology (KIT), Topic: *Neue Materialien und Konzepte für optoelektronische Anwendungen*
09/1996 – 06/2005	Grammar school, Klettgau-Gymnasium Tiengen, Waldshut-Tiengen, **Abitur** (06/2005)
09/1992 – 07/1996	Primary school, Grundschule Grießen

Scholarships

02/2012 – present	Karlsruhe School of Optics and Photonics (KSOP) PhD program
03/2009 – 03/2011	Careerbuilding Programm of Femtec.Network

8.2 Publications, Conference Contributions, Patents

Publications

- L. Bergmann, J. Friedrichs, M. Mydlak, T. Baumann, M. Nieger, S. Bräse, *Chem. Commun.* **2013**, *49*, 6501–6503. *Outstanding luminescence from neutral copper(I) complexes with pyridyl-tetrazolate and phosphine ligands.*

- D. Volz, L. Bergmann, D. M. Zink, T. Baumann, S. Bräse, *SPIE Newsroom* **2013**, Doi: 10.1117/2.1201308.005080. *Are copper(I) complexes tough enough to be processed from solution?*

- D. M. Zink, L. Bergmann, D. Ambrosek, M. Wallesch, D. Volz, M. Mydlak, *Transl. Mater. Res.* **2014**, *1*, 015003. *Singlet harvesting copper-based emitters: a modular approach towards next-generation OLED technology.*

- L. Bergmann, G. J. Hedley, T. Baumann, S. Bräse, I. D. W. Samuel, *Sci. Adv.* **2015**, *2*, e150088. *Direct observation of intersystem crossing in a thermally activated delayed fluorescence copper complex in the solid state.*

- L. Bergmann, M. Wallesch, D. M. Zink, S. Bräse, T. Baumann, D. Volz, *Top. Curr. Chem.* **2016**, accepted. *Metal-organic and Organic TADF-Materials: Status, Challenges and Characterization.*

Conference Contributions

- D. M. Zink, D. Volz, L. Bergmann, M. Nieger, S. Bräse, H. Yersin, T. Baumann, *Proc. SPIE* **2013**, *882907*, Doi:10.1117/12.2028619. *Novel oligonuclear copper complexes featuring exciting luminescent characteristics.*

- L. Bergmann, T. Baumann, J. Friedrichs, M. Mydlak, M. Nieger, S. Bräse, Large-area organic & printed electronics convention (LOPEC), 11.–13.06.2013, München. Poster contribution: *Highly luminescent, neutral copper(I) complexes for OLED applications.*

- L. Bergmann, T. Baumann, J. Friedrichs, M. Nieger, S. Bräse, 3MET I. International Workshop on Metal-to-Metal Communications in Oligonuclear Complexes, 08.–11.10.2012, Kaiserslautern. Poster contribution: *Novel luminescent copper complexes for OLED applications.*

- L. Bergmann, Karlsruhe Days of Optics & Photonics 2013, 05.11.2013, Karlsruhe. Poster contribution: *Novel solutions and materials for OLED applications.*

- L. Bergmann, T. Baumann, J. Friedrichs, M. Nieger, S. Bräse, G. J. Hedley, I. D. W. Samuel, International Exhibition and Conference for Printed Electronics (LOPEC), 26.–28.05.2014, München. Poster contribution: *Photophysical studies of the thermally-activated delayed fluorescence of a mononuclear copper(I) compound.*

- L. Bergmann, A. Unterreiner, C. Schweigert, S. Höfener, S. Bräse, Assessment of the SFB 3MET, 04.–05.06.2014, Karlsruhe. Poster contribution: *Electronic dynamics of luminescent 3MET systems (Part I).*

- L. Bergmann, T. Baumann, J. Friedrichs, M. Nieger, S. Bräse, G. J. Hedley, I. D. W. Samuel, 10th International Conference on Electroluminescence and Organic Optoelectronics (ICEL), 31.08.–03.09.2014, Köln. Poster contribution: *Photophysical studies of the thermally-activated delayed fluorescence of a mononuclear copper(I) compound.*

Patents

- T. Baumann, T. Grab, L. Bergmann, *Copper(I) Complexes for Optoelectronic Devices*, WO2013/007707 A1, PCT application.

8.3 Acknowledgements

A PhD is not made without the great help and encouragement of others and so I would like to thank all those people around me who provided their support and contribution to the success of this PhD thesis.

First of all, I would like to thank my supervisor, Prof. Dr. Stefan Bräse from the Karlsruhe Institute of Technology, for his excellent scientific support and for offering me the opportunity of a PhD in his research group. I would like to thank him for his guidance as well as the great freedom to carry out my own ideas.

My greatest thanks go to Dr. Thomas Baumann, Dr. Tobias Grab and Michael Bächle for offering me a PhD position at CYNORA GmbH with an exciting and up-to-date topic. Besides the outstanding scientific supervision, especially by Dr. Thomas Baumann, I was given a great thrust and support to play a part in CYNORA and adapt further competences.

I further would like to thank Prof. Dr. Ifor D. W. Samuel from the University of St. Andrews, for accepting me as a guest researcher in the Organic Semiconductor Optoelectronics Group and for his scientific supervision and expertise in all aspects of time-resolved spectroscopy. Many thanks also go to the whole OSO group for their warm welcome, assistance and encouragement. Especially Gordon J. Hedley provided his great knowledge for the success of this research project.

Furthermore, I would like to thank all my cooperation partners, without whose support this doctoral thesis would not have been possible up to this extent: Martin Nieger from the University of Helsinki, Christian Strassert from the University of Münster, and Markus Leitl and Alexander Schinabeck from the University of Regensburg.

I also want to express my gratitude to all the people in the Institute of Organic Chemistry of the KIT, especially Pia Lang, Tanja Ohmert, Angelika Mössle and Ingrid Roßnagel from the analytic department.

For fruitful scientific discussions, provision of compounds or measurements, encouragement and help, especially in the last months of this PhD, and also for the great working atmosphere, I would like to thank my colleagues from CYNORA GmbH, in particular Dr. Jana Friedrichs for DFT calculations, Dr. Georgios Liaptsis and Dr. Sebastian Dück for cyclovoltammetry, Dr. Harald Flügge for the fabrication of OLEDs, and Dr. Daniel Zink and Dr. Daniel Volz as great lab partners from the very beginning. And for correction of this manuscript I want to thank Daniel Zink, Daniel Volz, David Ambrosek and Gordon J. Hedley.

Last but not least, I want to thank my family and friends, and most notably Benjamin, for their support and encouragement in both challenging and interesting times of my PhD. Thank you!